COMPUTATIONAL MULTISCALE MODELING OF MULTIPHASE NANOSYSTEMS

Theory and Applications

Innovations in Chemical Physics and Mesoscopy

COMPUTATIONAL MULTISCALE MODELING OF MULTIPHASE NANOSYSTEMS

Theory and Applications

Alexander V. Vakhrushev, DSc

Apple Academic Press Inc.
3333 Mistwell Crescent
Oakville, ON L6L 0A2 Canada

Apple Academic Press Inc.
9 Spinnaker Way
Waretown, NJ 08758 USA

First issued in paperback 2021

ISBN 13: 978-1-77-463670-1 (pbk)
ISBN 13: 978-1-77-188528-7 (hbk)

Library and Archives Canada Cataloguing in Publication

Vakhrushev, Alexander V., author
Computational multiscale modeling of multiphase nanosystems : theory and applications / Alexander V. Vakhrushev, DSc.

Includes bibliographical references and index.
Issued in print and electronic formats.

ISBN 978-1-77188-528-7 (hardcover).--ISBN 978-1-315-20744-5 (PDF)
1. Nanotechnology--Mathematical models. 2. Multiscale modeling. I. Title.

T174.7.V35 2017 620'.5015118 C2017-903265-8 C2017-903266-6

Library of Congress Cataloging-in-Publication Data

Names: Vakhrushev, Alexander V., author.
Title: Computational multiscale modeling of multiphase nanosystems : theory and applications / Alexander V. Vakhrushev, DSc.
Description: Toronto ; New Jersey : Apple Academic Press, 2017. | Series: Innovations in chemical physics and mesoscopy | Includes bibliographical references and index.
Identifiers: LCCN 2017021263 (print) | LCCN 2017022317 (ebook) | ISBN 9781315207445 (ebook) | ISBN 9781771885287 (hardcover : alk. paper)
Subjects: LCSH: Nanotechnology--Mathematical models. | Multiscale modeling.
Classification: LCC T174.7 (ebook) | LCC T174.7 .V35 2017 (print) | DDC 620/.5015118--dc23
LC record available at https://lccn.loc.gov/2017021263

Apple Academic Press also publishes its books in a variety of electronic formats. Some content that appears in print may not be available in electronic format. For information about Apple Academic Press products, visit our website at **www.appleacademicpress.com** and the CRC Press website at **www.crcpress.com**

For Valery, Stepan, and Varvara

CONTENTS

LIST OF ABBREVIATIONS

A	angstroms
a.m.u.	atomic mass units
AMBER	Assisted Model Building with Energy Refinement
AO	atom orbitals
CGM	controlled gas media
CHARMM	chemistry at HARvard macromolecular mechanics
CL	contact layer
CM	continuum mechanics
CVFF	consistent valence force field
DEL	double electrical layer
DNS	direct numerical simulation
ECD	electrocodeposition
FE	finite element
FEM	finite element method
GROMOS	Groningen molecular simulation
IJE	impinging jet electrode
LAMMPS	large-scale atomic/molecular massively parallel simulator
LCAO	linear combinations of atomic orbitals
LES	large-eddy simulation
MD	molecular dynamics
MEAM	modified embedded atom method
MMEC	metal matrix composite electrochemical coatings
MMFF	Merck molecular force field
MO	molecular orbitals
MOFs	metal–organic frameworks
MSD	mesodynamics
NAMD	nanoscale molecular dynamics
ns	nanosecond
NW	nanowhisker
OPLS	optimized potential for liquid simulation
PDB	Protein Data Bank
PMs	powder materials
PZT	plumbum zirconate-titanate
QD	quantum dot

QM	quantum mechanics
RANS	Reynolds averaged Navier–Stokes
RCE	rotating cylinder electrode
SCF	self-consistent field method
SDS	stressed-deformed state
TFED	theory of the functional of electron density
VLS	vapor-liquid-solid method
VMD	visual molecular dynamics

LIST OF SYMBOLS

Ψ	Schrödinger wave function
$\hat{\mathbf{H}}$	Hamilton operator (an analog of classical Hamiltonian function)
$U_k\,(x_k, y_k, z_k, S_k, t)$	external field potential acting on *k*th particle
$U_{kj}\,(x_k, y_k, z_k, S_k, x_j, y_j, z_j, S_j, t)$	interaction potential between *j*th and *k*th particles
h	Planck's constant
N^a	number of atomic nuclei in nanosystems
N^{el}	number of electrons in nanosystems
m_i	*i*th atom mass
n	number of atoms composing nanosystems
$\vec{\mathbf{r}}_{i0}$	initial radius-vector of *i*th atom
$\vec{\mathbf{r}}_i$	current radius-vector of *i*th atom
$\vec{\mathbf{r}}_{\hat{j}}$	radius-vector of *j*th atom image, the distance from which to the *i*th atom is the least among the distances from the *i*th atom to the images of the *j*th atom
$\vec{\mathbf{F}}_{i\hat{j}}$	force acting on *i*th atom from the side of *j*th atom
$\vec{\mathbf{F}}_i(\vec{\mathbf{r}})$	total force acting on *i*th atom from the side of other atoms
$\vec{\mathbf{V}}_{i0}$	*i*th atom initial velocity
$\vec{\mathbf{V}}_i$	*i*th atom current velocity
Ω	region occupied by nanosystem
E	energy
$E(\vec{\mathbf{r}})$	potential describing bounded and unbounded interactions
E_b	potential describing the variation of bond length between two atoms
K_r	constant of extension-compression of bond
b_0	equilibrium length of bond
b	current length of bond
E_v	potential describing the variation of bond angle
K_θ	force constant of bending
θ_0	equilibrium magnitude of bong angle
θ	current magnitude of bond angle

$E\varphi$	potential describing the variation of torsion angles
χ	multiplicity of torsion barrier
φ_0	phase displacement
φ	dihedron
H	height of potential barrier of dihedrons φ
E_{ej}	potential describing the variation of flat groups
E_{LJ}	potential describing Van der Waals interactions
$\frac{e_j^*}{2}$	distance at which the interaction energy is half of the minimal energy of the separation of two atoms of the j type
ε_i	potential well depth for ith atom
R_{ij}	distance at which the interaction between atoms i and j takes place
$k_Б$	Boltzmann constant
W	volume of calculation cell
E_{sw}	interaction potential when the separation procedure is taken into account
E_{nsw}	interaction potential when the separation procedure is not taken into account
R_{in}	inner radius of separation
R_{out}	outer radius of separation
a	length of calculation cell
k_1, k_2, k_3	numbers of cells-images in all directions
$\vec{\mathbf{i}}_j$	displacement vector
$\vec{\mathbf{r}}_i^*$	radius-vector of ith atom after transport from one plane of periodic cell to another
$\vec{\mathbf{r}}_{i,im(k_1,k_2,k_3)}$	radius-vector of atom image
$\vec{\mathbf{x}}(x_1, x_2, x_3)$	coordinates of atoms
$\vec{\mathbf{V}}(V_{x_1}, V_{x_2}, V_{x_3})$	atom velocities
Ω	nanoparticle volume
S	nanoparticle surface
T	temperature
$\vec{\mathbf{F}}(F_{x_1}, F_{x_2}, F_{x_3})$	vector of forces
$\hat{\sigma}\ \hat{\varepsilon}, \hat{\varepsilon}, \hat{\varepsilon}$	tensors of stresses and deformations, respectively
$\vec{\mathbf{u}}$	displacement vector
$\hat{\dot{\varepsilon}}$	tensor of deformation rates
$\hat{\varepsilon}^e$	tensor of elastic deformations

$\hat{\varepsilon}^p$	tensor of plastic deformations
$\hat{\varepsilon}^R$	tensor of rheonomic deformations
$\hat{\varepsilon}^T$	tensor of temperature deformations
$\hat{\varepsilon}^w$	tensor of deformations due to material swelling
$\hat{\sigma}_0, \hat{\varepsilon}_0$	tensors of initial stresses and deformations, respectively
$\hat{\sigma}_r, \hat{\varepsilon}_r$	tensors of residual stresses and deformations, respectively
σ, τ	normal and tangential stress stresses, respectively
$\sigma_1, \sigma_2, \sigma_3$	main normal stresses
$\{\varepsilon\}, \{\sigma\}$	tensor of deformations and stresses in finite-element model, respectively
$\{\boldsymbol{\delta}\}$	vector of displacement of finite-element nodes
U	potential energy
ω	parameter of material fault probability
$\sigma_{eq}, \varepsilon_{eq}$	equivalent stresses and deformations, respectively
σ_p, σ_b	material yield strength and ultimate strength, respectively
σ_b	ultimate compression strength and tensile strength, respectively
Ω, S	volume and surface, respectively
ρ, θ	material density and porosity, respectively
E, ν	elastic modulus and Poisson ratio, respectively
T, $_P$	temperature and pressure, respectively
a_T	thermal diffusivity
λ	heat conductivity coefficient
χ	diffusion coefficient
α_T	coefficient of volumetric thermal expansion
c_p, c_v, R	specific heat capacity of gas and gas constant, respectively
q, α	heat flow and heat-transfer coefficient, respectively
$\hat{I}$	unit tensor
$\vec{\nabla} = (\frac{\partial}{\partial x_1}, \frac{\partial}{\partial x_2}, \frac{\partial}{\partial x_3})$ $\nabla^2 = (\frac{\partial^2}{\partial x_1{}^2}, \frac{\partial^2}{\partial x_2{}^2}, \frac{\partial^2}{\partial x_3{}^2})$	nabla operator and Laplace operator, respectively
t	time.

Symbols that are not indicated in the list are explained in the text.

PREFACE

Mathematical modeling is a powerful tool for investigation and designing in different areas of technology. It is especially important to use the mathematical modeling methods in new "pioneer" fields of science and technology, in which the operation experience and experimental data have not been accumulated yet. It relates in full to nanotechnologies associated with the atomic and molecular "assembly" nanoscale objects and their use.

The main difficulty in the mathematical description of these nanoscale objects is that because of their sizes they occupy an intermediate position between macroobjects, the physical processes of which are described by statistical physics and thermodynamics (consequently, by classical mechanics), and microobjects, the behavior of which is only described by quantum mechanics. By what sign are macro- and microobjects distinguished? It is thought that the size of macroobjects is much larger than some characteristic correlation length of the order of the distance between atoms (usually several angstroms). However, in some cases, especially in the region of phase transitions, this magnitude can grow up to tens of nanometers. As one goes from micro- to macroobjects, the characteristics of the system significantly change and not necessarily in a regular manner. It is an intermediate state sweeping the range to 100 nm (such size is called mesoscopic; the name was given by van Kampen in 1981, is described by mesoscopic physics, which includes classical and quantum mechanics at the boundaries of their application).

The specific feature of the physical processes in nanoscale systems is that the key phenomena determining the behavior of a nanoscale system in real time at the macroscale take place at small space and time scales. The properties of a nanoscale system depend not only on the properties of its constituent elements but also on the regularities of the spatial arrangement of the nanoscale system and the parameters of the nanoelements interaction. In this connection, the use of the continuum mechanics methods, which have been successfully applied for the mathematical modeling of macromechanics problems, is limited at the nanometer scale. While classical physical body mechanics is based on the hypothesis of continuum according to which all the properties of a body are determined by the mechanics of continuum, this hypothesis cannot be applied to nanoscale systems. In "Nanotechnology," the displacement of separate atoms, molecules, and their formations are considered rather than the movement of continuum described with macroparameters (density, viscosity, mechanical characteristics, etc.). In doing so, it is necessary to observe the influence of the properties and parameters of the nanoelements on the system behavior at the macroscale. All of these point to the need to develop multilevel modeling techniques applied to the problems of nanotechnology.

The book deals with a systematic description of the theory of multiscale modeling of nanotechnology application in various fields of science and technology. The problems of computing nanoscale systems at different structural scales are defined, and algorithms are given for their numerical solution by the quantum/continuum mechanics, molecular dynamics, and mesodynamics methods. Much consideration is given to the processes of the formation, movement, and interaction of nanoparticles, the formation of nanocomposites and the processes accompanying the application of nanocomposites. The book concentrates on different types of nanosystems: solid, liquid, gaseous, and multi-phase consisting of various elements interacting with each other, with other elements of the nanosystem and with the environment. The book includes a large number of examples of numerical modeling of nanosystems.

The information presented can be useful to engineers, researchers, and postgraduate students engaged in the design and research in the field of nanotechnology.

ABOUT THE AUTHOR

Alexander V. Vakhrushev, DSc, is a Professor at the M.T. Kalashnikov Izhevsk State Technical University in Izhevsk, Russia, where he teaches theory, calculating, and design of nano- and microsystems. He is also the Chief Researcher of the Department of Information-Measuring Systems at the Institute of Mechanics of the Ural Branch of the Russian Academy of Sciences, and Head of Department of Nanotechnology and Microsystems of Kalashnikov Izhevsk State Technical University. He is a Corresponding Member of the Russian Engineering Academy. He has over 400 publications to his name, including monographs, articles, reports, reviews, and patents. He has received several awards, including an Academician A.F. Sidorov Prize from the Ural Division of the Russian Academy of Sciences for significant contribution to the creation of the theoretical fundamentals of physical processes taking place in multi-level nanosystems and Honorable Scientist of the Udmurt Republic. He is currently a member of the editorial boards of journals *Computational Continuum Mechanics, Chemical Physics and Mesoscopia* and *Nanobuild.* His research interests include multiscale mathematical modeling of physical–chemical processes into the nano-hetero systems at nano-, micro-, and macro-levels, static and dynamic interaction of nanoelements, and basic laws relating to the structure and macro characteristics of nano-hetero structures.

ABOUT THE SERIES INNOVATIONS IN CHEMICAL PHYSICS AND MESOSCOPY

The Innovations in Chemical Physics and Mesoscopy book series publishes books containing original papers and reviews as well as monographs. These books and monographs will report on research developments in the following fields: nanochemistry, mesoscopic physics, computer modeling, and technical engineering, including chemical engineering. The books in this series will prove very useful for academic institutes and industrial sectors round the world interested in advanced research.

BOOKS IN THE SERIES

Multifunctional Materials and Modeling
Editors: Mikhail. A. Korepanov, DSc, and Alexey M. Lipanov, DSc
Reviewers and Advisory Board Members: Gennady E. Zaikov, DSc, and A. K. Haghi, PhD

Mathematical Modeling and Numerical Methods in Chemical Physics and Mechanics
Ali V. Aliev, DSc, and Olga V. Mishchenkova, PhD
Editor: Alexey M. Lipanov, DSc

Applied Mathematical Models and Experimental Approaches in Chemical Science
Editors: Vladimir I. Kodolov, DSc, and Mikhail. A. Korepanov, DSc

Computational Multiscale Modeling of Multiphase Nanosystems: Theory and Applications
Alexander V. Vakhrushev, DSc

INTRODUCTION

At the end of the 20th century, a new interdisciplinary field in science and technology appeared, which was named "Nanotechnology"; it is related to the investigation and control of groups of atoms and molecules, the characteristic sizes of which (or the size in one or two directions) do not exceed 100 nm (1 nm = 10^{-9} m). The objects were termed "nanoelements". At present, a large number of nanoelements are known: nanoparticles, nanotubes, fullerenes, nanofibers, etc. The main specific feature of nanoelements is that electrons, atoms, and molecules form new structures in them, which have been unknown in science earlier. As experimental investigations show, mechanical, electromagnetic, chemical, and other properties of the nanoelements are superior by orders of magnitude to those of the materials in which atoms and molecules are organized in an ordinary manner known in classical physics for a long time. In this connection, the application of nanotechnologies allowing to obtain new advanced materials and to organize physicochemical processes in a new manner is quite promising in many fields of technique. Let us consider several examples, which illustrate what kind of effects can be obtained when nanotechnologies are used in machine-building in the power industry.

To increase the heating capacity and combustion efficiency of solid and liquid fuels, nanoparticles are utilized. For example, when nanoparticles are added into solid propellant, the combustion rate of the latter noticeably increases. Nanoparticles based on titanium and cerium oxides are capable of decomposing dangerous for the human health nitrogen and carbon oxides, which are found in the automobile exhaust gases. Therefore, they are added into fuel and pyrotechnic compositions to decrease the content of dangerous substances in exhaust gases and to prevent the formation of undesired combustion products. The use of nanoparticles as a catalyst leads to the effective transformation of animal fats into bio-diesel fuel.

Nanocomposites with high specific strength are used in the aircraft and space industry. The application of nanoparticles makes it possible to change the structure and to improve the service parameters of constructional materials used in power plants, for example, the strength properties of powder metal composites (Fig. 1). When such composite is formed from powders with ordinary particle sizes, that is with the particle diameter of the order of hundreds of microns (Fig. 1b), an unfilled space (2) appears between the structural elements (1), which is the source of "initial" flaws (pores, microcracks) noticeably decreasing both the strength properties and the service life of the materials. When nanoparticles in small amounts are added (3) (Fig. 1a) filling the inter-element space in the composite, they decrease the sizes of the "initial" flaws and creates additional bonds between the structural elements of the composite. It leads to an essential improvement of the service and strength properties of the material.

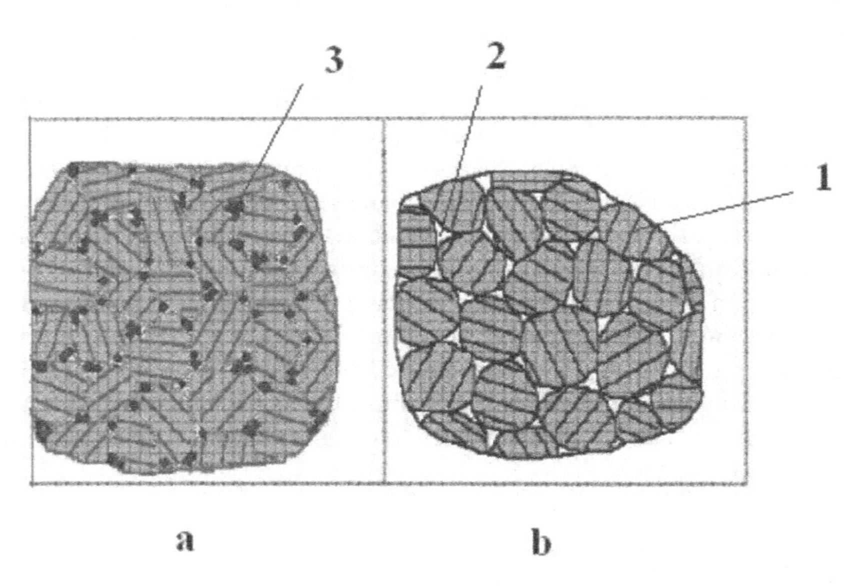

FIGURE 1 The structure of powder composite with nanoparticle additives (a) and without them (b): 1—structural elements of powder composite; 2—pores in the structure of the material; 3—nanoparticles filling pores and voids.

Nanoparticles are also used for the formation of the fine-grained structure of materials. The size of a crystal grain can be controlled by doping molten metal with nanoparticles (Fig. 2). Nanoparticles (2) are the centers of crystallization. The larger is the amount of nanoparticles, the larger is the number of such centers, and, consequently, the smaller is the characteristic size of the structural "grain" of the material (1). One can easily see that the structure of the material, which has been crystallized with the nanoparticle additives (Fig. 2a), differs from that of the material crystallized without such additives (Fig. 2b).

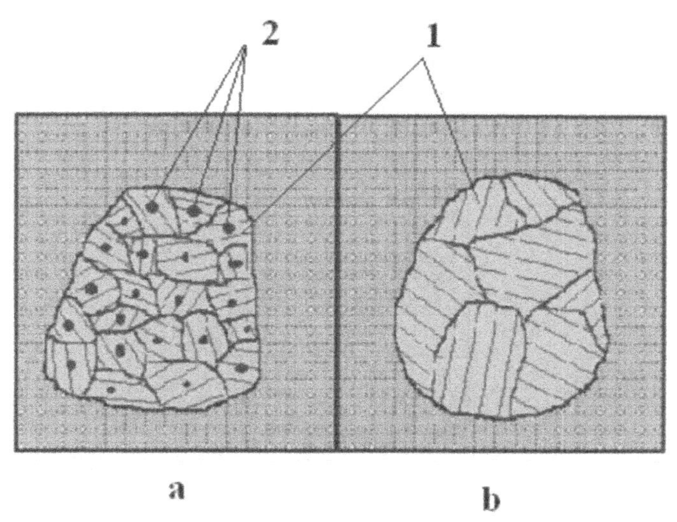

FIGURE 2 The structure of the material, in which nanoparticles are used as the crystallization centers (a), and the material structure without nanoparticles (b): 1—a "grain" of the material structure; 2—nanoparticles.

Figure 3 shows an example of the nanoparticles utilization for the improvement of the properties of glues and sealants for healing defects in materials. When crack (2) appears in material (1), it is filled with an adhesive composition containing nanoparticles as filler. Doping with the above additives improves the strength of adhesive materials in narrow cracks due to the penetrability of nanopowders and their capability to form additional strong bondings.

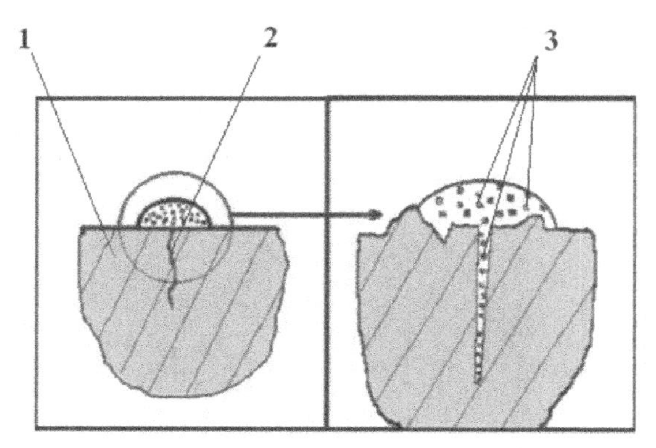

FIGURE 3 Nanoparticle additives for the improvement of the adhesion of glues and sealants: 1—material; 2—crack; 3—nanoparticles in adhesive liquid.

The application of a thin layer of nanopowder (Fig. 4(2)) upon filter element (1) leads to an essential improvement of the efficiency of its operation due to a considerable reduction of pores in the material near-surface layer, which provides fine cleaning of the liquid filtered.

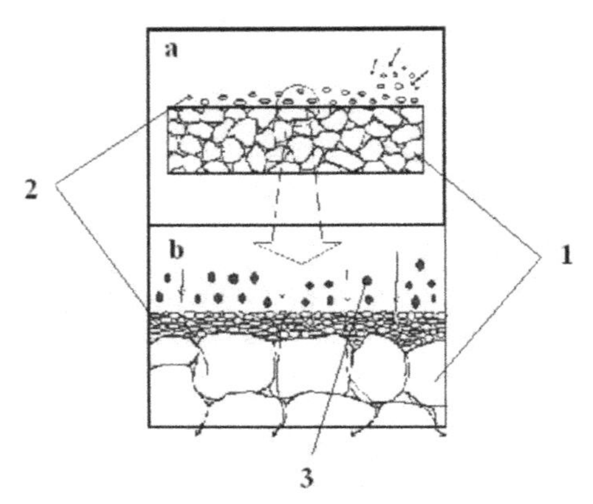

FIGURE 4 The improvement of the operating efficiency of a filter element (1) by applying nanoparticles (2) upon its surface.

The doping of fuels and oils with nanoparticles is widely used (e.g., boric acid nanoparticles improve lubricating properties of motor oils). Figure 5 demonstrates the use of nanoparticles as additives for oils. In the process of the interaction of the "working" faces of components (1) (in Fig. 5, one of the working contacting faces is shown), nanopowder (4) added into oil (3) fills microirregularities (2) forming a strong metal-composite surface layer. Friction in the contact diminishes substantially, the wear-resistance of the interacting components increases and the heat losses on the interacting surfaces decrease.

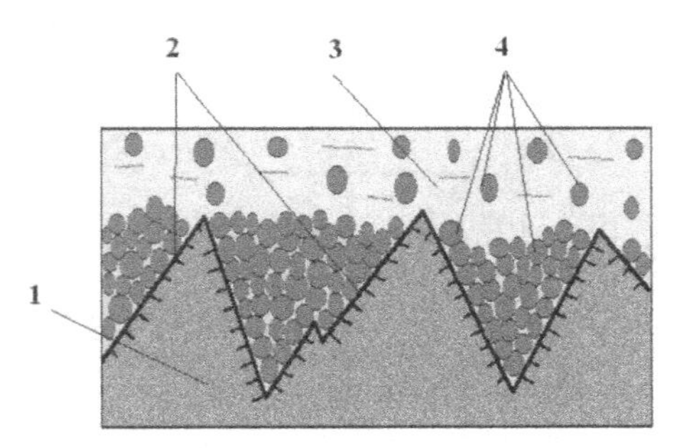

FIGURE 5 The decrease of the friction by filling of micro irregularities with nanoparticles: 1—"working" face of a component; 2—micro irregularities; 3—lubricant; 4—nanoparticles.

It should be noted that thermal propulsion systems can be the source of nanoelements. Solid propellant composition have been developed, the combustion of which leads to the formation of special controlled gas media (CGM) containing nanoparticles. Figure 6 demonstrates the use of a large-size gas-generator for the CGM formation in an industrial greenhouse. CGMs are complex media with specially selected concentrations of inorganic compounds in the form of nanoparticles for protecting living systems. They are used for the foliar nutrition of plants with macro- and microelements, the control of living system diseases, the protection of plants against frosts, the seed treatment, etc. The method comprises the creation of a controlled gas medium containing nanoparticles which can easily penetrate into foliage and stems of plants providing their metabolic activity. The results of the application of the above technology show the absence of carcinogens, heavy metals, and other dangerous substances both in foliage and fruits.

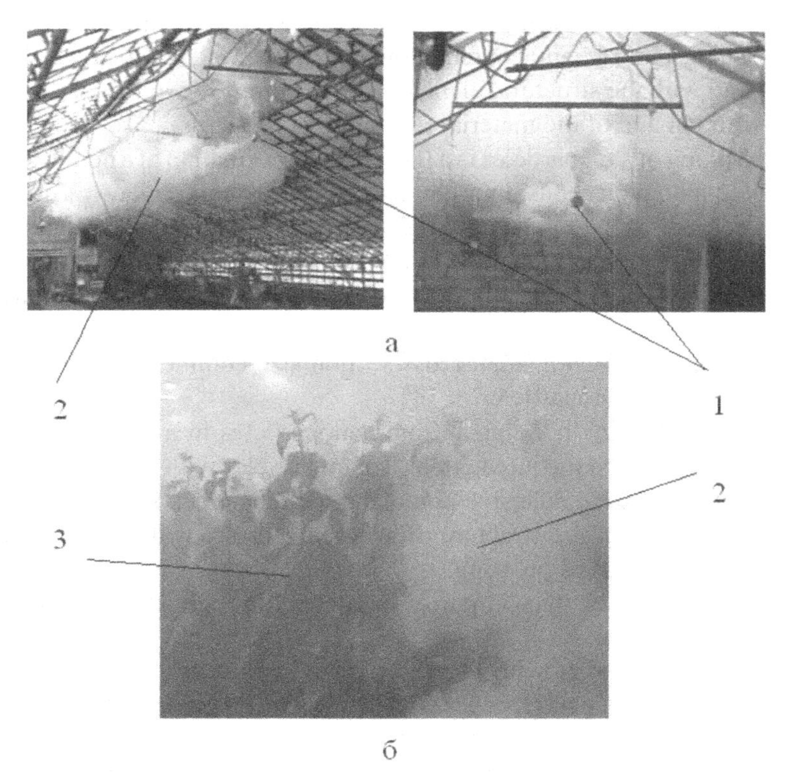

FIGURE 6 The diagram of the use of gas-generator (1): a) the general view of an industrial greenhouse; b) nano-aerosol gas medium (2) surrounding greenhouse plants (3).

Other examples can be given for the illustration of the constantly extending use of nanotechnologies for machine-building in the power industry for improving service properties of constructional nanoscale materials, increasing the service life of interacting surfaces of heat engines, improving the effectiveness of energy processes, etc.

It is necessary to underline that nanotechnological processes are complex and diverse, and the typical scale in which they take place is small and does not exceed 100 nanometers. It requires deep understanding of physicochemical processes taking place at the nanoscale; however, the nanometer scale of processes complicates their investigation by experimental methods and leads to the necessity of using mathematical modeling.

The above-mentioned examples give rise to a whole number of new problems, the solution of which is necessary for understanding the behavior characteristics of nanotechnological processes:

1. Nanoparticles mixing, and mixing of nanoparticles with macroparticles.
2. Static and dynamic interaction of nanoparticles with one another and with the structural elements of the composite in the processes of mixing and of the material formation.
3. Mixing of nanoparticles in a liquid and the formation of a homogeneous (or inhomogeneous, in accordance with the required law of nanoparticles distribution in a liquid) suspension "liquid – nanoparticles."
4. Stability of a nanoparticle in a liquid at the temperature of mixing.
5. The growth of a crystalline "grain" starting on the nanoparticle surface.
6. Movement of nanoparticles together with a high-viscous liquid in the channels with small cross-section size commensurable with the sizes of a nanoparticle.
7. Formation of "bondings" from nanoparticles in a crack, which are formed due to the interaction of nanoparticles with one another, with a polymerizing adhesive liquid and with the crack walls.
8. Formation of the structure of a nanodisperse powder-filtering layer in the process of pressing.
9. Movement of the impurity particles through the nanodisperse filtering layer.
10. Formation and growth of structures from nanoparticles in microcavities.
11. Interaction of nanoparticles with high-viscous liquids at high pressures.

From the above-listed problems, it is obvious that modeling in nanotechnologies requires the use of different physical models and mathematical methods.

This book is based on the lectures for the "Theory and practical application of nanotechnologies" course developed for the following specialties: Nanotechnology and Microsystems, Aircraft Building and Rocket Production; Rocket Motors; Reciprocating Motors and Combined Motors. These lectures the author is reading for the last 10 years in the Kalashnikov Izhevsk State Technical University.

The book deals with a systematic description of the theory of modeling the nanotechnology application in in various fields of science and technology. The problems of computing nanoscale systems at different structural scales are defined, and algorithms are given for their numerical solution by the quantum/continuum mechanics, molecular dynamics, and mesodynamics methods. Much consideration is given to the processes of the formation, movement, and interaction of nanoparticles, the formation of nanocomposites and the processes accompanying the application of nanocomposites.

The book concentrates on different types of nanosystems: solid, liquid, gaseous, and multi-phase consisting of various elements interacting with each other, with other elements of the nanosystem and with the environment. The book summarizes of research results of the author and his colleagues in the field of modeling of various nanosystems performed for 25 years. The results of these studies are presented in various articles and books.[7,145,154–172]

THE BOOK COMPRISES FIVE CHAPTERS

Chapter 1 contains the analysis of physical phenomena and mathematical models describing the processes of the formation of nanoscale systems from nanoparticles at different structural scales using the methods of quantum/continuum mechanics, molecular dynamics, and mesodynamics. The problems dealing with setting and solving the tasks of nanoelement formation, production of nanosystems from them, calculation of nanoelement and nanosystem properties, prognosticating the change in the structure and properties of nanoelements and nanosystems with time are formulated. The main calculation methods of nanoelements and nanosystems are considered.

In the chapter, the setting of the task of nanoelement structure and shape formation with quantum mechanics methods are considered. The main equations of the method are given and task boundary conditions are described. A detailed description of molecular dynamics method currently widely used for the computation of nanosystems is presented. The system of deferential equations of molecular dynamics method is given, the techniques for the calculation of interaction interatomic forces are described, and the boundary conditions for various computation tasks are demonstrated.

Also, this chapter is dedicated to the formulation of method, which is called in modern literature "mesodynamics" and targeted at the calculation of interaction, motion, and self-organization of nanoelements. The method equation system for symmetric and asymmetric interacting nanoelements for their "isotropic" and "anisotropic" interaction is given, the methods for calculating forces and moments defining the nanoelement interaction are described, the boundary conditions are considered, the method of successive enlargement of rated cell giving the possibility to transit from nano- to micro-scale during the calculation process is described.

The application of methods of continuous medium mechanics to the tasks of nanoparticle formation in "up down" processes, nanocomposite formation by pressing method, and calculation of stress fields and deformations in nanocomposites during their use are described. The systems of differential equations are given and boundary conditions for the indicated class of tasks are presented.

Chapter 2 presents numerical algorithms for the solution of the problems of nanotechnologies for different structural scales of modeling. The main algorithms of the numerical realization of the multiscale mathematical model of the formation of nanoparticles and the formation and loading of nanocomposites are given. The methods of numerical modeling of the tasks of quantum mechanics as applied to the calculation of atomic structure of nanoelements are presented. Numerical solution methods as applied to the problem indicated are classified. Hartree–Fock method and electron density functional method are described in detail.

This chapter is dedicated to the detailed description of various numerical schemes for integrating the equations of molecular dynamics method. One-step and multi-step difference schemes are considered; their accuracy, convergence, and computation time are evaluated. The methods of numerical integration of the equations of mesodynamics methods are presented. The acceleration of the integration of nanoelement motion equations is analyzed with the methods of mesodynamics as compared with the method of molecular dynamics. The algorithms of finite element method as applied to the tasks of powder dispersing, nanocomposite pressing, and calculation of the parameters of stressed-deformed state of nanocomposites during their operation are given. The convergence and stability of integration of the equations of finite element method with different class of problems are evaluated. The software currently applied for the calculation of nanosystems is reviewed and the structure of typical software for multi-level modeling of nanosystems with various methods is given.

In Chapter 3, the examples of the numerical simulation of nanoparticle formation are described. In first part of chapter, the examples of the calculation of nanoparticle formation with the method of condensation from the gaseous phase are given. The regularities of composite nanoparticle formation are considered, their structure and shape formation processes are analyzed, and the probabilistic regularities of nanoparticle parameter distribution are discussed. The influence of various parameters on the processes of nanoparticle formation in gaseous phase is demonstrated. The second part of chapter is dedicated to the investigation of nanoparticle formation kinetics in "up down" processes of powder dispersion under different types of their dynamic loading: fast high-pressure release, particle impact against an obstacle, and particle destruction in the impact wave. Physical processes of particle destruction are described in detail.

In Chapter 4, the problems of numerical simulation of nanosystem formation statistic and dynamic are considered. First, the interactions of nanoelements are described. Different types of nanoelement interaction are described: adhesion, fusion, and absorption. Stable and unstable forms of nanoparticle system equilibrium are investigated. The changes in the strength of powder nanomaterials depending upon the nanoparticle diameter are evaluated. The task for forming nanocomposites with different pressing schemes is presented. The complete pressing cycle is considered: loading and unloading of powder material. The evolution of technological stresses in the material being pressed and changes in the material parameters are discussed in detail. It is demonstrated that based on numerical calculations, the pressing and unloading modes providing the uniform density of the material by volume and minimal residual stresses can be selected. Simulation of electrodeposition of metal matrix nanocomposite coatings is presented. Simulation of formation of superficial nano-hetero structures: nanowhiskers and quantum dots are described in detail.

Chapter 5 is dedicated to the numerical calculation of nanosystem properties. It is demonstrated that the liquid structure considerably changes under different current conditions and changes with time. The dependence of water viscosity on nanochannel diameter is obtained. The dynamics of complex nanosystem consisting of a nanotube and fullerene and different types of gases as applied to the processes of gas accumulation, storage, and isolation in the given system is considered. The dynamics of fullerene motion is studied, thermodynamic parameters defining the processes indicated are determined, and limiting characteristics of the given system on gas absorption are evaluated. The task for calculating the elastic properties of metal nanoparticles is given. It is demonstrated that with the diameter decrease, the elasticity modulus of metal nanoparticles increases nonlinearly.

In Epilogue, further perspectives for applying the modeling results to develop nanotechnologies are considered.

ACKNOWLEDGMENTS

The author expresses the sincerest gratitude to Academician A. M. Lipanov for fruitful cooperation, counseling, constant support, and attention to the work.

I would also like to express gratitude for the pleasure to work with Prof. V. I. Kodolov, Prof. A. V. Aliev, Prof. V. N. Alikin, Prof. P. V. Trusov, Cand. Tech. Sc. V. B. Golubchikov.

The author is grateful to his colleagues Cand. Phys.-Math. Sc. A. A. Vakhrushev, Cand. Tech. Sc. M. V. Suetin, Cand. Phys.-Math. Sc. A. Yu. Fedotov, Cand. Tech. Sc. A. A. Shushkov, Cand. Phys.- Math. Sc. A. V. Severyukhin, Cand. Phys.-Math. Sc. O. Yu. Severyukhina, Cand. Tech. Sc. A. V. Suvorov, Cand. Tech. Sc. L.L. Vakhrusheva, Cand. Tech. Sc. Ig. A. Shestakov, graduate students E. K. Molchanov, Givotkov A. V., and A. A. Andreev for active participation in the development of the software complex, in conducting numerous calculations and analyzing their results.

The work was supported by the grants of the Russian Foundation for Basic Research, various research programs of the Ural Branch of the Russian Academy of Sciences RAS, and programs of the Russian Academy of Sciences Presidium, by the Kalashnikov Izhevsk State Technical University.

The calculations were performed at the Joint Supercomputer Center of the Russian Academy of Sciences.

CHAPTER 1

PHYSICAL AND MATHEMATICAL MODELS FOR NANOSYSTEMS SIMULATION

CONTENTS

ABSTRACT

This chapter contains the analysis of physical phenomena and mathematical models describing the processes of the formation of nanoscale systems from nanoparticles at different structural scales using the methods of quantum/continuum mechanics, molecular dynamics, and mesodynamics. The problems dealing with setting and solving the tasks of nanoelement formation, production of nanosystems from them, calculation of nanoelement and nanosystem properties, prognosticating the change in the structure and properties of nanoelements and nanosystems with time are formulated. The systems of differential equations are given and boundary conditions for the indicated class of tasks are presented.

1.1 THE PROBLEM STATEMENT

The specific feature of the physical processes in nanoscale systems is that the key phenomena determining the behavior of a nanoscale system in real time at the macroscale take place at small space and time scales.[32–34,45,97] The properties of a nanoscale system depend not only on the properties of its constituent elements but also on the regularities of the spatial arrangement of the nanoscale system and the parameters of the nanoelements interaction.[4,5,15,25,35,41,44,51,52,66,67,94,134] In this connection, the use of the continuum mechanics methods, which have been successfully applied for the mathematical modeling of macromechanics problems, is limited at the nanometer scale. While classical physical body mechanics is based on the hypothesis of continuum according to which all the properties of a body are determined by the mechanics of continuum, this hypothesis cannot be applied to nanoscale systems. In "Nanotechnology," the displacement of separate atoms, molecules, and their formations are considered rather than the movement of continuum described with macroparameters (density, viscosity, mechanical characteristics, etc.). In doing so it is necessary to observe the influence of the properties and parameters of the nanoelements on the system behavior at the macroscale.

The main difficulty in the mathematical description of nanoelements is that because of their sizes they occupy an intermediate position between macroobjects, the physical processes of which are described by statistical physics and thermodynamics (consequently, by classical mechanics), and microobjects, the behavior of which is only described by quantum mechanics. By what sign are macro- and microobjects distinguished? It is thought that the size of a macroobject is much larger than some characteristic correlation length of the order of the distance between atoms (usually several angstroms). However, in some cases, especially in the region of phase transitions, this magnitude can grow up to tens of nanometers. As one goes from micro- to macroobjects, the characteristics of the system significantly change and not necessarily in a regular manner. It is an intermediate state sweeping the range to 100 nm (such size is called mesoscopic; the name was given by van Kampen in 1981[77]) is described by mesoscopic physics,[73] which includes classical and quantum mechanics at the boundaries of their application.

Thus, for the modeling of nanoelements and nanoscale systems, methods are needed based both on classical mechanics and quantum mechanics depending on purposes and desired precision of modeling.

Let us consider in detail the main physical and mathematical models for nanoscale systems formed from nanoparticles.

The atomic structure and form of a nanoparticle depend on both its size and the methods of its formation, which can be divided into two main groups:

1. The formation of a nanoparticle in the process of the consolidation of atoms by their "assembling" and stopping (fixation) the process when a desired size of a nanoparticle is reached ("bottom up" processes). Stopping the process of the nanoparticle growth is realized by changing physical and chemical conditions of the nanoparticle creation or by limiting the space in which the nanoparticle is being formed.
2. The fabrication of nanoparticles by dispersion: grinding, crushing, and destruction of more massive (coarse) formations to fragments of the desired size (the above methods are classified as "up down" processes).

Let us consider "bottom up" processes and formulations of the modeling techniques using the problem of the nanoparticles formation and movement in a gaseous phase as an example. The main steps of the nanoparticles formation and movement processes are as follows (Fig. 1.1):

1. Burning of a highly condensed system and the movement of the gaseous mixture from the source of its generation over the space in which there is a working object.
2. The origination of molecular formations containing the above-indicated elements.

3. The integration of atoms and molecules into nanoparticles at cooling of the gaseous mixture down to normal temperature.
4. The nanoparticles movement in the gaseous mixture.
5. Deposition of the nanoparticles on the working object.
6. The nanoparticles penetration into the working object inner volume.
7. The nanoparticles' movement inside the working object through its microchannels and pores.
8. The decomposition of the nanoparticles into constituent molecules and atoms inside the working object.

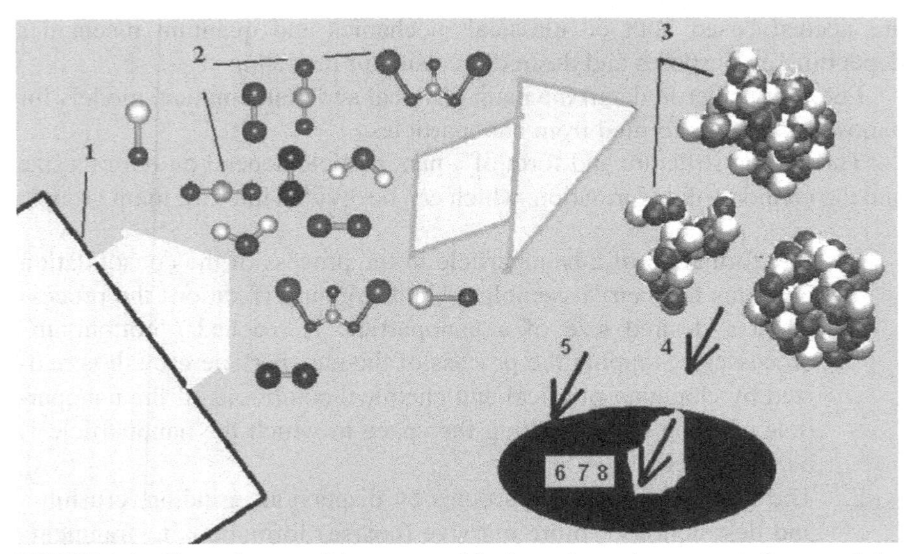

FIGURE 1.1 The main steps of the process of the formation and movement of nanoparticles during "bottom up" processes.

At fabricating materials from nanoparticles, the last four steps are replaced by the processes of nanoparticles compacting or integrating.

Let us consider "up down" processes using the dispersion of powder materials (PMs) to nanoparticles as an example, and point out the main structural scales in the modeling of physico-mechanical processes in nanoparticles (Fig. 1.2) and corresponding methods of calculation.

The first modeling scale is the macrostructural scale (Fig. 1.2a), which corresponds to the particles sizes more than 1 μm and the sizes of structural elements of a material (grain, crystal, etc.) by one or two orders of magnitudes smaller than the particle size. In this case despite the anisotropy of separate structural object and the difference of strength properties of structural elements and intergranular layers of the material, the description of the particle deformation, in general, can be performed on the basis of phenomenological equations of the isotropic medium owing to the statistical distribution of the orientation of the crystals anisotropy axes.

When the particle sizes are reduced to 100 nm, which is comparable with the characteristic size of a structural element (mesoscale), the influence of the inner structure of the material, that is, the grain shape, the direction of the anisotropy axis, the thickness and strength properties of the intermediate layers, and so forth, becomes significant (Fig. 1.2b). However, even at this scale, the modeling of the processes of the particle dispersion can be performed within the framework of the phenomenological equations of continuum mechanics since the sizes of the cells of a crystal lattice are much smaller than the grain sizes.

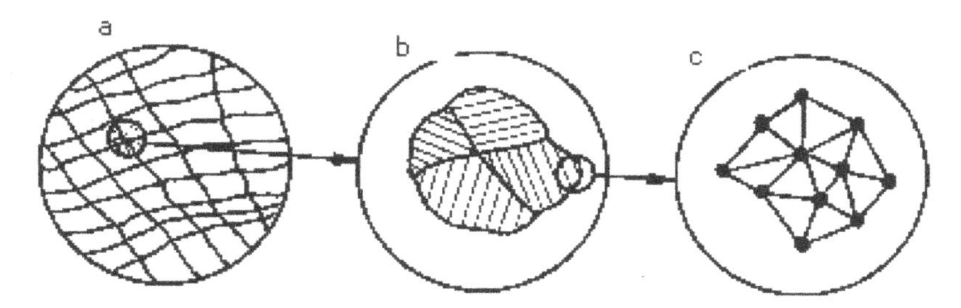

FIGURE 1.2 Structural scales of the modeling of the formation of nanoparticles at dispersion: (a) macro-, (b) meso-, and (c) microstructural scales.

The physical properties of a grain and corresponding equations of the material state are based on the transition from the crystal lattice theory to the continuum theory. At further decrease in the particle sizes to the magnitudes smaller than 100 nm, the material crystal lattice structure (Fig. 1.2c) gains in importance. The particle structure, physical properties, shape, sizes, and the stability of its existence in time depend directly on the forces of the atomic interaction. In this case, the description of the dispersion process by the continuum mechanics techniques does not reflect the essence of the physical phenomena taking place at the microstructural scale considered. Here, the mathematical model of the particles formation process should be based on the atomic interaction theory.

It should be noted that the division into the above-indicated scales is conditional to a certain extent and can be supplemented if necessary. In addition, the pointed-out scales are connected with one another so that calculation results at one scale can be the initial conditions for the other. Sometimes it is necessary to perform modeling at different scales simultaneously (e.g., when a particle with a nanometer size is being separated from a particle with a "large" size).

These problems are the subject of intense research.

Normally, the modeling of these tasks is performed using the molecular dynamics technique. However, the solution of the task within such framework requires much time and is computationally intensive. For example, because of the small mass of the interacting atoms it is necessary to select a time step of the order of 10^{-15} s to provide the stability of the model, which leads to the large time of calculation at the time integration of molecular dynamics equations. In addition, the collective behavior of atoms, molecules, and nanoparticles is observed. It gives rise to a multiscale character of the problems of modeling, which requires different physical and mathematical approaches.[22,29,65,70,81,95,100,103–106,121,124,129,135,137,147,152,178,179,180]

However, in general, the problem is still not solved, because of its difficulty and complexity, which makes the high relevance of research in the field of statics and dynamics of nanoparticles, since the structure and properties of nanoparticles determine all the necessary parameters of the material.

Changing the original shape and size of nanoparticles in these processes can lead to a radical restructuring of the structure and the substantial change in the properties of materials and mediums containing nanoparticles.

To summarize, let us note that both the "up down" and the "bottom up" processes of the formation and motion of nanoparticles and the formation of materials based on nanoparticles require successive and coordinated use of the above methods of modeling (from quantum mechanics to continuum mechanics) at different structural scales.

The main problems of such modeling are as follows:

- Multiscale nature and connectedness of problems.
- Large number of variables.
- Variation of scales both over space and in time.
- Characteristic times of processes at different scales differ by orders of magnitude.
- Variation of the problem variables at different scales of modeling.
- Matching of boundary conditions at the transition from one modeling scale to another when the problem variables are changed.
- Stochastic behavior of nanoscale systems.

Figure 1.3 displays the main methods of modeling and indicates their state-of-the-art capabilities regarding time, space, and the number of calculated atoms.

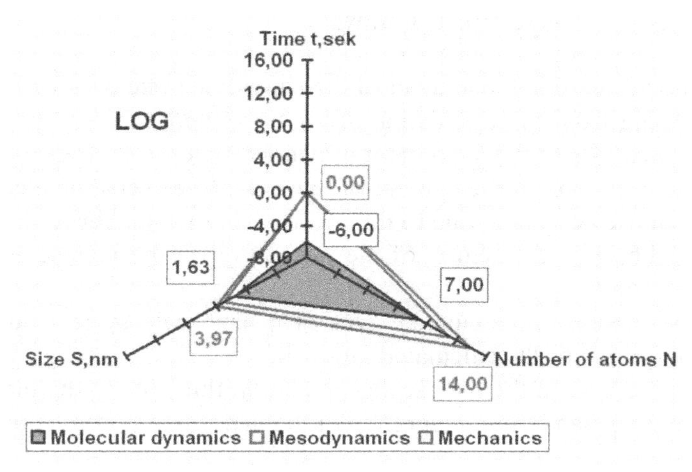

FIGURE 1.3 Methods for the solution of the nanoparticles formation problem.

The calculation of the configurations of a molecular formation, which are the constituents of a nanoparticle, is based on the quantum mechanics ("ab initio") methods of modeling.[12,13,16,27,43,48,50,54,87,90,99,101,105,106,122,128,132,138,140,185] These methods give the most complete and precise presentation of nanoobjects and take into account quantum effects, but they are very intensive computationally. At present the use of the quantum mechanics methods for the calculation of nanoscale systems is limited to 1000–2000 atoms composing a nanoscale system.

The modeling of the coalescence of molecules into nanoparticles can be performed by the molecular dynamics method.[2,6,20,26,68,71,96,110,118,119] The method allows us to consider systems containing up to 10 million atoms and more than that, but it does not take into account quantum phenomena.

The calculation of the movement of nanoparticles and their coalescence is the task for mesodynamics.[65,155,159,160] The characteristic feature of mesodynamics is the simultaneous use of the methods of molecular and classical dynamics.

It also should be noted that a number of phenomena, in particular the phenomena taking place at the final stages of the nanoscale system formation, can be considered within the framework of continuum mechanics.

Each of the above methods has its own advantages and limitations. The use of any of the above methods of modeling or their combination for specific nanotechnology problems depends on the calculation accuracy required. The main reasons and purposes for the transition from one method to the other at different modeling stages are as follows:

1. The decrease in the number of the nanoscale system variables due to passing from the calculation of the movement of separate atoms to the analysis of the movement of nanoparticles or their groups.
2. The reduction of the number of bonds between separate elements of a nanoscale system due to the decrease in the number of the nearest neighbors in the region of the interaction of the nanoscale system unit cells.
3. The increasing calculated sizes of a nanoscale system due to the coarsening of the calculated unit cell.
4. An increase in the nanoscale calculation time due to the increase in the time step of the integration of the nanoscale system equations.

The next sections of this chapter present a detailed consideration of the above methods of modeling and give complex analysis of the "history" of the nanoscale materials existence at different structural scales.

Main attention is given to nanoscale systems formed from nanoparticles. The following interconnected problems are considered:

- construction of the atomic structure and shape of a nanoparticle;
- movement of nanoparticles;
- interaction of nanoparticles;
- organization of nanoparticles in nano-, micro-, and macrostructures.

1.2 QUANTUM-CHEMICAL METHODS OF MODELING OF NANOSYSTEMS

Quantum-chemical methods of modeling involve mathematical tools of quantum mechanics and they are based on the solution of Schrödinger equation.[99] At the utilization of these methods, the full electronic and atomic structure of objects (atoms, molecules, and ions) is considered and the detailed configuration of all electron clouds is taken into account. In this case, the information about the behavior of the system comprising N particles (nuclei of atoms and electrons) in the x_1, x_2, x_3 coordinate system is determined by the wave function ψ, which depends on $3N$ coordinates of all the particles of the atomic system and the projections of their spins s on the x_3 axis and time t

$$\Psi = \Psi\left(x_{11}, x_{21}, x_{31}, s_{x_{31}}, x_{12}, x_{22}, x_{32}, s_{x_{32}}; \ldots;, x_{1N}, x_{2N}, x_{3N}, s_{x_{3N}}, t\right) \tag{1.1}$$

The variation of the wave function ψ in space and time determines the wave Schrödinger equation

$$\mathbf{i}h\frac{\partial \Psi}{\partial t} = \hat{\mathbf{H}}\Psi \tag{1.2}$$

where h is Planck constant; $\mathbf{i} = \sqrt{-1}$

$$\begin{aligned} \hat{\mathbf{H}} = \sum_{k=1}^{N}\left\{-\frac{h^2}{2m_k}\nabla_k^2 + U_k\left(x_{1k}, x_{2k}, x_{3k}, s_{x_{3k}}, t\right)\right\} + \\ + \sum_{k=1}^{N}\sum_{j \neq k=1}^{N} U_{kj}\left(x_{1k}, x_{2k}, x_{3k}, s_{x_{3k}}\, x_{1j}, x_{2j}, x_{3j}, s_{x_{3j}}, t\right) \end{aligned} \tag{1.3}$$

$\hat{\mathbf{H}}$ is Hamilton operator (an analog of the classical Hamiltonian function) for the considered atomic system; $U_k\left(x_{1k}, x_{2k}, x_{3k}, s_{x_{3k}}, t\right)$ is the potential of the external field acting on the kth particle; $U_{kj}\left(x_{1k}, x_{2k}, x_{3k}, s_{x_{3k}}\, x_{1j}, x_{2j}, x_{3j}, s_{x_{3j}}, t\right)$ is the potential of the interaction between particles j and k.

Equation 1.3 is valid if the following two conditions are fulfilled:

- in the process of the nanoscale system evolution, existing elementary particles do not disappear and new elementary particles do not appear;
- the velocity of the elementary particles is small in comparison with the velocity of light.

For a nanoparticle comprising N^a atomic nuclei and N^{el} electrons, Hamilton operator (in the stationary case, it is possible not to take into account electron spins) has the form

$$\hat{\mathbf{H}} = -\hat{\mathbf{K}}_{N^a} - \hat{\mathbf{K}}_{N^{el}} + U_{N^a N^a} + U_{N^a N^{el}} + U_{N^{el} N^{el}} \tag{1.4}$$

where

$$\hat{\mathbf{K}}_{N^a} = -\frac{h^2}{2m_i}\sum_{i=1}^{N^a}\nabla_k^2 \tag{1.5}$$

is the operator of the kinetic energy of atomic nuclei,

$$\hat{\mathbf{K}}_{N^{el}} = -\frac{h^2}{2m_{el}}\sum_{i=1}^{N^{el}}\nabla_k^2 \tag{1.6}$$

is the operator of the kinetic energy of electrons,

$$U_{N^a N^a} = e^2(\sum_{l=1}^{N^a}\sum_{\substack{k=1\\k\neq l}}^{N^a}\frac{Z_k Z_l}{r_{kl}}) \tag{1.7}$$

is the potential energy of the interaction of atomic nuclei,

$$U_{N^a N^{el}} = e^2(\sum_{k=1}^{N^a}\sum_{i=1}^{N^{el}}\frac{Z_k}{r_{ik}}) \tag{1.8}$$

is the potential energy of the interaction of atomic nuclei and electrons,

$$U_{N^{el} N^{el}} = e^2(\sum_{i=1}^{N^{el}}\frac{1}{r_{ij}}) \tag{1.9}$$

is the potential energy of the interaction of electrons;
e is the electron charge, m_{el} is the electron mass, Z_k, Z_l are the numbers of protons in an atomic nucleus, r_{kl}, r_{ik}, r_{ij} are the distances between the nuclei of atoms, between the nuclei of atoms and electrons, and between electrons, respectively.

Taking into account (1.5)–(1.9), Hamiltonian operator will obtain the form

$$\begin{gathered}\hat{\mathbf{H}} = -\frac{h^2}{8\pi m_{el}}\sum_{i=1}^{N^{el}}\nabla_k^2 - \\ -\frac{h^2}{8\pi m_i}\sum_{i=1}^{N^a}\nabla_k^2 - e^2(\sum_{k=1}^{N^a}\sum_{i=1}^{N^{el}}\frac{Z_k}{r_{ik}} - \sum_{i=1}^{N^{el}}\frac{1}{r_{ij}} - \sum_{l=1}^{N^a}\sum_{\substack{k=1\\k\neq l}}^{N^a}\frac{Z_k Z_l}{r_{kl}})\end{gathered} \tag{1.10}$$

In the general case, Schrödinger equation does not have an analytical solution and normally is solved with the use of numerical methods. In this case, the main peculiarity of Schrödinger equation is that the wave function exists only for the whole system. A separate particle (atomic nucleus or electron) cannot be in the state which can be described by the wave function for separate particle, and the general wave function cannot be presented as the product of the wave functions of separate particles. Therefore, for the direct solution of Schrödinger equation a computing machine with great computational power is necessary. For the considered nanoscale system consisting of N^a atomic nuclei and N^{el} electrons, Schrödinger function should be determined in the configuration space containing $3N^aN^{el}$ dimensions. When the number of the points of the integration over each dimension is 10^n, summing up should be carried out by 10^f elements of the configuration space volume where $f = 3nN^aN^{el}$. It is obvious that it is a great number even for a small object. For example, for a nanoparticle containing 100 atoms and 100 electrons, for 100 points of the integration over each coordinate the number of the elements of the configuration space volume will be 10^{60000}. It explains why at present the main efforts of scientists are directed to the development of the approximate methods of calculation, which will reduce the number of dimensions of a problem.

1.3 MOLECULAR DYNAMICS METHODS FOR MODELING OF NANOSYSTEMS

1.3.1 SYSTEM OF EQUATIONS

The molecular dynamics method had been developed by T.L. Hill, Dostovsky, Hughes, Ingold, Westheimer, Mayer, B. Alter, Alder Winniyard, and other scientists in 1946–1960, and it was intensively developed by many scientists concerning various problems of the modeling of processes in condensed, liquid and gaseous media at the atomic scale. The method is based on the Born–Oppenheimer concept of force surface which is a multi-dimensional space describing the system energy as the place function of the atomic nuclei forming the system. Thus, the molecular dynamics method can be used for the calculation of the motion of the atomic nuclei of a nanoscale system and not for the consideration of the electron motion.

The atomic nuclei motion is determined from Hamilton equations

$$\frac{d\vec{\mathbf{x}}_i}{dt} = \frac{\partial \mathbf{H}}{\partial \vec{\mathbf{p}}_i}, \tag{1.11}$$

$$\frac{d\vec{\mathbf{p}}_i}{dt} = -\frac{\partial \mathbf{H}}{\partial \vec{\mathbf{x}}_i}, \tag{1.12}$$

where

$$\mathbf{H} = \sum_{k=1}^{N} \left\{ \frac{p_k^2}{2m_k} + U_k\left(x_{1k}, x_{2k}, x_{3k}, t\right) \right\} + \\ + \sum_{k=1}^{N} \sum_{j \neq k=1}^{N} U_{kj}\left(x_{1k}, x_{2k}, x_{3k}, x_{1j}, _{2j}, x_{3j}, t\right) - \sum_{k=1}^{N} \alpha_k \vec{\mathbf{p}}_k \vec{\mathbf{x}}_k \tag{1.13}$$

is Hamiltonian function; $\overline{\mathbf{x}}_i$ is the vector of the coordinates (x_{1i}, x_{2i}, x_{3i}); $\overline{\mathbf{p}}_i$ is the momentum vector ($m_i \frac{dx_{1i}}{dt}, m_i \frac{dx_{2i}}{dt}, m_i \frac{dx_{3i}}{dt}$).

Substituting (1.13) in eqs 1.11 and 1.12 we obtain the equations of the motion of the nanosystem atoms in the form

$$m_i \cdot \frac{d\vec{\mathbf{V}}_i}{dt} = \sum_{j=1}^{N^a} \vec{\mathbf{F}}_{ij} + \vec{\mathbf{F}}_i(t) - \alpha_i m_i \vec{\mathbf{V}}_i, \qquad i = 1,2,..,N^a, \\ \frac{d\vec{\mathbf{x}}_i}{dt} = \vec{\mathbf{V}}_i, \tag{1.14}$$

where m_i is the *i*th atom mass; $\vec{\mathbf{F}}_{ij}$ are the forces of the interatomic interaction; α_i is the coefficient of "friction" in the atomic system; $\vec{\mathbf{F}}_i(t)$ are external forces; *t* is the time.

A stochastic force $\vec{\mathbf{F}}_i^g(t)$ is used as $\vec{\mathbf{F}}_i(t)$ in eq 1.14, which acts on the *i*th atoms and is given by Gauss distribution δ-correlated over the time by Gauss stochastic process having the following properties:

- the mean value of the stochastic force is 0

$$< \vec{\mathbf{F}}_i^g(t) >= 0 \tag{1.15}$$

- does not correlate with the velocity $\frac{d\vec{\mathbf{x}}_i}{dt}$ of the atom considered, therefore

$$< \vec{\mathbf{F}}_i^g(t) \frac{d\vec{\mathbf{x}}_i}{dt} >= 0 \tag{1.16}$$

$$< \overline{\mathbf{F}}_i^g(t) \overline{\mathbf{F}}_i^g(0) >= 2k_B T_0 \alpha_i m_i \delta(t) \tag{1.17}$$

where k_B is Boltzmann constant; δ(t) is Dirac function; T_0 is the temperature.

The above is the description of the interaction of the atomic-molecular system with the external medium (a heat reservoir), which consists of two parts, namely, the systematic force of friction

$$\sum_{j=1}^{N^a} \vec{\mathbf{F}}_i^f(t) = -\alpha_i m_i \vec{\mathbf{V}}_i \tag{1.18}$$

and the stochastic force $\vec{\mathbf{F}}_i^g(t)$ (noise).

In this case, the motion equations are called Langevin equations and the method of the calculation of the molecular dynamics in accordance with the above equations is called the Langevin dynamics method.

1.3.2 BOUNDARY AND INITIAL CONDITIONS

The formation of nanoparticles in the "bottom up" processes usually starts in the medium consisting of the atoms of different materials and molecules at the initial moment. When the medium is cooled due to condensation, additional nanoparticles appear in it. This process takes place in the macro-volume containing a great number of atoms and molecules. The modeling of such system by the molecular dynamics methods is impossible due to a great number of variables; therefore, it is conducted in a calculation cell occupying nano or microvolume (Fig. 1.4) with special boundary conditions on the surface. In this case, the molecular dynamics tools allow us to obtain the system general characteristics and to trace in detail the trajectories of each atom and each nanoparticle.

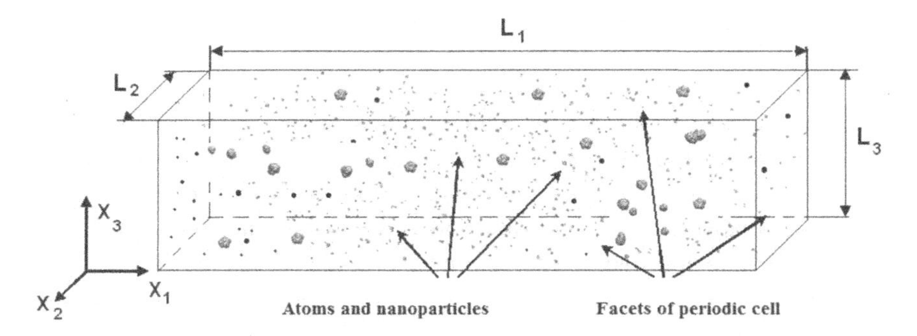

FIGURE 1.4 The calculation region for the problem of the formation of nanoparticles.

The boundary conditions got the name the Born–Karman boundary conditions. The essence of the indicated boundary conditions is explained by Figure 1.5.

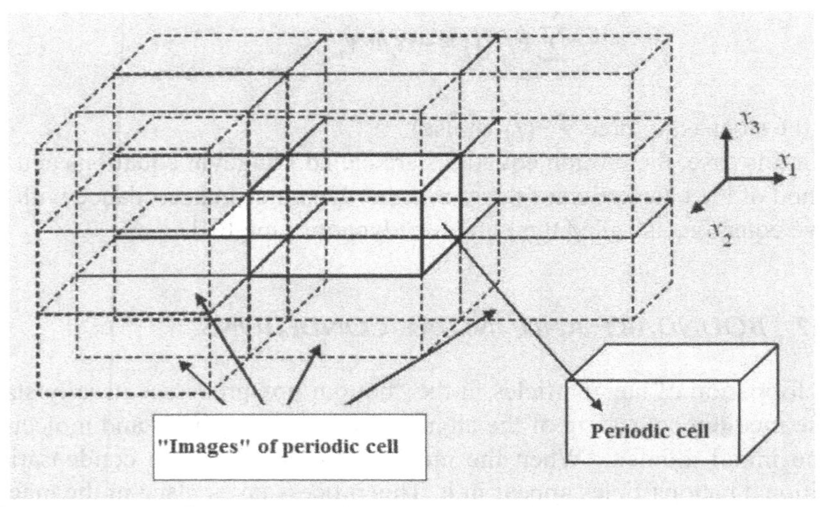

FIGURE 1.5 The cell with the periodic boundary conditions.

The space modeled is divided into 27 similar cells. The central cell is the calculation cell (it is displayed in Fig. 3.2), and the rest of the cells are its images. Modeling is conducted only in the central cell. The calculation cell and the rest of the cells-images contain the same set of atoms, molecules, and nanoparticles. The general number of the calculation cell images is 26.

The molecular dynamics equations are written only for the calculation cell. In the calculation cell images the trajectories of the atoms motion are specified by the similar trajectories of their motion in the calculation cell according to eq 1.14.

When an atom, molecule, or nanoparticle crosses any boundary within the inner space of the calculation cell, a similar atom, molecule, or nanoparticle with the same properties and velocity appears in the calculation cell from the outside from the opposite boundary. When an atom, a molecule, or a nanoparticle approaches the cell inner boundary, it is influenced by the atoms, molecules, and nanoparticles of the images. As a result, the entire calculation space is a set of equal calculation volumes, and the movement of microobjects in them is assumed to be identical, and the influence of the calculation cell faces is eliminated.

The hypothesis on the periodicity of the region modeled allows us to reduce the size of the calculation volume, which means the reduction of the calculation itself.

Let us consider the mathematical formulation of the periodic Born–Karman boundary conditions.

The set of images of the *i*th atom with the coordinate $\vec{\mathbf{r}}_i$ is specified in the following form:

$$\vec{\mathbf{r}}_{i,im(k_1,k_2,k_3)} = \vec{\mathbf{r}}_i + L_j \cdot \vec{\mathbf{i}}_j,\ k_1,k_2,k_3 = 1,2,3 \qquad (1.19)$$

where L_j is the length of the calculation cell along the corresponding coordinate, k_1,k_2,k_3 are the numbers of the cells-images in all the directions, $\vec{\mathbf{r}}_{i,im(k_1,k_2,k_3)}$ is the radius-vector of the *i*th atom, $\vec{\mathbf{i}}_j$ is the displacement vector:

$$\vec{\mathbf{i}}_j = \begin{pmatrix} i_1 \\ i_2 \\ i_3 \end{pmatrix},\ i_1,i_2,i_3\text{=-1,0,1},\ i_1 \neq i_2 \neq i_3 = 0 \qquad (1.20)$$

Eventually, we create all the 26 images of the *i*th atom. When the atom reaches the calculation cell boundary, it enters the cell from the opposite side and has the following coordinates:

$$\vec{\mathbf{r}}_i = \vec{\mathbf{r}}_i^* + L_j \cdot \vec{\mathbf{l}}_j \qquad (1.21)$$

where

$$\vec{\mathbf{l}}_j = \begin{pmatrix} l_1 \\ l_2 \\ l_3 \end{pmatrix},l_1,l_2,l_3\text{=-1,0,1},\ l_1 \neq l_2 = \pm 1,\ l_2 \neq l_3 = \pm 1,\ l_1 \neq l_3 = \pm 1 \qquad (1.22)$$

$\vec{\mathbf{r}}_i^*$ is the radius-vector of the *i*th atom on the calculation cell boundary, $\vec{\mathbf{l}}_j$ is the transfers vector.

The atom retains the motion parameters:

$$\vec{\mathbf{V}}(\vec{\mathbf{r}}_i) = \vec{\mathbf{V}}(\vec{\mathbf{r}}_i^* + L_j \cdot \vec{\mathbf{l}}_j) \qquad (1.23)$$

The above mathematical formulation of the periodic boundary conditions eliminates the influence of faces and allows us to describe precisely the interactions taking place in the calculation region.

To start the calculation of the problem of the formation of nanoparticles in the calculation cell with the use of the molecular dynamics method, let us use the structure of molecules of the gaseous mixture discussed at the first step of modeling by the quantum mechanics method and let us specify the coordinates and velocities of the atoms of all the molecules at the moment $t = 0$:

$$\vec{\mathbf{x}}_i = \vec{\mathbf{x}}_{i0}, \frac{d\vec{\mathbf{x}}_i}{dt} = \vec{\mathbf{V}}_i = \vec{\mathbf{V}}_{i0}, \ t = 0, \ \vec{\mathbf{x}}_i \subset \Omega \tag{1.24}$$

where Ω is the volume of the calculation cell.

The initial coordinates of atoms are chosen based on their uniform distribution in the gaseous mixture and their stochastic mixing within the calculation cell.

The moduli of the initial velocities of atoms are calculated in accordance with Maxwellian distribution[71] taking into account the gaseous mixture temperature T_o.

The connection between the initial temperature T_o and the initial velocities $\vec{\mathbf{V}}_{i0}$ is determined by the relation

$$T_0 = \frac{1}{3Nk_B} \sum_{i=1}^{N_a} m_i \left(\vec{\mathbf{V}}_{i0}\right)^2, \tag{1.25}$$

where N is the total number of the degrees of freedom of the gaseous mixture atoms, and N_a is the number of atoms in the gaseous mixture.

Maxwellian distribution for the velocity vector $\vec{\mathbf{V}}_0 = \left(\vec{\mathbf{V}}_{x_1 0}, \vec{\mathbf{V}}_{x_2 0}, \vec{\mathbf{V}}_{x_3 0}\right)$ is the product of distributions for each of the three directions of the coordinates:

$$f_V\left(\vec{\mathbf{V}}_{x_1 0}, \vec{\mathbf{V}}_{x_2 0}, \vec{\mathbf{V}}_{x_3 0}\right) = f_V\left(\vec{\mathbf{V}}_{x_1 0}\right) f_V\left(\vec{\mathbf{V}}_{x_2 0}\right) f_V\left(\vec{\mathbf{V}}_{x_3 0}\right) \tag{1.26}$$

where the distribution in one direction is determined by the normal law of distribution:

$$f_V\left(\vec{\mathbf{V}}_{x_1 0}\right) = \sqrt{\frac{m}{2\pi k_b T_0}} \exp\left(-\frac{m(\vec{\mathbf{V}}_{x_1 0})^2}{2k_b T_0} \right) \tag{1.27}$$

Upon integrating system of equation (1.14) with respect to the time with the use of the above initial conditions we obtain the main parameters at the moment t:

The mean kinetic energy of the system:

$$E(t) = \frac{\sum_{i=1}^{N} m_i \left(\vec{\mathbf{V}}_i(t)\right)^2}{2} \tag{1.28}$$

the instantaneous value of temperature

$$T(t) = \frac{1}{3Nk_B} \sum_{i=1}^{N} m_i \left(\vec{\mathbf{V}}_i(t)\right)^2 \tag{1.29}$$

The temperature is obtained by the averaging of the instantaneous values T (t) over a certain time interval

$$T(t) = \frac{1}{3Nk_B\tau} \int_{t_0}^{t_0+\tau} \sum_{i=1}^{n} m_i \left(\vec{\mathbf{V}}_i(t)\right)^2 dt \tag{1.30}$$

In real conditions, normally a molecular system exchanges its energy with the environment. For taking into account such energy interactions, special algorithms are used, namely, thermostats. The use of the thermostat allows us to calculate at the medium constant temperature or vice versa to change the medium temperature according to a certain law.

In the general case the thermostat temperature does not coincide with the temperature of the molecular system. At the fixed temperature of the thermostat, the molecular system temperature can vary due to various reasons. In the case of the established thermodynamic equilibrium, the thermostat temperature and the average temperature of the molecular system should coincide.

The simplest way to maintain the thermostat temperature constant is the scaling of velocities. The scaling is carried out according to the relation:

$$\vec{\mathbf{V}}_j^{new}(t) = \vec{\mathbf{V}}_j^{old}(t) \cdot \sqrt{3Nk_BT_0 \Big/ \left\langle \sum_{i=1}^{N} m_i \left(\vec{\mathbf{V}}_i^{old}(t)\right)^2 \right\rangle} \tag{1.31}$$

where T_o is the thermostat temperature. In the angle parenthesis, the averaging of the value of the total pulse over the time between the scaling of the velocities is shown.

There is another important point, namely, the influence of the initial temperature on the molecule velocity distribution. As shown above, in the molecular dynamics problem, the field of velocities at the initial moment is usually chosen according to Maxwellian distribution. This distribution has the form of normal distribution. As can be expected for gas at rest, the average velocity is zero in any direction.

It is also important to know the distribution of the velocities of molecules or atoms by the absolute values of velocities rather than by the projections. The velocity modulus V is determined as:

$$V = \sqrt{V_{x_1}{}^2 + V_{x_2}{}^2 + V_{x_3}{}^2} \tag{1.32}$$

Therefore the velocity modulus is always larger than zero or is zero. Since all the velocity components are distributed in accordance with the normal law, V^2 will have the xi-square distribution with three degrees of freedom. The probability density for the velocity modulus $f(V)$ has the form

$$f(V)dV = 4\pi V^2 \left(\frac{m_i}{2\pi k_B T}\right)^{3/2} \exp\left(\frac{-m_i V^2}{2k_B T}\right) dV \tag{1.33}$$

For calculating the pressure in the calculation cell of a nanoscale system, the following expression is used

$$P(t) = \frac{1}{3W}\left[\sum_{i=1}^{N} m_i \left(\vec{\mathbf{V}}_i(t)\right)^2 - \sum_{\substack{i,j \\ i<i}} (\vec{\mathbf{r}}_{\tilde{j}}(t) - \vec{\mathbf{r}}_i(t))\vec{\mathbf{F}}_{i\tilde{j}}(t)\right] \tag{1.34}$$

where W is the volume occupied by the nanoscale system.

The first summand in (1.33) depends on the energy of the motion of atoms or molecules, and the second summand is determined by the pairwise interaction of atoms. Along with the pair of atoms i and j, all the images of the jth atom are considered and the interaction of the ith atom with nearest of them $\tilde{j}$ is calculated. The function $\overline{\mathbf{F}}_{i\tilde{j}}(t)$ characterizes the value of the interaction between the atoms.

The use of the algorithms of barostats makes it possible to model the system behavior at a constant pressure. The simplest of them is Berendsen barostat, in which the pressure value is maintained constant by the scaling of the calculation cell. The position of the system particles at each time step is modified according to the scaling coefficient μ of Berendsen barostat:

$$\vec{\mathbf{r}}_i(t) \to \mu \vec{\mathbf{r}}_i(t), i = 1, 2, ..., N \tag{1.35}$$

The scaling coefficient is determined by the relation

$$\mu = \sqrt[3]{1 - \frac{\Delta t}{\tau_P}\left(P - P_0\right)} \tag{1.36}$$

where Δt is the integration step, τ_P is the time of the use of the barostat, P is the current pressure value, P_0 is the barostat pressure value.

The transformation of the position of the particles according to formula (1.34) leads to the change of the sizes and volume of the calculation cell, and, consequently, to the change in the pressure value.

1.3.3 THE ATOMIC INTERACTION POTENTIALS

The calculation of the forces of the interactions between atoms (molecules) occupies the main place in the molecular dynamics method.[71] As follows from (1.3), the forces of this interaction are potential and they are determined form the relation

$$\vec{\mathbf{F}}_i = -\sum_{i=1}^{N} \frac{\partial U(\vec{\mathbf{r}})}{\partial \vec{\mathbf{r}}_\mathbf{i}} \tag{1.37}$$

where $\vec{\mathbf{r}} = \left\{\vec{\mathbf{r}}_1, \vec{\mathbf{r}}_2, \ldots, \vec{\mathbf{r}}_N\right\}$ is the radius-vector of the ith atom; $U(\vec{\mathbf{r}})$ is the potential of the intramolecular interaction depending on the mutual arrangement of all the atoms.

The method is based on the use of the concept of Born–Oppenheimer force surface which is a multidimensional space describing the energy of a system as the function of the position of the atomic nuclei forming it.

In the general case, the potential $U(\vec{\mathbf{r}})$ is given as the sum of several components corresponding to different types of interaction:

$$U(\vec{\mathbf{r}}) = U_b + U_\theta + U_\varphi + U_{ej} + U_{LJ} + U_{es} + U_{hb} \tag{1.38}$$

where the summands correspond to the following types of interactions: U_b is the change of the bond length, U_θ is the change of the bond angle, $U\varphi$ are torsion angles, U_{ej} are flat groups, U_{LJ} are Van der Waals interactions, U_{es} are electrostatic interactions, U_{hb} are hydrogen bonds.

The above summands have different functional forms.

The potential of chemical bonds U_b defines the interaction between atoms depending on the distance between them (Fig. 1.6).

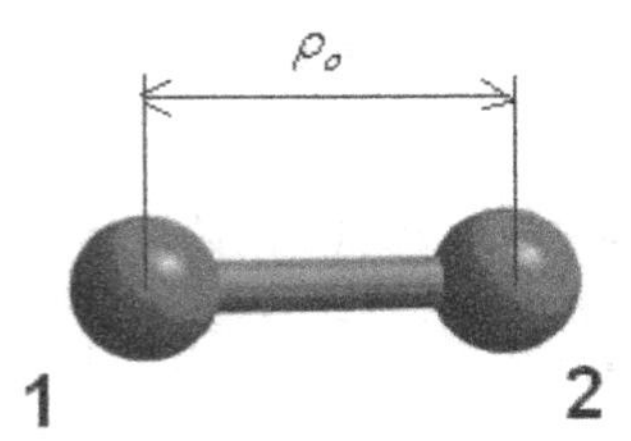

FIGURE 1.6 The change of the distance between atoms (1) and (2): ρ_o is the equilibrium length of bond.

This potential can be written in different forms. Let us consider some of them.

The harmonic potential is widely used, which is written as follows:

$$U_b = 1/2\sum_b D_b(\rho_{ij} - \rho_o)^2 \tag{1.39}$$

For different material Morse potential is used

$$U_b = D_b\,(\exp(-2\alpha_b(\rho_{ij} - \rho_0\,))\text{ -}2\exp(-\alpha_b(\rho_{ij} - \rho_0\,))) \tag{1.40}$$

Lennard-Jones potential is also used

$$U_b = 4\,D_b\left(\rho_{ij}^{-12} - \rho_{ij}^{-6}\right) \tag{1.41}$$

This potential is two-parameter, therefore it has limited possibilities for the variation of the macroscopic parameters of a material modeled with it. Lennard-Jones potential quite precisely describes Van der Waals forces playing an important role in solids. The unquestionable advantage of Lennard-Jones potential is its computational simplicity. It is widely used as a classical model potential.

Let us consider other potentials of chemical bond:

corrected harmonic potential

$$\begin{aligned} &U_b = 1/2D_b(\rho_{ij} - \rho_o)^2, \quad |\rho_{ij} - \rho_o| \le \rho_c; \\ &U_b = 1/2D_b\rho_c^{\,2} + D_{rh}\rho_c(|\rho_{ij} - \rho_o| - \rho_c); \quad |\rho_{ij} - \rho_o| > \rho_c; \end{aligned} \tag{1.42}$$

the fourth-order potential

$$U_b = \frac{D_b}{2}\left(\rho_{ij} - \rho_o\right)^2 + \frac{D_{b1}}{3}\left(\rho_{ij} - \rho_o\right)^3 + \frac{D_{b2}}{4}\left(\rho_{ij} - \rho_o\right)^4 \tag{1.43}$$

modified Buckingham potential

$$U(\vec{\rho}_{ij})_m = \frac{\varepsilon}{1-\frac{6}{\lambda}}\left(\frac{6}{\lambda}\exp(\lambda\,(1-\frac{|\vec{\rho}_{ij}|}{\rho_0})-(\frac{|\vec{\rho}_{ij}|}{\rho_0})^{-6}\right) \tag{1.44}$$

where summing-up is conducted over the number of the chemical bonds b of atoms in a nanoparticle; ρ_o is the equilibrium length of the bond; ρ_{ij} are current bond lengths; D_b,D_{b1},D_{b2} are the constants; λ is the constant characterizing the steepness of the exponential repulsion in modified Buckingham potential.

It should be noted that modified Buckingham potential is one of the main in the method of atom–atomic potentials which approximate the potential of the intermolecular interactions by the sum of the atom–atomic interactions.

The valence angles (Fig. 1.7) are specified by the potential

$$U_\theta = 1/2\sum_b D_\theta(\theta - \theta_o)^2 \tag{1.45}$$

where θ_o are the equilibrium angles; θ are the current values of the angles; D_θ is the constant.

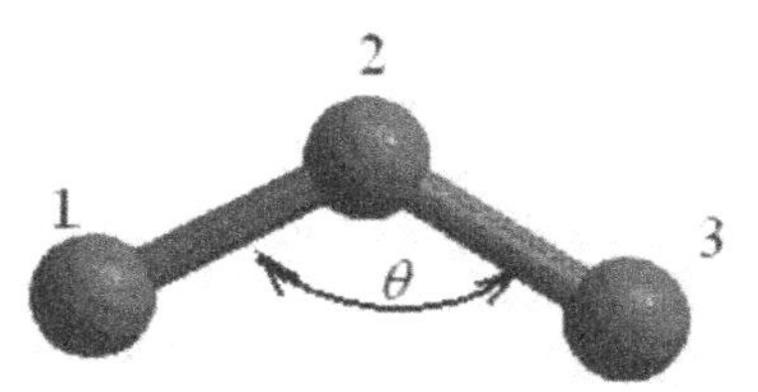

FIGURE 1.7 The change of the angle between atoms (1) and (3): θ is the angle of the bond.

The potential of the torsion interactions (Fig. 1.8) has the form

$$U_\varphi = \sum_{tr} D_\varphi(\cos(n\varphi - \varphi_o) + 1) \tag{1.46}$$

where the order of the torsion barrier n, the phase shift φ_o, the constant D_φ determine the heights of the potential barriers of the dihedral angles φ; tr is the number of torsion bonds.

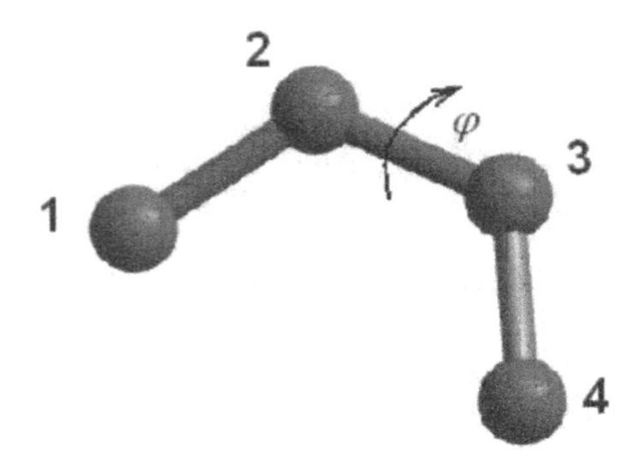

FIGURE 1.8 The change of the torsion angle between atoms (1) and (2) and (3) and (4): φ is the value of the dihedral angle.

The potential corresponding to the flat groups has the form similar to that in (1.36)

$$U_{eg} = \sum_{tp} D_{tp}(\cos(\chi\Theta - \Theta_o) + 1) \tag{1.47}$$

where χ, Θ_o, D_{tp} are the constants determining the heights of the potential barriers of flat angles Θ; tp is the number of flat groups.

Van der Waals interactions of atoms are written with the use of Lennard-Jones potentials:

$$U_{vv} = 4\varsigma\left(r_{ij}^{-12} - r_{ij}^{-6}\right) \tag{1.48}$$

where ς is the depth of the potential well; $r_{ij} = R_j / \rho_0$ is the reduced interatomic distance; $\mathbf{R}_{ij}$ is the vector connecting the centers of atoms i and j; ρ_0 is the interatomic distance corresponding to zero potential.

Electrostatic interactions are specified by the Coulomb potential

$$U_{es} = \sum_{N} {q_i q_j}\big/{er_{ij}} \tag{1.49}$$

where q_i, q_j are the partial charges of atoms; e is the dielectric permeability of the medium.

The functional form of the hydrogen bond potential is similar to that of the potential of Van der Waals interactions:

$$U_{hb} = 4\varepsilon_h\left(r_{ij}^{-12} - r_{ij}^{-6}\right) \tag{1.50}$$

Sometimes, when the interaction of carbon is calculated (e.g. in the force field Charmm22[16,101]), Urey-Bradley potential is additionally taken into account, which characterizes the interaction between two carbon atoms bound with the third situated between them (Fig. 1.9):

$$E_{ub} = \sum_{an} K_{ub}(s - s_0)^2 \tag{1.51}$$

where K_{ub} is the force constant of the distance between two atoms in relation to the central atom; s_0 is the equilibrium value of the distance; s is the current value of the distance.

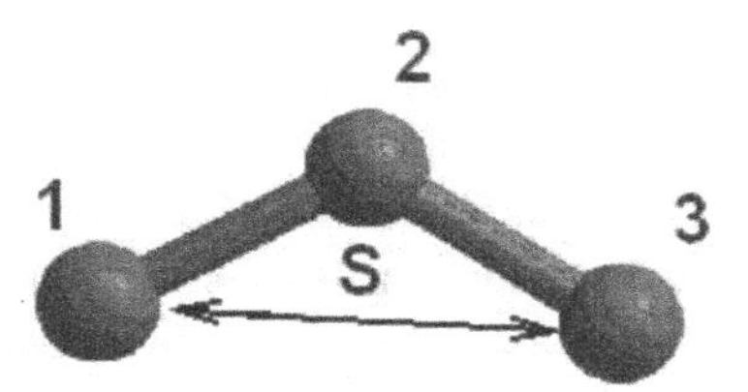

FIGURE 1.9 The distance s between atoms (1) and (3).

Equations 1.39–1.51 describe the main types of atomic interactions. It should be noted that there exist other forms for writing of the indicated potentials.

Owing to the simplicity of its mathematical notation, Lennard-Jones potential is widely in calculations and has a number of modifications: Mason-Stamp potential, Klein-Henley potential, and Kihara potential.

Like Lennard-Jones potential, Buckingham potential also has a number of modifications: Buckingham–Korner potential, Alrihs-Penko-Scholes potential, Kara-Konovalov potential, Barker-Pomp potential, and Smirnov potential.

Barker developed the potential for the atomic interactions of gases krypton and xenon, which gives a better accuracy compared to Lennard-Jones and Buckingham potentials

$$U(\vec{\rho}_{ij}) = \varepsilon\,(U_0(\vec{\rho}_{ij}) + U_1(\vec{\rho}_{ij}) + U_2(\vec{\rho}_{ij})) \tag{1.52}$$

where ε is the constant; $U_0(\vec{\rho}_{ij}), U_1(\vec{\rho}_{ij}), U_2(\vec{\rho}_{ij})$ are complex functions of the interatomic distance, which include additional 18 constants.

There are other potential used, the parameter values of which are determined experimentally (crystallographic, spectral, calorimetric, and other experiments) and on the basis of quantum-chemical calculations. For a more precise description of the properties of nanosystems are using many-particle potentials.[99]

1.4 MESODYNAMICS METHODS FOR MODELING OF NANOPARTICLE INTERACTION AND MOTION

The calculations of the nanosystem formation by the molecular dynamics method are effective at the initial stage of the nanoparticle appearance. However, at the atomic scale, the modeling of processes taking place in a gas medium using the molecular dynamics methods requires large computational resources and much time. In this connection, the developments of economical computational methods are very important. In this section, a method is offered, which is based on the mesodynamics theory. Note that mesodynamics[65,155] is the development of the particle method.

The essence of the method is as follows. When atoms and molecules are coalescing in nanoparticles, more and more of them reveal collective behavior. Atoms and molecules constituting a nanoparticle move together making only small fluctuations near their equilibrium position in the nanoparticle structure. It allows us to diminish the number of the objects modeled and to use a different modeling method—mesodynamics. Mesodynamics is based on the collective behavior of atoms and employs force parameters calculated by the molecular dynamics methods, and the motion of nanoparticles is studied by classical mechanics. Let us consider the main stages of the mesodynamics realization.

The first stage is the computation of the interaction of two nanoparticles. Let us give the statement of the problem for symmetrical nanoparticles and then for nanoparticles of arbitrary shape.

With this in mind, consider a nanoparticle at the moment t = 0 consisting of N^a atoms occupying a region Ω (Fig. 1.10). The position of each ith atom of the nanoelement is given by the coordinates x_{i1}, x_{i2}, x_{i3}. The atoms interact with one another. Figure 1.10 shows the interaction forces $\vec{\mathbf{F}}_{ik}$ (the interaction force between the atoms i and k) and $\vec{\mathbf{F}}_{ij}$ (the interaction force between the atoms i and j. The interaction force of the two atoms is directed along the line connecting their centers. In addition, each ith atom experiences the action of the external force $\vec{F}_i^b$. The direction and magnitude of this force are determined by the type of the nanoelement interaction with the environment. Under the action of the set of the above forces, the nanoparticle atoms displace.

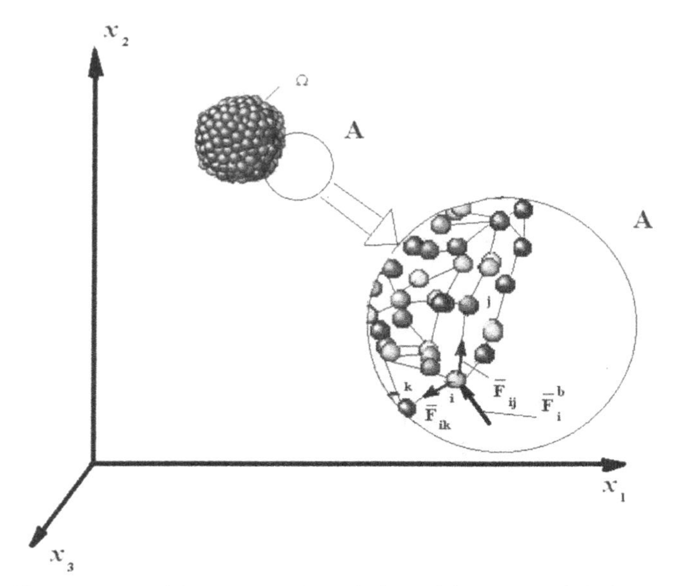

FIGURE 1.10 A nanoparticle: (A) an enlarged view of the nanoparticle fragment.

According to the molecular dynamics method, the motion of the atoms constituting the nanoparticle is described by the set of differential equation (1.19) into which the forces are added caused by the interaction of the atoms with the environment,

$$\vec{\mathbf{F}}_{bi} = -\sum_{1}^{N} \frac{\partial U_b(\vec{\rho}_{bi})}{\partial \vec{\rho}_{bi}} \vec{\mathbf{e}}_{\text{ib}} \tag{1.53}$$

where ρ_{bi} is the distance between the atom i and the atom b from the environment; $\bar{e}_{ib}$ is the unit vector directed from the atom i to the atom b; $U_b(\vec{\rho}_{bi})$ is the potential of the interaction of the nanoparticle atoms with the atoms of the environment.

1.4.1 INTERACTION OF NANOPARTICLES

Let us consider two symmetrical nanoparticles situated at the distance S from one another (Fig. 1.11).

In this case, eq 1.14 will have the form

$$m_i \cdot \frac{d^2\vec{\mathbf{x}}_i}{dt^2} = \sum_{j=1}^{N_1+N_2} \vec{\mathbf{F}}_{ij} + \vec{\mathbf{F}}_i(t) - \alpha_i m_i \frac{d\vec{\mathbf{x}}_i}{dt}, \qquad i = 1,2,..,(N_1 + N_2), \quad (1.54)$$

at the initial conditions,

$$\vec{\mathbf{x}}_i = \vec{\mathbf{x}}_{i0}, \vec{\mathbf{V}}_i = \vec{\mathbf{V}}_{i0}, \ t = 0, \ \vec{\mathbf{x}}_i \subset \Omega_1 \bigcup \Omega_2 \qquad (1.55)$$

where N_1,N_2 are the number of atoms in the first and second nanoparticles, respectively; Ω_1, Ω_2 are the regions occupied by the first and second nanoparticles, respectively.

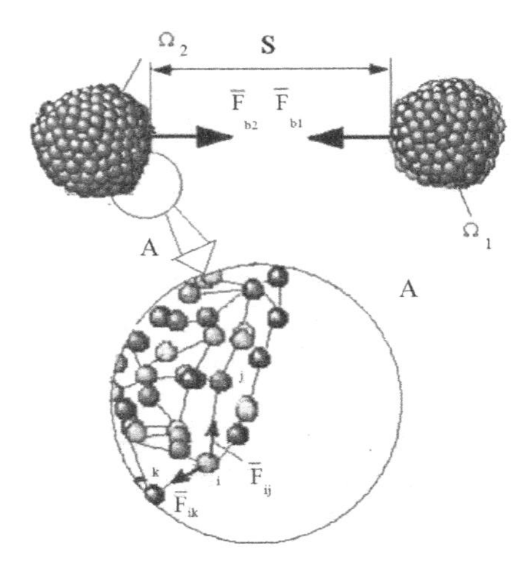

FIGURE 1.11 The scheme of the interaction of nanoparticles; A is an enlarged view of the nanoparticle fragment.

Solving eq 1.54 at the boundary conditions (1.55) allows us to calculate the trajectory of the motion of each nanoparticle, and consequently, nanoparticles in general. The total forces of the interaction between the nanoparticles can be calculated as

$$\vec{\mathbf{F}}_{b1} = -\vec{\mathbf{F}}_{b2} = \sum_{i=1}^{N_1} \sum_{j=1}^{N_2} \vec{\mathbf{F}}_{ij} \qquad (1.56)$$

where *i* and *j* are the atoms of the first and second nanoparticles, respectively.

In the general case the force of the nanoparticles interaction $\vec{\mathbf{F}}_{bi}$ can be written as the product of functions depending on the size of the nanoparticles and the distance between them:

$$\vec{\mathbf{M}}_{bi} = \Phi_{11}(S_{\tilde{n}}) \cdot \Phi_{12}(D) \tag{1.57}$$

The direction of the vector $\vec{\mathbf{M}}_{bi}$ is determined by the direction cosines of the vector connecting the centers of mass of the nanoparticles. Of course, the forces of the interaction between the nanoparticles change in time fluctuating near the mean value. Therefore, eq 1.57 gives the mean value of the nanoparticles interaction force.

Let us consider two interacting asymmetrical nanoparticles situated at the distance S between their centers of mass and oriented relative to one another under certain specified angles (Fig. 1.12).

In contrast to the previous problem, the interaction of the atoms constituting the nanoparticles leads not only to the relative displacement of the latter but their rotation of as well. Consequently, in the general case the set of all the forces of the interaction of the nanoelements atoms is reduced to the resultant vector of the forces $\vec{\mathbf{F}}_c$ and the resultant moment $\vec{\mathbf{M}}_c$:

$$\vec{\mathbf{F}}_c = \vec{\mathbf{F}}_{b1} = \vec{\mathbf{F}}_{b2} = \sum_{i=1}^{N_1} \sum_{j=1}^{N_2} \vec{\mathbf{F}}_{ij} \tag{1.58}$$

$$\vec{\mathbf{M}}_c = \vec{\mathbf{M}}_{b1} = -\vec{\mathbf{M}}_{b2} \tag{1.59}$$

where *i* and *j* are the atoms of the first and second nanoparticles, respectively.

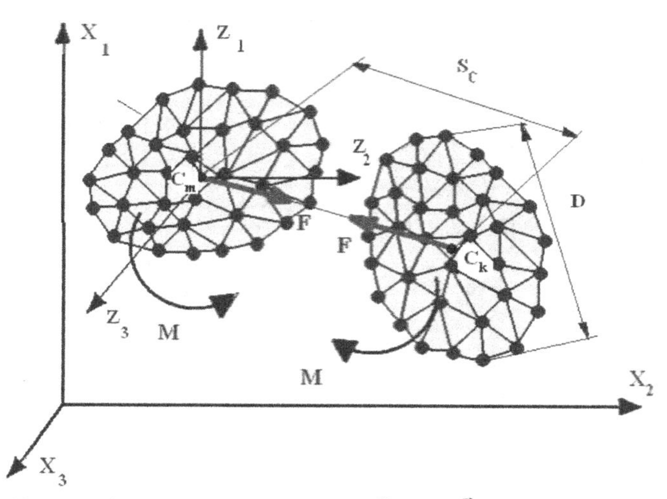

FIGURE 1.12 Two interacting nanoelements; $\vec{\mathbf{M}}_c$ and $\vec{\mathbf{F}}_c$ are the resultant moment and resultant vector of forces, respectively.

The main objective of this stage of computation is the construction of the dependences of the forces and moments of the interaction of the system elements on the distance S_c between the centers of mass of the nanostructure elements, the angles of mutual orientation of the nanoelements $\Theta_1,\Theta_2,\Theta_3$ (the shapes of the nanoelement) and the nanoelement characteristic size D.These dependences can be presented in the form

$$\vec{\mathbf{M}}_{bi} = \vec{\mathbf{\Phi}}_1(S\ ,\Theta_1,\Theta_2,\Theta_3,D) \tag{1.60}$$

$$\vec{\mathbf{M}}_{bi} = \vec{\mathbf{\Phi}}_2(S\ ,\Theta_1,\Theta_2,\Theta_3,D) \tag{1.61}$$

For spherical nanoelements the angles of mutual orientation do not influence the force of their interaction; therefore, in eq 1.61 the moment is identically zero.

The functions in (1.60) and (1.61) can be approximated similarly to those in (1.57) as the product of the functions S_0, Θ_1, Θ_2, Θ_3,D, respectively.

1.4.2 MOTION OF THE INTERACTING NANOPARTICLE SYSTEM

When the evolution of the system of the interacting nanoparticles is investigated, we consider the motion of each nanoparticle as that of the whole. The translational motion of the mass center of each nanoparticle is specified in the coordinate system X_1, X_2, X_3 and the rotation is described in the coordinate system Z_1, Z_2, Z_3 associated with the nanoparticle mass center (Fig. 1.13).

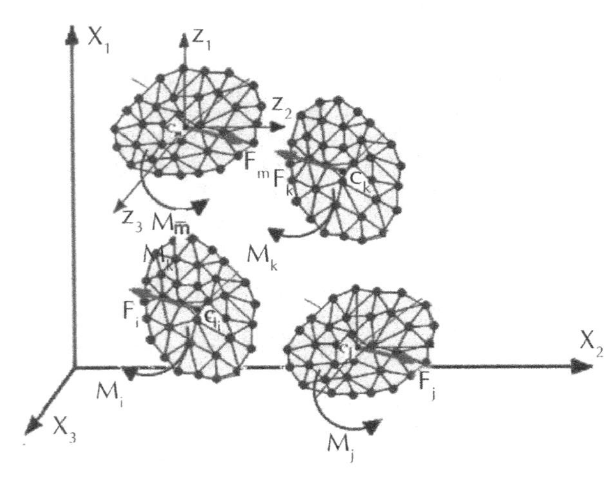

FIGURE 1.13 The system of the interacting nanoparticles.

It allows us to pass to other variables (from the coordinates and velocities of atoms to the coordinates and velocities of the nanoparticle mass centers) and to reduce their number. Compared to the molecular dynamics method, for a nanoscale system comprising N^{np} nanoparticles each of which contains N^{a} atoms the number of the mesodynamics variables decreases by k_x which is computed by the following formula:

$$k_x = \xi_x \frac{N^a N^{np}}{N^{np}} = \xi_x N^a \tag{1.62}$$

where $\xi_x = 1$ for the force interaction of nanoparticles; $\xi_x = 0.5$ when the forces and moments of the nanoparticles interaction are taken into account.

The next set of equations describing the motion of the interacting nanoparticles has the form

$$\begin{cases} M_k \dfrac{d^2 X_1^k}{dt^2} = \sum\limits_{j=1}^{N_e} F_{X_1}^{kj} + F_{X_1}^{ke}, \\ M_k \dfrac{d^2 X_2^k}{dt^2} = \sum\limits_{j=1}^{N_e} F_{X_2}^{kj} + F_{X_2}^{ke}, \\ M_k \dfrac{d^2 X_3^k}{dt^2} = \sum\limits_{j=1}^{N_e} F_{X_3}^{kj} + F_{X_3}^{ke}, \\ J_{Z_1}^k \dfrac{d^2\Theta_1^k}{dt^2} + \dfrac{d\Theta_2^k}{dt} \cdot \dfrac{d\Theta_3^k}{dt}(J_{Z_3}^k - J_{Z_2}^k) = \sum\limits_{j=1}^{N_e} M_{Z_1}^{kj} + M_{Z_1}^{ke}, \\ J_{Z_2}^k \dfrac{d^2\Theta_2^k}{dt^2} + \dfrac{d\Theta_1^k}{dt} \cdot \dfrac{d\Theta_3^k}{dt}(J_{Z_1}^k - J_{Z_3}^k) = \sum\limits_{j=1}^{N_e} M_{Z_2}^{kj} + M_{Z_2}^{ke}, \\ J_{Z_3}^k \dfrac{d^2\Theta_3^k}{dt^2} + \dfrac{d\Theta_2^k}{dt} \cdot \dfrac{d\Theta_1^k}{dt}(J_{Z_2}^k - J_{Z_1}^k) = \sum\limits_{j=1}^{N_e} M_{Z_3}^{kj} + M_{Z_3}^{ke}, \end{cases} \tag{1.63}$$

where X_i^k and Θ_i are the coordinates of the centers of mass and the angles of the orientation in the space of the principal axes Z_1, Z_2, Z_3 of nanoelements; $F_{X_1}^{kj}, F_{X_2}^{kj}, F_{X_3}^{kj}$ are the interaction forces of nanoelements calculated by formula (4.5) or (4.6); $F_{X_1}^{ke}, F_{X_2}^{ke}, F_{X_3}^{ke}$ are the external forces acting on nanoelements; N_k is the number of nanoelements; M_k is the nanoelement mass; $M_{Z_1}^{kj}, M_{Z_2}^{kj}, M_{Z_3}^{kj}$ are the moments of the interaction forces of nanoelements calculated by formulae (4.9); $M_{Z_1}^{ke}, M_{Z_2}^{ke}, M_{Z_3}^{ke}$ are external moments acting on nanoelements; $J_{Z_1}, J_{Z_2}, J_{Z_3}$ are the nanoelement inertia moments.

$$\vec{\mathbf{X}}^k = \vec{\mathbf{X}}_0^k,\ \Theta^k = \Theta_0^k, \vec{\mathbf{V}}^k = \vec{\mathbf{V}}_0^k, \frac{d\Theta^k}{dt} = \frac{d\Theta_0^k}{dt} \tag{1.64}$$

In the process of the motion of the system of nanoparticles the coalescence of a certain pair of nanoparticles may occur. In this case, the number of nanoparticles decreases by one and, consequently, the number of the equations in set (1.63) decreases by six. After that the calculation of the parameters of the new coalesced nanoparticle is carried out and the integration of the new reduced set of equations is conducted.

It should be noted that if the moments of the interaction of nanoparticles with one another and with the environment are zero, in equation set (1.63) only three first equations remain.

1.4.3 MOTION OF THE SYSTEM OF NO INTERACTING NANOPARTICLES

Equations of the motion of nanoparticles (1.63) take into account all the interactions of nanoparticles and their interaction with the environment. However, in a gaseous medium, in which nanoparticles are formed with a large distance between them and often with short-range interactions.

For this case equation set (1.63) can be written in the form

$$\begin{cases} M_k \dfrac{d^2 X_1^k}{dt^2} = F_{X_1}^{ke}, \\ M_k \dfrac{d^2 X_2^k}{dt^2} = F_{X_2}^{ke}, \\ M_k \dfrac{d^2 X_3^k}{dt^2} = F_{X_3}^{ke}, \\ J_{Z_1}^k \dfrac{d^2\Theta_1^k}{dt^2} + \dfrac{d\Theta_2^k}{dt} \cdot \dfrac{d\Theta_3^k}{dt} (J_{Z_3}^k - J_{Z_2}^k) = M_{Z_1}^{ke}, \\ J_{Z_2}^k \dfrac{d^2\Theta_2^k}{dt^2} + \dfrac{d\Theta_1^k}{dt} \cdot \dfrac{d\Theta_3^k}{dt} (J_{Z_1}^k - J_{Z_3}^k) = M_{Z_2}^{ke}, \\ J_{Z_3}^k \dfrac{d^2\Theta_3^k}{dt^2} + \dfrac{d\Theta_2^k}{dt} \cdot \dfrac{d\Theta_1^k}{dt} (J_{Z_2}^k - J_{Z_1}^k) = M_{Z_3}^{ke}. \end{cases} \tag{1.65}$$

If the moments and rotation of the nanoparticles is not taken into consideration, equation set (1.65) obtains the following form

$$\begin{cases} M_k \dfrac{d^2 X_1^k}{dt^2} = F_{X_1}^{ke}, \\ M_k \dfrac{d^2 X_2^k}{dt^2} = F_{X_2}^{ke}, \\ M_k \dfrac{d^2 X_3^k}{dt^2} = F_{X_3}^{ke} \end{cases} \tag{1.66}$$

The forces of the nanoparticles interaction with the gaseous medium can be presented as follows:

$$\overline{\mathbf{F}_{X_i}^{ke}}\ (t, \vec{\mathbf{r}}(t)) = -M_i g + \overline{\mathbf{f}}_i(t) - m_i b_i \frac{d\overline{X_i}(t)}{dt},\ i = 1,2,...,n \tag{1.67}$$

where $\overline{\mathbf{f}}_i(t)$ is the random force acting on the nanoparticle i from the gaseous medium; b_i is the coefficient of "friction" in the system nanoparticle—gaseous medium.

The random force $\overline{\mathbf{f}}_i(t)$ is similar to the random force in the Langevin dynamics. Note that the random force $\overline{\mathbf{f}}_i(t)$ reflects the action of the molecules of the gaseous medium on the nanoparticles moving in it, and it is found from the Gauss distribution law and has the following properties: the mean value of the random force $\overline{\mathbf{f}}_i(t)$ is zero, and the force $\overline{\mathbf{f}}_i(t)$ does not correlate with the velocity $\overline{\mathbf{V}}_i(t)$ of the nanoparticle considered, that is why

$$< \overline{\mathbf{f}}_i(t) \overline{\mathbf{V}}_i(t) >= 0 \tag{1.68}$$

$$< \overline{\mathbf{f}}_i(t) \overline{\mathbf{f}}_i(0) >= 2 k_B T_0 b_i m_i \delta(t) \tag{1.69}$$

To represent the random force $\overline{\mathbf{f}}_i(t)$ in eq 1.69 we employ the Box–Muller transformation. Unlike, for example, the methods based on the central limit theorem, this method is accurate.

Let x and y be independent random quantities uniformly distributed on the line element $[-1, 1]$. Let us define $R = x^2 + y^2$. If $R > 1$ or $R = 0$, the values x and y should be generated again. When the condition $0 < R \leq 1$ is fulfilled, z_0 and z_1 are calculated, which will be random quantities satisfying standard normal distribution

$$z_0 = x \cdot \sqrt{\frac{-2 \ln R}{R}} \tag{1.70}$$

$$z_1 = y \cdot \sqrt{\frac{-2\ln R}{R}} \tag{1.71}$$

When the standard normal random quantity z is obtained, we can easily pass to the quantity $\xi \sim N\left(\mu, \sigma^2\right)$ which is normally distributed and has mathematical expectation μ and standard deviation σ

$$\xi = \mu + \sigma \cdot z \tag{1.72}$$

In eq 1.69, the random force $\overline{\mathbf{f}}_i(t)$ has mathematical expectation $\mu = 0$ and standard deviation

$$\sigma = \sqrt{2k_B T_0 b_i m_i \delta(t)} \tag{1.73}$$

The processes of the condensation of nanoparticles in the gas medium are associated with the presence of the interaction between the atoms. In the gas medium containing nanoparticles the processes of coalescence are determined by two main factors:

- the distance between the interacting nanoparticles;
- the direction and magnitude of the velocities of the nanoparticles' motion.

Let us consider two nanoparticles at the arbitrary moment (Fig. 1.14), which are at the distance R_{ij}. When R_{ij} is small, the condition of "agglutination" of nanoparticles is fulfilled.

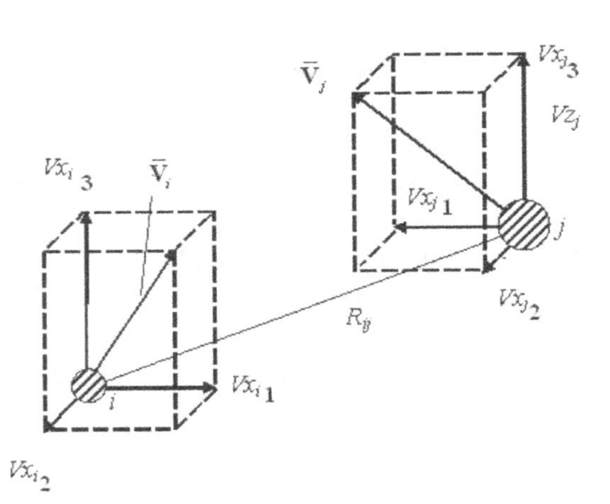

FIGURE 1.14 The relative position of the ith and jth nanoparticles.

The second factor influencing the condensation of nanoparticles is determined by the magnitudes and direction of the velocities. It is obvious that very "fast" nanoparticles can go past one another even at mutual osculation. The angle α between the vectors of the velocities (Fig. 1.15) showing the direction of the nanoparticles' motion is also essential for the nanoparticles coalescence.

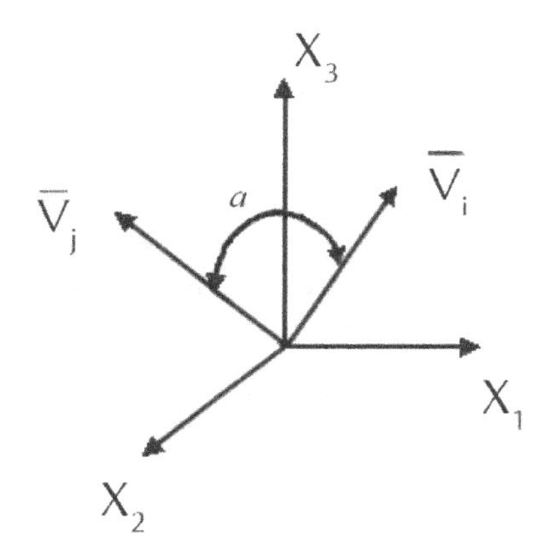

FIGURE 1.15 Relative position of the velocity vectors of the *i*th and *j*th nanoparticles.

The processes of the formation of new nanoparticles and the dynamics of their motion depend on the choice of an adequate condition of the nanoparticles agglutination. The criterion of the nanoparticles agglutination can be presented in the form

$$\Phi(|R| - R_{ij}; \bar{\mathbf{V}}_i - \bar{\mathbf{V}}_j) = 0 \qquad (1.74)$$

In the problem of the coalescence of the interacting nanoparticles the main role is played by the correct choice of the time integration step for eq 1.66 for both the analytical solution and the numerical solution. The choice of a small time step leads to an increase in the time of the problem calculation. Too large integration step can lead to the fact that nanoparticles will go past one another despite the condition (1.74). Thus, the problem of an optimal time step arises.

Figure 1.16 shows an arbitrary couple of nanoparticles at a certain moment. The nanoparticles have the velocities $\bar{\mathbf{V}}_i$ and $\bar{\mathbf{V}}_j$ and the position is determined by $\bar{\mathbf{r}}_i$ and $\bar{\mathbf{r}}_j$ which are current radius-vectors, respectively.

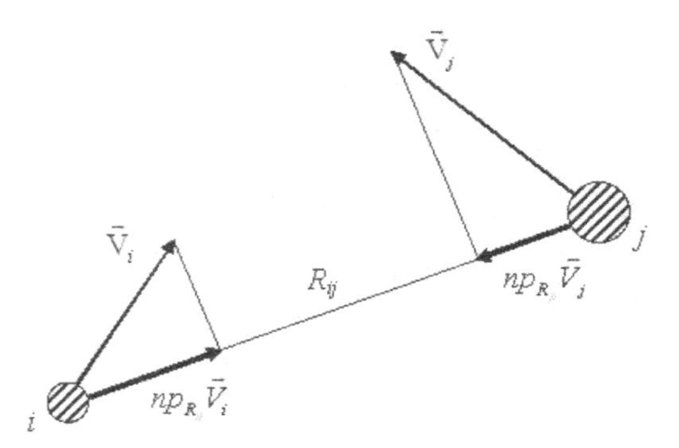

FIGURE 1.16 Two nanoparticles moving with the velocities $\bar{\mathbf{V}}_i$ and $\bar{\mathbf{V}}_j$.

The projections of the velocities of the *i*th and *j*th nanoparticles are defined according to the formulae

$$np_{R_{ij}}\bar{\mathbf{V}}_i = \bar{\mathbf{V}}_i \cdot \bar{\mathbf{R}}_{ij} / R_{ij} \tag{1.75}$$

$$np_{R_{ij}}\bar{\mathbf{V}}_j = \bar{\mathbf{V}}_j \cdot \bar{\mathbf{R}}_{ji} / R_{ij} \tag{1.76}$$

The vectors $\bar{\mathbf{R}}_{ij}, \bar{\mathbf{R}}_{ji}$ and the distance between the nanoparticles R_{ij} are calculated from the following relations:

$$\bar{\mathbf{R}}_{ij} = \bar{\mathbf{r}}_j - \bar{\mathbf{r}}_i, \bar{\mathbf{R}}_{ji} = \bar{\mathbf{r}}_i - \bar{\mathbf{r}}_j \tag{1.77}$$

$$R_{ij} = \left|\bar{\mathbf{R}}_{ij}\right| = \left|\bar{\mathbf{R}}_{ji}\right| = \left|\bar{\mathbf{r}}_j - \bar{\mathbf{r}}_i\right| \tag{1.78}$$

The interval when the collision of the *i*th and *j*th nanoparticles becomes possible is directly proportional to the distance between the nanoparticles and inversely proportional to the projections of the velocities:

$$\Delta t_{ij} = \frac{R_{ij}}{\ddot{\imath}\,\eth_{R_{ij}}\bar{\mathbf{V}}_i + \ddot{\imath}\,\eth_{R_{ij}}\bar{\mathbf{V}}_j} \tag{1.79}$$

Going over all the possible couples of nanoparticles the smallest positive value Δt_{ij} is found and the integration step is calculated with the use of formula (1.79):

$$\Delta t = \frac{1}{m} \cdot \min_{i,j,\, dt>0} \left(\Delta t_{ij}\right); i = 1, 2, ..., n; j = i, i+1, ..., n, \tag{1.80}$$

where m is the integral number determining the value of the step of the time integration.

Thus, the integration step value is influenced first of all by the motion velocity of the fastest particles. In order not slow down solving the problem by additional calculations, it is more practical to make the choice of the integration step in a particular number of iterations rather than for each one.

1.4.4 METHOD FOR THE ENLARGEMENT OF THE CALCULATION CELL

The important aspect of the multiscale modeling of the process of the nanoparticles formation is the gradual enlargement of the spatial scale in the course of calculation. This possibility arises due to the fact that during the formation of nanoparticles the number of atoms and molecules decreases in the gas mixture. They combine forming nanoparticles and after that they exhibit collective behavior. Consequently, the number of variables of the problem of modeling decreases.

The above process takes place at the stage of the modeling of a gas system by the molecular dynamics method. When nanoparticles are sufficiently large, their concentration quickly decreases in the calculation cell considered. The gas phase is not the source of nanoparticles; their enlargement is due to the agglutination of nanoparticles with a smaller size. Therefore, from here on it is unpractical to conduct the calculation of a nanoscale system by the molecular dynamics method; one should employ the mesodynamics methods instead. At further enlarging of nanoparticles, the situation can arise when within the calculation cell the nanoparticles do not almost interact since their trajectories do not intersect. However, if the influence of nanoparticles from the neighboring cells is taken into account, the condensation process will continue. Thus, for the adequate investigation of the problem of the nanoparticles condensation it is necessary to enlarge the spatial scale combining several calculation cells into one in proper time.

Since the problem is solved with the use of the periodic boundary conditions, the space scale can be enlarged by the symmetrical imaging of atoms, molecules, and nanoparticles onto the neighboring calculation cells.

Figure 1.17 displays the example of a four-fold enlargement of the calculation cell volume. In this case, linear sizes of the cell are changed which requires the correction of the parameters in the mathematical formulation with the Born–Karman periodic conditions. These conditions are represented above by relations (1.24)–(1.27). The notation of the conditions for the enlarged cell remains valid. New sizes of the cell are fitted into them and new images of modeled objects are constructed in the new enlarged "images" of the calculation cell.

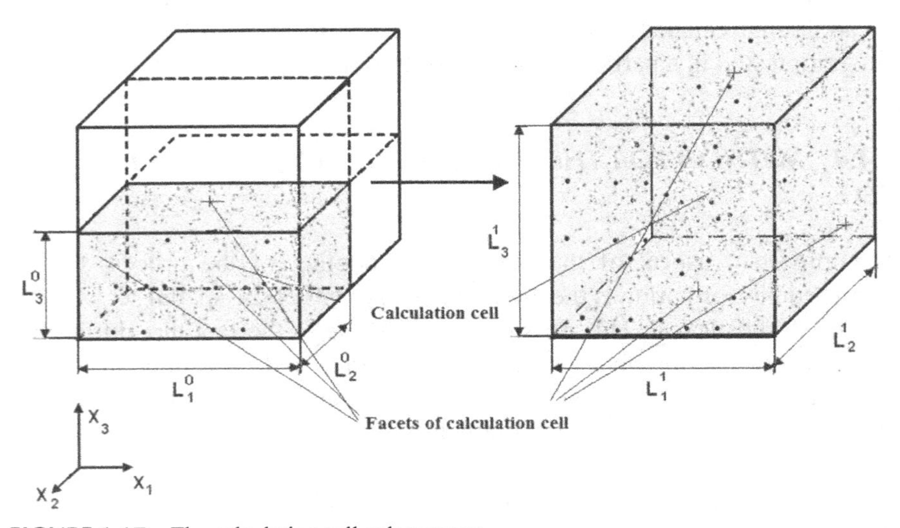

FIGURE 1.17 The calculation cell enlargement.

The enlargement of the calculation volume can happen repeatedly as nanoparticles are being enlarged.

Based on the results of the molecular dynamics calculation, modeling by the mesodynamics methods allows us to calculate linear and angular velocities of the nanoparticles motion, the coordinates of the mass center of the nanoparticles $\overline{\mathbf{V}}_i, \Theta_3^k, \overline{\mathbf{x}}_i$ and the variation of the indicated parameters in time. Based on these data, the shape and spatial structure of nanostructures formed from nanoparticles are calculated.

For the example shown in Figure 1.18 the following relations are fulfilled:

$$L_1^1 = L_1^0; L_2^1 = 2L_2^0; L_3^1 = 2L_3^0 \tag{1.81}$$

It should be noted that the above-discussed calculation methods are not isolated. For the solution of a number of problems they can be used simultaneously. For example, the calculation of the trajectories of the motion of atoms and molecules can be carried out by the molecular dynamics methods, while the forces of the interaction between the atoms and molecules at each calculation step in time can be calculated by the quantum mechanics methods. The motion of nanoparticles in the gas phase can be considered as the motion of super molecules, the trajectory of which is calculated by the mesodynamics methods, and the motion of atoms and molecules of the gas mixture—with the use of the molecular dynamics methods.

1.5 CONTINUUM MECHANICS METHOD

1.5.1 DISPERSION OF NANOPARTICLE

In this section, the physics and mathematical models of the dispersion processes for powders are considered at impulse temperature–power action. The occurring physical phenomena are analyzed and the boundary problems describing the indicated processes are formulated. First, let us consider the physical processes of dispersion of powder particles.

The physical essence and kinetics of the dispersion of a PM are determined by the parameters of loading of the PM particle, namely, the environment pressure and temperature, the rate of the change of these magnitudes in time and physical-mechanical characteristics of the material particle. Depending on the relation of the indicated parameters, it is possible to observe various types of the PM particle destruction:[115] slabbing destruction of the solid phase of the material of the particle, fusion of particles, simultaneous disintegration of the solid and liquid phases, evaporation of the material from the particle surface, and so forth.

Let us consider a solid particle with an arbitrary shape in the medium with pressure and temperature varying with time. If the combination of the indicated parameters does not cause a change in the phase state of the material (its melting or evaporation), the analysis of the particle destruction process can be carried out within the framework of the theory of deformed solid. When the phase state of the material is changed, it is necessary to consider the combined behavior of the solid, liquid, and gaseous phases. Below the main mechanisms of the destruction of the particle in the solid and multiphase states are presented.

At the force action of the external pressure P_0 the particle under consideration with the volume Ω_o and the surface S_0 is deformed accumulating elastic energy (Fig. 1.18). Let us assume that at a point A of the particle the equivalent stress at the given moment reaches the value of the ultimate stress of its material. It will lead to the fragmentation of the particle over the surface S^* going through the point A into two parts with the volumes Ω_1 and Ω_2 and the surfaces S_1 and S_2, respectively. In the particle fragments formed, the pulsation of the stress fields continues, which can cause their destruction. The process is repeated till the maximal equivalent stress exceeds the material ultimate stress. The level of the equivalent stress is conditioned by content of the deformation elastic energy remained in each newly formed fragment after the particle destruction.

In the dispersion process, an initial particle loses a portion of energy for inelastic deformations of destruction and for the separation of the fragments formed. Apparently, the moment will come when the remained energy is not sufficient for further fragmentation of the particle fragments and the dispersion process will stop.

The kinetics of the solid phase dispersion can be influenced by the presence of initial defects: pores, microcracks, foreign inclusions, from which destruction normally starts. In addition, the interaction of stress waves with different types of defects can essentially change spatial and time distribution of stresses in the particle. However, the process physics will not change in this case.

Thermal action (both the external action caused by the environment and the internal one due to the dissipation of the energy at the destruction and inelastic deformations) also influences the kinetics and topography of the particle destruction. An increase or decrease in temperature leads to the formation of additional temperature stresses in the particle and to the change in the material elastic and strength properties. In particular, the Young modulus and Poisson ratio, which determine the velocity of the stress wave propagation and the material ultimate stress, that is, the beginning of destruction, significantly depend on temperature. In addition, with increasing temperature the manifestation of the material inelastic properties increases.

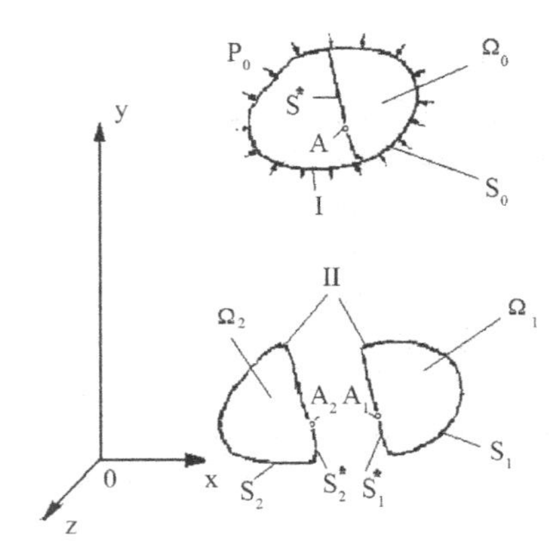

FIGURE 1.18 A solid particle before (I) and after (II) destruction.

At high-speed loading, intensive thermal flows and small heat capacity of the material, the particle temperature can exceed the temperature of melting of its material, which will cause the phase transition "solid–liquid." Different mechanisms of the particle destruction are possible, which correspond to the formation of a liquid phase on the surface and within the material (Fig. 1.19a). The latter is observed when the particle is inhomogeneous and the temperature of the material melting on the surface is higher than that within the particle.

As the result of the phase transition "solid–liquid," the compressibility of the material and its other physical-mechanical characteristics are changed, and additional stresses appear associated with the change of the material volume (the increase of the volume of the liquid in comparison with that of the solid). When the internal regions of the phase transition "solid–liquid" are formed, additional tensile stresses appear in the particle. Let us consider in turn the fragmentation mechanisms of the particle covered with a liquid layer and the particle containing liquid in its internal region.

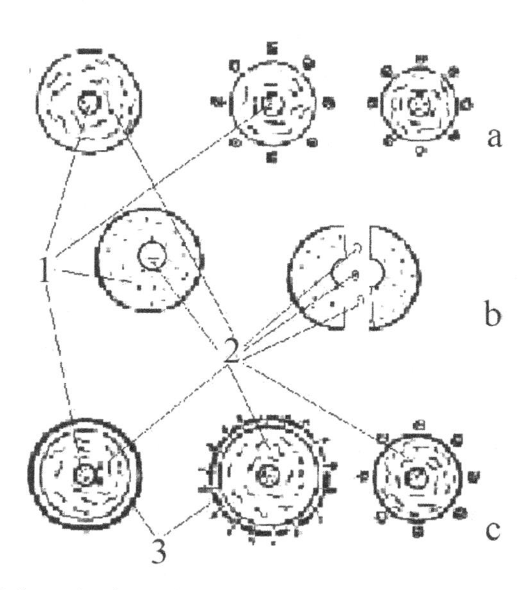

FIGURE 1.19 Schematic view of the destruction of a particle containing: (a, b) solid (1) and liquid (2) phase; (c) solid and liquid phases contained in a solid shell (3).

In the first case, after the dynamic removal of the external pressure, the liquid surface layer and the solid nucleus begin to fluctuate simultaneously. The destruction of such system caused by the propagation of stress waves in the particle can happen as the result of the solid nucleus destruction as well as of the destruction of the destruction of the external layer, the strength of which is conditioned by the action of the forces of the liquid surface tension. It should be noted the velocities of the motion of waves in the liquid and the solid nucleus are not the same; therefore, the reflection of them from the solid–liquid phase interface is possible.

If the liquid phase was formed within the particle, the first stage of its destruction will be the splitting of its solid part (Fig. 1.19b). Further the fragmentation process is developed in the order described in the above paragraph.

Many metals are covered with oxides, the melting temperature and strength of which are much higher than those of the materials themselves. The presence of the oxide film can significantly change the physics of the PM destruction.

At loading by external pressure, the system "solid nucleus–oxide cover" (Fig. 1.19c) is compressed and due to the heat exchange with the environment it is warmed up. The surface oxide film does not melt. Melting or even evaporation starts in the base material under the film. It causes the appearance of additional stresses compressing the nucleus and stresses stretching the external oxide cover, which can lead to its destruction when the volume of the melted material is greatly changed and to the subsequent destruction of the liquid layer.

The particle outer cover can also be destructed due to the external pressure in the process of fluctuations of the complex system: solid nucleus, liquid layer, gas interlayer, and surface solid film. After the cover destruction, the gas being within the particle spreads into the environment. The surface liquid layer is torn off forming drops and the solid nucleus is destructed into smaller parts if there is sufficient potential energy. As the environment cools down, the evaporated substance is condensed on the solid and liquid particles formed. In this case, the latter harden as they cool down.

Thus, the analysis of the physical processes taking place during PM dispersion shows that for its description it is necessary to take into account the whole set of interconnected phenomena such as: dynamic deformation and destruction of a solid, heat exchange and thermal conduction, and melting and evaporation of a substance. The determining parameter is the temperature of a process, the action of which causes a significant change in the material properties and the phase transitions such as solid–liquid, liquid–gas, and it also influences the kinetics and topography of the destruction. Taking into account the above it is practical to consider the construction of the mathematical model of the dispersion process in isothermal and non-isothermal conditions. Below are presented such models for a single elastic and inelastic particle at constant and variable temperatures.

For the formulation of mathematical models, let us consider a single particle from the material that is nonuniform in the general case, which is in the medium with the initial pressure P_0, temperature T_0 and occupies the area Ω_0 (Fig. 1.18). The position of the particle is determined by the coordinates $\overline{x}(x, y, z)$. Owing to the compression over the surface S_0, the particle accumulates the deformation elastic energy U_0, to which the initial fields of the displacement vectors $\overline{u}_0$, tensors of deformation $\hat{\varepsilon}_o$ and stresses $\hat{\sigma}_o$ correspond.

The value of the elastic energy accumulated by the particle is determined by the work of the pressure forces on its surface

$$U_o = \frac{1}{2} \int_{S_o} P_o \, \vec{\mathbf{n}} \circ \vec{\mathbf{u}}_o \, dS_o, \tag{1.82}$$

which is equal to the work of stresses $\hat{\sigma}$ on the deformations $\hat{\varepsilon}$ in the volume Ω_0

$$A_o = \frac{1}{2} \int_{\Omega_o} \hat{\sigma}_o \circ \hat{\varepsilon}_o \, d\Omega_o, \tag{1.83}$$

where $\overline{u}_0$ is the vector of displacements on the surface S_0; $\overline{n}$ is normal to the surface S_0; $\circ$ is the sign of the internal product.

At the moment t_0 the pressure U_0 changes and the potential energy U_0 accumulated in the particle passes into kinetic energy of its elastic vibrations. In the particular conditions, at the moment $t = t_*$ the destruction of the particle takes place and two new particles are formed having the volumes Ω_1 and Ω_2 and the surfaces S_1 and S_2. These particles can also be destructed with time forming smaller particles till there is the energy sufficient for destruction in them.

The vibrations and destruction of particles can be accompanied by inelastic scleronomic or rheonomic deformations owing to which a portion of the energy accumulated at compression dissipates and also by the formation of various defects. Taking into account the indicated phenomena considerably complicates the modeling of the dispersion process. Therefore, at the first stage, in order to obtain the solution allowing to perform the preliminary analysis of the dispersion process, we assume that the material is elastic. The non-linear properties of the material we will take into account later, at the statement of the problem of non-isothermal loading.

The set of equations describing the change of the shape of an elastic solid and its destruction with time in the isothermal conditions includes:

- differential equations of motion

$$\vec{\nabla} * \hat{\sigma} = \rho(\vec{\mathbf{x}}) \frac{\partial^2 \vec{\mathbf{u}}}{\partial t^2} \tag{1.84}$$

- equations of coupling of the deformation tensor $\hat{\varepsilon}$ and the displacement vector $\bar{u}$

$$\hat{\varepsilon} = \frac{1}{2}\left(\vec{\mathbf{u}} \otimes \vec{\nabla} + \vec{\nabla} \otimes \vec{\mathbf{u}}\right) \tag{1.85}$$

determining equation for elastic body

$$\hat{\sigma} = \mathbf{D}(\vec{\mathbf{x}})\,\hat{\varepsilon} \tag{1.86}$$

where $\mathbf{D}(\vec{\mathbf{x}})$ is the matrix of the material elastic constants.

The boundary conditions for the system of equations (1.84)–(1.86) contain:

- force conditions on the surface S_o of the initial particle and the surfaces S_1 and S_2 of the newly formed particles including the destruction surfaces S_1^* and S_2^*:

$$\left.\begin{array}{l} \tau_n = 0,\ \sigma_n = -P_o,\ \ t = 0 \\ \tau_n = 0,\ \sigma_n = -P(t),\ 0 < t \le t_* \end{array}\right\} \vec{\mathbf{x}} \subset S_0 \tag{1.87}$$

$$\tau_n = 0,\ \sigma_n = P(t),\ t > t_*,\ \vec{\mathbf{x}} \subset S_1 \bigcup S_2 \bigcup S_1^* \bigcup S_2^*$$

where σ_n and τ_n are the normal and tangential stresses acting on the surfaces of the initial particle and on the surfaces of the particle fragments formed as the result of the destruction; t_* is the moment of the appearance of the destruction;

- the force and deformation conditions within the initial particle Ω_o and the newly formed particles Ω_1 and Ω_2:

$$\hat{\sigma} = \hat{\sigma}_o,\ \hat{\varepsilon} = \hat{\varepsilon}_{o,}\ t = 0,\ \vec{\mathbf{x}} \subset \Omega_o$$

$$\left.\begin{array}{ll} \vec{\mathbf{u}} = \vec{\mathbf{u}}_1^*, & \vec{\mathbf{x}} \subset \Omega_1 \\ \vec{\mathbf{u}} = \vec{\mathbf{u}}_2^*, & \vec{\mathbf{x}} \subset \Omega_2 \end{array}\right\} \; t = t_* \tag{1.88}$$

where $\overline{u}_1^*$ and $\overline{u}_2^*$ are the displacement vectors within the initial particle at the moment of destruction t_*;

- the condition of the destruction

$$\sigma_{eq} = \sigma_b, \;\; \overline{x} \subset A \tag{1.89}$$

where A is the starting point of the destruction; σ_{eq} is the equivalent stress; σ_b is the ultimate stress of the particle material;

- the condition determining the geometry of the destruction surface S^*, which can be written with the assumption that the unit vector $\overline{n}_{S^*}$ perpendicular to the destruction surface S^* is parallel to the maximal principal stress $\overline{\sigma}_1$

$$\vec{\mathbf{n}}_{\sigma_1} \circ \vec{\mathbf{n}}_{S^*} = 1 \tag{1.90}$$

where $\overline{n}_{\sigma_1}$ is the unit vector determining the direction of the principal stress $\overline{\sigma}_1$.

The construction of the surface S^* starts at the point A, in which the condition of the destruction (1.8) is fulfilled. In this case, we assume that the time of the surface S^* formation is small and the particles of the material located on different sides from the surface S^* move away from one another for an infinitely small distance.

Thus, the system of equations (1.84)–(1.86), in view of the conditions (1.87)–(1.90), composes the statement of the problem on the deformation and destruction of an elastic particle in the isothermal conditions.

Next, consider a mathematical model for the fragmentation of a PM particle at time-varying temperature is presented with regard to the scleronomic and rheonomic properties of the material (here, the phase transitions of the types "solid-liquid" and "liquid-gas" are not taken into consideration).

Let us consider a particle from an elastovisco-plastic material occupying an area Ω and having a surface S in the medium with the time-varying pressure $P(t)$ and temperature $T(t)$. With a change in the above parameters, the particle can deform and break down remaining in the solid state.

The mathematical statement of this problem is formulated based on equation of motion (1.84), Cauchy equation (1.4), determining equations for a viscoelastic body with consideration for temperature deformations (for the region of the viscoelastic deformation of a particle)

$$\hat{\varepsilon} = \mathbf{D}(\vec{\mathbf{x}})^{-1}\left(\hat{\sigma} + \mathbf{R}(\vec{\mathbf{x}}) * \hat{\sigma}\right) + \hat{\varepsilon}^{T} \tag{1.91}$$

determining equations for inelastic deformation (in the region of the plastic deformations of a particle)

$$\hat{\sigma} = \hat{\Psi}\left(\hat{\sigma}, \hat{\varepsilon}, \hat{\dot{\varepsilon}}, \mathbf{D}(\vec{\mathbf{x}}), \mathbf{R}(\vec{\mathbf{x}}), \mathbf{G}(\vec{\mathbf{x}}), T, t, \ldots\right) \tag{1.92}$$

and equations of thermal conductivity

$$\lambda \nabla^2 T + q_p = \frac{\lambda}{a_T} \frac{\partial T}{\partial t} \tag{1.93}$$

where

$$\hat{\varepsilon}^{T} = \alpha_T (T(t) - T_0)\hat{I} \tag{1.94}$$

Here, in the general case $\hat{\Psi}$ is the functional, which determines the coupling between the stress tensors $\hat{\sigma}$ and deformations $\hat{\varepsilon}$ and takes into account the influence of the deformation rate $\hat{\dot{\varepsilon}}$, the temperature T, the elastic $\mathbf{D}(\vec{\mathbf{x}})$, rheonomic $\mathbf{R}(\vec{\mathbf{x}})$, and plastic parameters $\mathbf{G}(\vec{\mathbf{x}})$ of the material; a_T is the temperature conductivity coefficient; λ is the thermal conductivity coefficient; $\hat{\varepsilon}^T$ is the temperature deformation; α_T is the coefficient of the volume temperature expansion; T_0 is the initial temperature; q_p is the heat flow from internal heat sources (as a result of plastic deformations); $\hat{I}$ is the unit tensor.

The system of equations (1.84), (1.85), and (1.91)–(1.94) is supplemented with the conditions of plasticity and destruction

$$\sigma_{eq} = \sigma_p, \vec{\mathbf{x}} \subset S^p \tag{1.95}$$

$$\sigma_{eq} = \sigma_b \text{ or } \omega = 1, \vec{\mathbf{x}} \subset A \tag{1.96}$$

where σ_p is the material plastic limit; ω is the parameter of the destruction rate; S^p is the surface separating the plastic deformation regions from the viscoelastic deformation regions.

To calculate the parameter of the destruction rate, the kinetic equation of damage accumulation is used[72]

$$\omega = \varsigma(\hat{\sigma}, \hat{\varepsilon}, T, ...) \tag{1.97}$$

The boundary conditions for this problem includes

$$\left.\begin{array}{l} \hat{\sigma} = \hat{\sigma}_0, \hat{\varepsilon} = \hat{\varepsilon}_0, T = T_0, \vec{\mathbf{x}} \subset \Omega \\ \sigma_n = -P_0, \tau_n = 0, \vec{x} \subset S \end{array}\right\} t = 0 \tag{1.98}$$

$$\left.\begin{array}{l} q_n = \alpha(T_S - T(t)) \\ \sigma_n = -P(t),\ \tau_n = 0 \end{array}\right\} \vec{\mathbf{x}} \subset S,\ t > 0 \tag{1.99}$$

$$\left.\begin{array}{l} q_n = \alpha(T_S - T(t)) \\ \sigma_n = -P(t), \tau_n = 0 \end{array}\right\} \vec{\mathbf{x}} \subset S^*, t > t_* \tag{1.100}$$

$$\vec{\mathbf{u}}^{ve} = \vec{\mathbf{u}}^{ep}, \sigma_{\mathrm{n}}^{\mathrm{ve}} = \sigma_{\mathrm{n}}^{\mathrm{ep}}, \vec{\mathbf{x}} \subset S^P \tag{1.101}$$

where q_n is the heat flow normal to the initial surface S of the particle; T_S is the particle surface temperature; $T(t)$ is the temperature of the environment; S^* is the surface of the destruction; $\vec{\mathbf{u}}^{ve}, \vec{\mathbf{u}}^{ep}$ are the displacements on the surface S^P; $\sigma_{\mathrm{n}}^{\mathrm{ve}}, \sigma_{\mathrm{n}}^{\mathrm{ep}}$ are the normal stresses perpendicular to the surface S^P, respectively; α is the heat-transfer coefficient.

The surface S^P geometry of the loaded particle is determined by the regularities of the development of plastic deformations in accordance with eq 1.92, and the surface S^* profile of the destruction depends directly on the condition of destruction (1.89). Thus, the system of equations (1.84), (1.85), and (1.91)–(1.93) together with the conditions (1.95)–(1.101) determines the statement of the problem on the deformation and destruction of the particle from the elastovisco-plastic material at the varying temperature.

To conclude the above, it should be noted that a rigorous description of the destruction processes with regard to the phase transitions of a substance from a solid state to a liquid or gaseous state is a very complicated task. The more detailed consideration of these processes is presented in Ref. 160.

1.5.2 FORMATION OF NANOCOMPOSITES

In this section, the physical phenomena are considered and the mathematical models of the main steps of the compaction of powder nanocomposites are constructed, namely, the travel of a press-tool, removal of the pressing force, and re-pressing and squeezing a finished item from the die block.

The consideration is also given to the processes of the rearrangement of the fields of stresses and deformations with time in powder nanocomposites caused by the rheological properties of the materials.

The formation of mechanical energy and other properties of compacted nanocomposites is realized in the process of their conversion into finished items by the methods of pressing. In this case, the technological process of the composite fabrication is often conducted simultaneously with the item fabrication and is accompanied by a significant change in the stressed-deformed state (SDS) of the powder being pressed.

As an example, Figure 1.20 shows a typical scheme of the process for fabricating a nanocomposite: 1—active pressing (in the present case, double pressing with hydroshaping); 2—instantaneous unloading at the puncheon exit or die splitting; and 3—a relaxation period observed during the item "rest" or other operations. Note that the powder compaction during pressing is characterized by significant volume deformations and a continuous change in the parameters such as density, elasticity, plasticity, rheonomic, and strength properties.

Inhomogeneity of SDS of the compacted material leads to the inhomogeneity of its plastic deformation both over the volume and in the direction. It results in the continuously changing nonuniformity of the material physical-mechanical characteristics during the process of pressing and in the appearance of residual (technological) stresses in the pressed item. The composite physical-mechanical properties change significantly. Before pressing, it is a powder mixture with high porosity and low density. During the puncheon stroke, it consolidates, which results in the formation of an inhomogeneous compact (in the general case).

At the step of the repeated pressing, after some material powder has been added to the matrix, an item is formed consisting of two parts with different physical-mechanical properties. It is explained by that one part of the material is pressed repeatedly and the second one is just in the formation process, namely, a powdered body is converting into a compact. An adhesive bond appears at the interface of the materials, which provides the connection of the successively compacted parts of the PM into a single whole.

In the process of the squeezing of the finished item from the die block, significant elastic unloading is observed in it, which changes its shape and sizes, and in some cases (e.g. at the "overpress") it leads to a destruction of the item. Hydroshaping can be a final operation in the technological process leading to the essential decrease of the inhomogeneity of the material and the significant removal or decrease of the technological stresses.

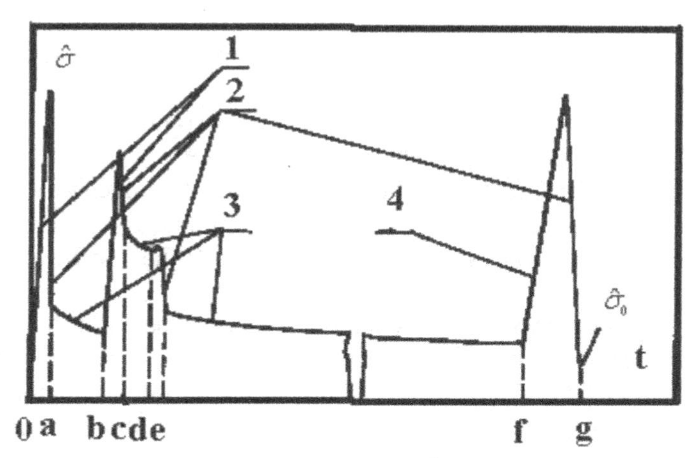

FIGURE 1.20 The scheme of the change of stresses $\hat{\sigma}$ in the pressing with time during the process flow: o–a and b–c are the steps of the pressing process; a–b and c–d–e are unloading with an intermediate "rest;" e–f are the item storage; and f–g are the item loading during an operation.

At the step of the repeated pressing, after some material powder has been added to the matrix, an item is formed consisting of two parts with different physical-mechanical properties. It is explained by that one part of the material is pressed repeatedly and the second one is just in the formation process, namely, a powdered body is converting into a compact. An adhesive bond appears at the interface of the materials, which provides the connection of the successively compacted parts of the PM into a single whole.

In the process of the squeezing of the finished item from the die block, significant elastic unloading is observed in it, which changes its shape and sizes, and in some cases (e.g. at the "overpress") it leads to a destruction of the item. Hydroshaping can be a final operation in the technological process leading to the essential decrease of the inhomogeneity of the material and the significant removal or decrease of the technological stresses.

There are some other physical processes accompanying pressing that influence the parameters of the condition of the material pressed and the SDS of the item formed. First, it is the gas release from the pressed item (which is especially characteristic of super disperse systems containing a great initial amount of gas), which can lead to the growth of internal pressure within the pressed item and the appearance of internal hollows and microcracks in it.

Both the external friction occurring due to the interaction of the powder mixture and the press-tool and the internal friction resulting from the displacement of particles relative to one another also essentially influence the pressing process. The value of the frictional forces depends on the sizes and shapes of particles, the tool surface quality, and they are controlled with various technological additions.

Thus, pressing is a set of interconnected physical processes taking place in a material on the background of the essential change of its aggregative state, from a loose powder-like material to a compacted inhomogeneous body.

The selection of an optimal mode of pressing and the investigation of the kinetics of the accompanying processes require the construction of the complete mathematical model of process of pressing.

Let us consider the mathematical models of the process of the nanocomposite compaction. Figure 1.21 shows the schematic views of the above technological steps and the typical regions of the system "material being formed– press-tool." Here, Ω_1 is the area occupied by the powder; $\Omega_{2,}$ Ω_3, and Ω_4 are the areas of the die, base, and puncheon, respectively; Ω_5 is the volume of an elastic element that is usually present in the system indicated and improves the conditions of pressing. The area Ω_1 is subdivided into the areas $\Omega^1_{1,}$ $\Omega^2_{1,}$ and $\Omega^k_{1,}$ which are formed during material shaping during "k" of the press-tool throws.

The system of equations describing the problem statement includes the following:

equilibrium equations

$$\overline{\nabla} * \hat{\sigma} = 0 \tag{1.102}$$

geometrical equations

$$\hat{\varepsilon} = \frac{1}{2}\left(\overline{u} \otimes \overline{\nabla} + \overline{\nabla} \otimes \overline{u} + \overline{\nabla} \otimes \overline{u} * \overline{u} \otimes \overline{\nabla}\right) \tag{1.103}$$

determining equations describing the mechanical properties of the contacting objects: the press-tool (1.104) and the composite mass compacted during pressing (1.105) and unloading (1.106)

$$\hat{\sigma} = 2\mu_n \hat{\varepsilon} + \lambda_n \varepsilon_0 \hat{I} \tag{1.104}$$

$$N = 2, 3, 4, 5, \ \overline{\mathrm{x}} \subset \Omega_2, \ \Omega_3, \ \Omega_4, \ \Omega 5,$$

$$\hat{\sigma}_m = \hat{\psi}(\hat{\sigma}, \hat{\varepsilon}, \theta, h_i, \ldots) \tag{1.105}$$

$$M = 1, 2, \ldots, k; \ \overline{\text{x}} \subset \Omega_1^1, \Omega_1^2, \ \ldots, \Omega_1^k,$$

where μ_n and λ_n are the Lamé constants; λ_0 is the volume deformation; $\overline{\text{x}}$ is the coordinates; $\hat{\psi}$ is the functional of the plasticity of a porous body; θ is the current porosity; h_i is the structural parameters of a porous material. For writing eq 1.105, the theory of plasticity of porous bodies is used.[55]

For the regions of the ultimately compacted material, when the nonuniform distribution of density over the volume is taken into account, the functional from (1.24) degenerates into the equations of state of an inhomogeneous elastic body with the variable characteristics $\mu_m(\overline{x})$ and $\lambda_m(\overline{x})$:

$$\hat{\sigma} = 2\mu_m(\overline{x})\hat{\varepsilon} + \lambda_m(\overline{x})\varepsilon_0 \hat{I} \tag{1.106}$$

Note that it is sufficient to preserve the linear connection for the areas Ω_2, … and Ω_5 in eq 1.3 because the deformations in the press-tool are usually small.

Equations (1.102)–(1.106) for the calculation of the SDS of the system of the compacted and elastic bodies should be supplemented with the equations of heat conduction and the gas filter equations in the mixture being compacted.

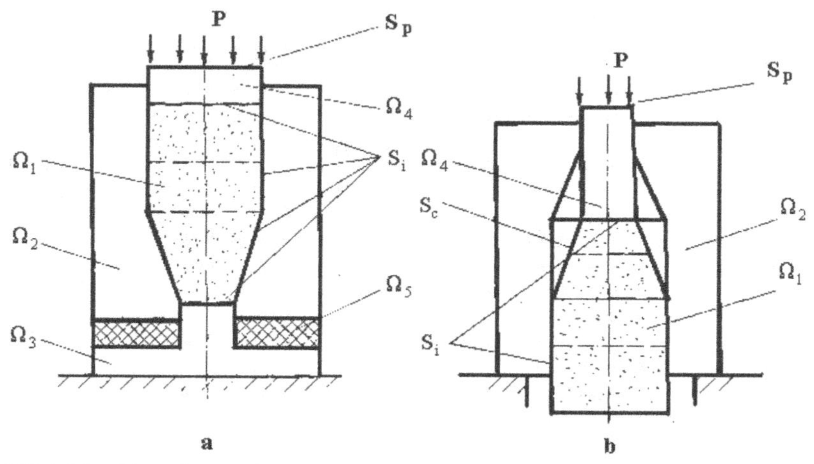

FIGURE 1.21 The scheme of pressing of powders: (a) is the active travel of the puncheon and unloading; (b) is the pressing-out of the finished item.

The heat conduction equation for an inhomogeneous porous body pressed taking into account heat release during pressing has the following form

$$\frac{\partial}{\partial x}(\lambda(\theta)\frac{\partial}{\partial x}(T))+\frac{\partial}{\partial y}(\lambda(\theta)\frac{\partial}{\partial y}(T))+\frac{\partial}{\partial z}(\lambda(\theta)\frac{\partial}{\partial z}(T))+$$
$$+q_P=\frac{\lambda(\theta)}{a_T(\theta)}\frac{\partial T}{\partial t} \tag{1.107}$$

where q_p is the power of heat sources, which depends on the plastic deformation levels; $a_T(\theta)$ and $\lambda(\theta)$ are the coefficients of temperature conductivity and heat conduction, which depend on the porosity of a PM;

$$q_P=\Phi(\hat{\varepsilon}^P) \tag{1.108}$$

For the press-tool, the heat conduction equation can be written in the form

$$\nabla^2 T=\frac{1}{a_T}\frac{\partial T}{\partial t} \tag{1.109}$$

For the derivation of the equation of the gas or liquid filtering through a powder composite, we use Darcy law for porous medium

$$\bar{V}=-\chi\bar{\nabla}P \tag{1.110}$$

where $\bar{V}$ is the gas movement velocity; P is the gas pressure; χ is the filtration coefficient depending on the parameters of the gas and powder media. The equation of continuity for the porous medium can be written in the form

$$\bar{\nabla}*(\rho\bar{V})=-\frac{\partial(\theta\rho)}{\partial t} \tag{1.111}$$

Substituting eq 1.9 in eq 1.10 and taking into account the equations of the ideal gas state

$$P=\rho RT \tag{1.112}$$

we obtain the differential equation determining the change of the pressure in the gas medium by the volume of the powder body during the process of pressing

$$\overline{\nabla} * (\frac{P}{T} \chi \overline{\nabla} * P) = \frac{\partial}{\partial t}(\theta \frac{P}{T}) \tag{1.113}$$

Equation (1.113) is nonlinear because θ and χ change with time and are not the same at different points of the composite due to the nonuniform pressing of its separate volumes.

The system of the above equations is closed by the boundary conditions, which represent the initial state of the areas under study at each technological step, the loading conditions and the interaction of the system of the bodies on the surfaces S_1, S_2,... of the contact (the non-interpenetration of the contacting surfaces, their unilateral connection, and the presence of frictional forces) and the conditions of gas and heat exchange with the environment.

The variable boundary conditions in parallel with the variable mechanical properties of the PM pressed determine the peculiarities of the deformation behavior of the system of the bodies at each step of the process of shaping.

Let us consider the boundary conditions in relation to the schemes of the process of shaping displayed in Figure 1.21.

A downward puncheon throws (Fig. 1.21a):

$$\left.\begin{matrix} \hat{\sigma}_l^m = \left(\hat{\sigma}_l^m\right)^{\mathrm{I}}, \ \theta_m = (\theta)^I{}_m, \\ \overline{x} \subset \Omega_l \quad m = 1, 2, \ldots \\ \hat{\sigma}_k = \left(\hat{\sigma}_k\right)^{\mathrm{I}}, \ \overline{x} \subset \Omega_k \\ k = 2, \ldots, 5 \end{matrix}\right\} h = 0 \quad \left.\begin{matrix} u_{n_j} = u_{n_i}, \\ \tau_{S_i} = \mathrm{sign}(u_{S_i}) f \sigma_{n_i}, \\ if \quad \sigma_n < 0: \\ \tau_{S_i} = 0, \quad \sigma_n = 0 \end{matrix}\right\} \begin{matrix} \overline{x} \subset S_i, \\ 0 \le h \le 1 \end{matrix} \quad \left.\begin{matrix} u_{n_j} = u_{n_i}, \\ \tau_{S_i} = \mathrm{sign}(u_{S_i}) f \sigma_{n_i}, \\ if \quad \sigma_n < 0: \\ \tau_{S_i} = 0, \quad \sigma_n = 0 \end{matrix}\right\} \begin{matrix} \overline{x} \subset S_i, \\ 0 \le h \le 1 \end{matrix}$$

$$\left.\begin{matrix} \sigma_n = 0, \quad \tau_S = 0, \quad \overline{x} \subset S_c \\ P = P_0 / h, \qquad \overline{x} \subset S_p \end{matrix}\right\} h \uparrow \tag{1.114}$$

$$T = T_0, \ \overline{x} \subset \Omega_m, \ m = 1, 2, \ldots, k; \ h = 0$$

$$T = T_c, \ \overline{x} \subset S_c, \ 0 \le h \le 1$$

where S_c is the free surface of the system "item being pressed–press-tool;" T_c is the environment temperature; h is the parameter of proportional loading.

It should also be kept in mind that if pressing is carried out in different directions, <<S_p could vary at each step of pressing.

Unloading (Fig. 1.21a)

$$\left.\begin{array}{l} \hat{\sigma}_1^m = \left(\hat{\sigma}_1^m\right)^{\mathrm{II}},\ \theta_m = (\theta)_m^{II}, \\ \bar{x} \subset \Omega_1,\ m = 1,2,..., \\ \hat{\sigma}_k = \left(\hat{\sigma}_k\right)^{\mathrm{II}},\ \bar{x} \subset \Omega_k, \\ k = 2,...,5 \end{array}\right\} h = 1$$

$$\left.\begin{array}{ll} \sigma_n = 0, \quad \tau_S = 0, & \bar{x} \subset S_c, \\ P = P_0 / h, & \bar{x} \subset S_p \end{array}\right\} h \downarrow\ (0 \le h \le 1)$$

$$\left.\begin{array}{l} u_{n_j} = u_{n_l}, \\ \tau_{S_i} = \mathrm{sign}(u_{S_i}) f \sigma_{n_i}, \\ \text{ï ðè} \quad \sigma_n < 0: \\ \tau_{S_i} = 0 \text{ ï ðè} \quad \sigma_n = 0 \end{array}\right\} \begin{array}{c} \bar{x} \subset S_i, \\ h \downarrow (0 \le h \le 1) \end{array}$$

$$T = T_1,\ \bar{x} \subset \Omega_m,\ m = 1,2,...;\ h = 0$$

$$T = T_c, \bar{x} \subset S_c, 0 \le h \le 1, \tag{1.115}$$

where the index I designates the variable parameters of the process, corresponding to the end of the puncheon throw.

The index II designates the variable parameters of the system state at pressing repeated many times after unloading. For single pressing and the first step of pressing repeated many times, the initial stresses in the system are zero and the porosity corresponds to the bulk porosity of the powder composite mixture.

Pressing-out (Fig. 1.21b):

$$\left.\begin{array}{l}\hat{\sigma}_1^m = \left(\hat{\sigma}_1^m\right)^{\mathrm{I}}, \theta_m = (\theta)_m, \\ \overline{x} \subset \Omega_1,\ m = 1,2,... \\ \hat{\sigma}_k = \left(\hat{\sigma}_k\right)^{\mathrm{I}}, \overline{x} \subset \Omega_k \\ k = 2,...,5\end{array}\right\} h = 0$$

$$\left.\begin{array}{l}u_{n_j} = u_{n_i}, \\ \tau_{S_i} = \mathrm{sign}(u_{S_i}) f\, \sigma_{n_i}, \\ if \quad \sigma_n < 0: \\ \tau_{S_i} = 0, \quad \sigma_n = 0\end{array}\right\} \begin{array}{l}\overline{x} \subset S_i, \\ 0 \le h \le 1\end{array}$$

$$\left.\begin{array}{ll}\sigma_n = 0, \quad \tau_S = 0, & \overline{x} \subset S_c(h) \\ P = P_1 / h, & \overline{x} \subset S_p\end{array}\right\}$$

$$T = T_1, \overline{x} \subset \Omega_m \qquad m = 1,2,...;\ h = 0$$

$$T = T_{\hat{n}}, \overline{x} \subset S_c, 0 \le h \le 1 \tag{1.116}$$

where S_i is the contacting surfaces; u_n is the normal displacement on the surface of contact; τ_s and σ_n are the tangential and normal stresses on the surface of contact; P is the pressure of pressing; P_0 is the maximal pressure of pressing; f is the friction coefficient.

Here, P_1 is the maximal pressure of pressing-out. Note that in the process of pressing-out, the free surface of the pressing essentially increases and the surface of the contact interaction of the composite and the press-tool decreases. When the item is leaving the instrument, the loading parameter can either increase or decrease in the range from 0 to 1.

1.5.3 EXPLOITATION OF NANOCOMPOSITES

In this section the physics of the processes taking place during long-term use of pressed composites in an aggressive medium is considered and the mathematical formulation of the problem is presented.

Powder composites are exposed to the action of various physical and force fields during their utilization. The characteristics of the materials undergo the appreciable influence of temperature, the change of which during the materials use leads to the appearance of the fields of temperature stresses due to the nonuniform warming of an item and restricted conditions of its deformation. In addition, an increase and decrease in temperature significantly influence the material mechanics and the rate of physicochemical processes accompanying the storage and use of items. When the item life is assessed, it is also necessary to take into account the ambient influence causing the material aging, particularly its surface layers, the surface drying at low humidity and its swelling at high humidity of the ambient. It is especially important for the technologies of the fabrication of components from nanopowders (e.g. high-capacity capacitors) since the discussed materials actively interact with moisture.

During storage some powder composites, in particular high-energy ones, that are used as solid fuels graphitize and interact with coating, which change their properties as well.

In general, in the course of time the net effect of the service conditions leads to the change of the shape and geometry of an item, the material properties, the coating properties, the degradation of mechanical characteristics, the appearance of cracks, and so forth. Complex assessment of the indicated phenomena and reliable prediction of the life of an item made from a composite material require the development of mathematical models allowing quick evaluation of the influence of the main service conditions (temperature, humidity, force actions, internal technological stresses, etc.) on physics-mechanics of powder composites with the use of computing machines.

Consider a powder body occupying a region Ω, bounded with a surface S (Fig. 1.22). The body is in a space with the temperature A T_1 and the concentration n of gaseous substances $C_1^1, C_2^1, ..., C_n^1$ at the moment $t=0$. Part of the body surface $S^k \subset S_1^k \bigcup S_2^k \bigcup S_3^k ... \bigcup S_m^k$ is fixed or is in the contact with a deformable solid body. The body surface $S^c \subset S_1^c \bigcup S_2^c \bigcup S_3^c ... \bigcup S_l^c$ is not fixed and interacts with the ambient. The rest part of the surface $S^\Pi \subset S / S^k / S^c \subset S_1^L \bigcup S_2^L ... \bigcup S_i^L$ is coated which prevents the penetration of gaseous or liquid substances from the ambient into the bulk of the body. The initial concentration of the above substances inside the body is $C_1^0, C_2^0, ..., C_n^0$.

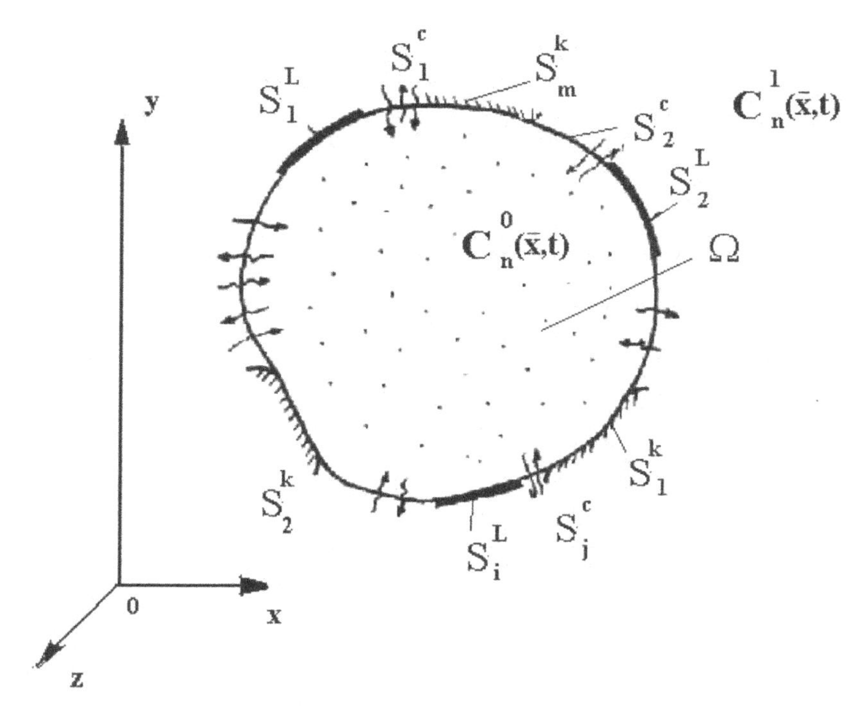

FIGURE 1.22 A powder composite exchanging gaseous and liquid substances with the ambient: S^c—free surfaces; S^L—isolated surfaces; and S^k—surfaces contacting with other bodies.

In addition, in the body there is the initial field of technological residual stresses $\hat{\sigma}_r$, to which the field of residual deformations $\hat{\varepsilon}_r$ corresponds. Due to diffusion processes, heat exchange and heat conductivity, the body temperature and the concentration of liquid and gaseous substances vary. It results in the volume deformation in the form of swelling or shrinking $\hat{\varepsilon}^w(\overline{x},t)$ and the gas release within the volume $\hat{\varepsilon}^g(\overline{x},t)$. It leads to the appearance of stresses due to the inhomogeneity of the volume deformation field and the restricted conditions of deformation on the surface S^k.

The system of differential equations describing the set of the indicated physical processes (diffusion, swelling, and the formation of the fields of stresses) includes the following equations:

equations of the diffusion of liquid and gaseous substances into the body bulk from the ambient

$$\begin{cases} \frac{\partial}{\partial x}(\chi_1(\theta)\frac{\partial}{\partial x}(C_1)) + \frac{\partial}{\partial y}(\chi_1(\theta)\frac{\partial}{\partial y}(C_1)) + \frac{\partial}{\partial z}(\chi_1(\theta)\frac{\partial}{\partial z}(C_1)) = \frac{\partial C_1}{\partial t}, \\ \frac{\partial}{\partial x}(\chi_2(\theta)\frac{\partial}{\partial x}(C_2)) + \frac{\partial}{\partial y}(\chi_2(\theta)\frac{\partial}{\partial y}(C_2)) + \frac{\partial}{\partial z}(\chi_2(\theta)\frac{\partial}{\partial z}(C_2)) = \frac{\partial C_2}{\partial t}, \\ \dots\dots\dots\dots\dots\dots \\ \frac{\partial}{\partial x}(\chi_n(\theta)\frac{\partial}{\partial x}(C_n)) + \frac{\partial}{\partial y}(\chi_n(\theta)\frac{\partial}{\partial y}(C_n)) + \frac{\partial}{\partial z}(\chi_n(\theta)\frac{\partial}{\partial z}(C_n)) = \frac{\partial C_n}{\partial t}, \end{cases} \quad (1.117)$$

equation of thermal conductivity

$$\frac{\partial}{\partial x}(\lambda(\theta)\frac{\partial}{\partial x}(T)) + \frac{\partial}{\partial y}(\lambda(\theta)\frac{\partial}{\partial y}(T)) + \frac{\partial}{\partial z}(\lambda(\theta)\frac{\partial}{\partial z}(T)) = \frac{\lambda(\theta)}{a_T(\theta)}\frac{\partial T}{\partial t} \quad (1.118)$$

equation of release and diffusion of gases

$$\chi_g \nabla^2 C_g + \beta = \frac{\partial C_g}{\partial t} \quad (1.119)$$

equation of body swelling

$$\frac{\partial \varepsilon_0^w}{\partial t} = \varphi(C_1, C_2, \dots, C_n, T, t) \quad (1.120)$$

and the group of equations determining the SDS of an item; differential equation of equilibrium (1.101) and Cauchy equation (1.84). For writing the above equations, let us consider the total effect of the above-mentioned conditions of service. Taking into account that the ultimate strains of the destruction of compacted PMs are usually small, the tensor of total deformations can be presented as the sum of the components resulting from the action of each separate condition

$$\hat{\varepsilon} = \hat{\varepsilon}^F + \hat{\varepsilon}^w + \hat{\varepsilon}^g + \hat{\varepsilon}^T + \hat{\varepsilon}^r \quad (1.121)$$

where $\hat{\varepsilon}$ is the tensor of total deformations; $\hat{\varepsilon}^F$ is the tensor of the deformations caused by the action of external loadings; $\hat{\varepsilon}^w$ is the tensor of "swelling" caused by the absorption of gases from the ambient; $\hat{\varepsilon}^g$ is the tensor of deformations caused by the processes of gas release within a powder composite; $\hat{\varepsilon}^T$ is the tensor of temperature deformations; $\hat{\varepsilon}^r$ is the tensor of deformations appearing due to the action of technological stresses.

Let us consider the structure of each tensor separately.

Temperature is one of the main factors influencing the physics-mechanics of compacted powder composites since the properties of the materials are significantly determined by its magnitude. An increase in temperature intensifies diffusion processes, enhances rheonomic properties of the material, decreases its modulus of elasticity, and so forth. In addition, a change in temperature directly influences the volume and shape of an item due to respective temperature deformation. The latter is determined by the spherical tensor of temperature deformations

$$\hat{\varepsilon}^{T} = \alpha_{T}\ \Delta T\ \hat{I} \tag{1.122}$$

where α_T is the temperature coefficient of the material linear expansion; ΔT is the change of temperature relative to the temperature of the item fabrication; $\hat{I}$ is the unit spherical tensor.

Being porous materials, compacted powder nanocomposites have the capability to absorb gases from the ambient. As it takes place, two main processes occur, namely, adsorption, that is, the absorption of gases by the item surface, and absorption, that is, the absorption of gases by the inner content. These processes are simultaneous and lead to the change in the item volume and its swelling. In addition, the capillary condensation and capillary contraction of a liquid are essential for porous bodies since they lead to the significant volume deformation.

Physical equations describing the above phenomena are rather complicated; therefore, their combined effect can be evaluated based on the phenomenological equations. The tensor of the deformations occurring due to the material swelling can be written in the form

$$\hat{\varepsilon}^{w} = \varepsilon_0^{W}\ \hat{I} \tag{1.123}$$

where ε_0^W is the swelling of the material volume unit, the rate of the enlargement of which with time is determined according to (1.120).

In powder composites that are utilized in an aggressive medium and at elevated temperature, gas release is observed. The pressure of the gases released can reach a significant magnitude causing the item volume deformation. The connection of the deformation with the gas pressure within the material P^g can be given in the form

$$\hat{\varepsilon}^{g} = \frac{P^{g}}{3K}\hat{I} \tag{1.124}$$

where K is the modulus of the material volume compaction, which depends on temperature and humidity in the general case.

Let us present the deformation caused by external forces as the sum

$$\hat{\varepsilon}^F = \varepsilon^e + \hat{\varepsilon}^p + \hat{\varepsilon}^c \tag{1.125}$$

where $\hat{\varepsilon}^e$ are the elastic deformations, $\hat{\varepsilon}^p$ are the plastic deformations, and $\hat{\varepsilon}^c$ are the rheonomic deformations of a compacted metal composite.

The elastic deformations are linearly connected with the stresses acting in the item

$$\hat{\varepsilon}^e = \mathbf{D}\hat{\sigma} \tag{1.126}$$

The stress dependence of the plastic and rheonomic deformations can be written in the general form

$$\hat{\varepsilon}^p = \mathbf{R}_p * \hat{\sigma} \tag{1.127}$$

$$\hat{\varepsilon}^c = \mathbf{R}_c * \hat{\sigma} \tag{1.128}$$

where $\mathbf{D}$ is the matrix of the material elasticity; $\mathbf{R}_p$ and $\mathbf{R}_c$ are the matrices-operators determining the material plastic and rheonomic properties, respectively. In the general case, the matrix and the operators are temperature-dependent.

Technological (residual) stresses appear at the stage of the item fabrication. Their magnitude can be significant, therefore, after the item has been fabricated, its sizes do not match the sizes of the forming tool.[151] During the storage and service of the item, the technological stresses relax due to its rheonomic properties. In this case, two processes take place. First, because of the relaxation of the residual stresses and a decrease in their general level, the elastic aftereffect that appears right after the item fabrication decreases, which leads to a change in the item sizes. On the other hand, the technological stresses cause the material creep, which may result in an increase in the item sizes.

The relations determining the magnitude of the residual stresses have the form

$$\hat{\varepsilon}^r = -\mathbf{D}^{-1}\hat{\sigma}^r \tag{1.129}$$

where $\hat{\sigma}^r$ is the tensor of the residual technological stresses.

At the simultaneous action of the above-indicated service conditions on the item, in accordance with eqs 1.41–1.48 the deformation of the latter can be given in the form

$$\hat{\varepsilon} = \alpha_T \Delta T \hat{I} + \varepsilon_0^w \hat{I} + \frac{P^g}{3K}\hat{I} - \mathbf{D}^{-1}\hat{\sigma}^r + \\ + \mathbf{D}^{-1}\hat{\sigma} + \mathbf{R}_p * \hat{\sigma} + \mathbf{R}_c * \hat{\sigma} \tag{1.130}$$

Combining the summands in equation (1.130), which correspond only to the volume deformation, we obtain

$$\hat{\varepsilon} = \mathbf{D}^{-1}(\hat{\sigma} - \hat{\sigma}^r) + \mathbf{R}_p * \hat{\sigma} + \mathbf{R}_c * \hat{\sigma} + \gamma_w \hat{I} \tag{1.131}$$

where

$$\gamma = \alpha_T \Delta T + \varepsilon_0^w + \frac{P^g}{3K} \tag{1.132}$$

is the phenomenological function taking into account the influence of temperature, swelling processes, and gas release on the item deformation during its service and storage.

Let us express the stress using equation (1.130)

$$\hat{\sigma} = \tilde{D}(\hat{\varepsilon} - \tilde{P} * \hat{\sigma} - \tilde{R} * \hat{\sigma} - \gamma_w \hat{I}) + \hat{\sigma}^r \tag{1.133}$$

Equation 1.132 closes the system of equations (1.84), (1.101), and (1.116)–(1.19), which is supplemented by the boundary conditions for solving the diffusion and force problems included in it:

$$C_1(0,\bar{x}) = C_1^0(\bar{x}),\ C_2(0,\bar{x}) = C_2^0(\bar{x}), \ldots C_n(0,\bar{x}) = C_n^0(\bar{x}), \bar{x} \subset \Omega$$

$$\frac{\partial C_i}{\partial n} = 0 \ \bar{x} \subset S^k \subset S_1^k \bigcup S_2^k \bigcup S_3^k \ldots \bigcup S_m^k$$

$$\frac{\partial C_i}{\partial n} = 0 \ \bar{x} \subset S^L \subset S_1^L \bigcup S_2^L \ldots \bigcup S_i^L$$

$$C_1(0,\bar{x}) = C_1^1(\bar{x})\, C_2(0,\bar{x}) = C_2^1(\bar{x}), \ldots C_n(0,\bar{x}) = C_n^1(\bar{x})\, \bar{x} \subset S^c$$

$$\hat{\sigma}(0,\bar{x}) = \hat{\sigma}_0(\bar{x})$$

$$\hat{\varepsilon}(0,\bar{x}) = \hat{\varepsilon}_0(\bar{x}), \bar{x} \subset \Omega;\ u_n = 0, \tau_n = \sigma_n f, \bar{x} \subset S^k;$$

$$\sigma_n(t,\bar{x})\, l_n = 0, \bar{x} \subset S^{\tilde{n}} \cup S^L; \tag{1.134}$$

where $\hat{\sigma}_0(\bar{x})$, $\hat{\varepsilon}_0(\bar{x}), C_1^0(\bar{x}), C_2^0(\bar{x}),..., C_n^0(\bar{x})$ are the initial distribution of tensors, stresses, deformations, and gaseous substances in the powder body; l_n are the direction cosines of the normal to the surface; u_n, τ_n, and σ_n are the displacements over the normal, tangential, and normal stresses on the contacting surface, respectively; f is the friction coefficient; n is the normal to the surface; Ω is the body volume; S^c is the free surfaces; S^L is the isolated surfaces; S^k is the surfaces contacting with other bodies.

Note that for the body free surface, the boundary conditions of the first type are given since the processes of the adsorption of substances from the ambient are much more intensive than their diffusion, and their concentration on the free surface can be taken as equal to their concentration in the ambient.

KEYWORDS

- **physics models**
- **mathematical models**
- **quantum-chemical method**
- **molecular dynamic method**
- **mesodynamics method**
- **continuum mechanics method**

CHAPTER 2

NUMERICAL METHODS FOR MODELING OF NANOSYSTEMS

CONTENTS

ABSTRACT

This chapter presents numerical algorithms for the solution of the problems of nanotechnologies for different structural scales of modeling. The methods of numerical modeling of the tasks of quantum mechanics as applied to the calculation of atomic structure of nanoelements are presented. The detailed description of various numerical schemes for integrating the equations of molecular dynamics method and mesodynamics are described. The algorithms of finite element method as applied to the tasks of powder dispersing, nanocomposite pressing, and calculation of the parameters of stressed-deformed state of nanocomposites during their operation are given. The software currently applied for the calculation of nanosystems is reviewed and the structure of typical software for multi-level modeling of nanosystems with various methods is given.

2.1 NUMERICAL ALGORITHMS FOR THE SOLUTION OF QUANTUM-CHEMICAL PROBLEMS

2.1.1 TASK DEFINITION

The main task of quantum mechanics with regard to the problem of the nanoparticles formation is the construction of the basis of the particle wave function similar to the molecular wave function using Schrödinger equation (1.3) for stationary states in the adiabatic approximation. In this case, nuclear and electronic wave functions (1.1) can be separated, and the equations for the motion of nuclei and electrons can be solved separately

$$\hat{\mathbf{H}}\Psi = E_e \Psi , \tag{2.1}$$

where $\hat{\mathbf{H}}$ is the Hamilton operator determined by relation (1.3); E_e is the total electron energy

$$E_e = \mathbf{K}_{N^{el}} + U_{N^a N^{el}} + U_{N^{el} N^{el}} . \tag{2.2}$$

As it has been mentioned in the previous chapter, the analytical solution of eq 2.1 can be obtained only for the simplest systems; therefore, at present the approximated methods of solving the Schrödinger equations are used. There are two main groups of these methods, namely, the methods of perturbation theory and variational methods.[53]

In the first group, all the interactions in a system are divided into "principal" and "secondary." It allows us to introduce changes in the Hamilton operator (1.3), which simplify the initial problem to the problem having an exact solution. In accordance with the perturbation theory, corrections, which take into account the perturbation of the Schrödinger equation (2.2) solution at minor change of the Hamiltonian (1.2), can be added to the exact analytical solution.

In contrast to the perturbation theory, variational methods for finding solutions of the Schrödinger equation are based on the simplification of the wave function rather than on the simplification of the Hamiltonian of the system. The main idea of the variational method is that any system tends to the state with minimal energy, which, in its turn, depends on the wave function. If we substitute approximated or "trial" functions instead of a wave function (which is unknown a priori) in the integral for the system total energy, that function which will give the minimal value of the energy can be considered to be the solution of the Schrödinger equation in a certain approximation. It is obvious that it is impossible to select such trial function arbitrarily; it is necessary to use optimization methods. Therefore the Schrödinger equation solution is replaced by the search of the extreme points of the functional of the system energy with regard to the variations (the virial) of the trial function. The task indicated above is reduced to two main problems:

- the selection of the class of trial functions;
- the selection of such trial function in the selected class, which provides the minimal total energy of the system.

It is obvious from the formulation of the problem that the variational method provides the systematic refinement of the solution since the energy minimal value can only decrease. The rest solutions are not considered. In addition, there is always the upper bound of the system total energy in the variational method.

For the construction of molecules the method of molecular orbitals (MO) is normally used, which is based on the assumption that in a molecule as well as in an atom it is possible to construct the set of allowed discrete energy levels and wave functions (MO) corresponding to them, which describe the behavior of an electron in the molecule.

It should be noted that there is a sufficient number of programs for quantum-chemical calculations. These programs, as a rule, are also accompanied by the description of the theoretical foundations of quantum-chemical methods.[36,53] The detailed description of the MO theory can be found in special literature.[12] For applied quantum-chemical calculations only general propositions of the MO theory are necessary; therefore, below we will give only a short description of this theory and indicate the methods most frequently employed in calculations.

First, let us consider more thoroughly the variational methods and after that the methods of the perturbation theory.

2.1.2 THE HARTREE–FOCK METHOD

The most routine variational method is the Hartree–Fock method[129] (the self-consistent field method (SCF)). It is believed that every electron in the molecular system moves in the field of the atomic nuclei that are fixed in space and in the effective (averaged) field of the other electrons of the molecular system. In this case, the Schrödinger wave function for all the electrons of the system (many-electron wave function) is sought in the form of the antisymmetrized product of spin-orbitals, that is, the single electron MOs $\phi_i(v)$ multiplied by spin wave functions α or β of the respective electron:

$$\Psi = C_n \det\left[\phi_1(1)\alpha(1)\phi_1(1')\beta(1')...\phi_n(n')\beta(n')\right] \tag{2.3}$$

where n is the half of the quantity of electrons in a molecule; v is the number of an electron; C_n is the constant determined from the condition $|\Psi|^2 = 1$.

In the Hartree–Fock approximation, the Schrödinger equation is re-arranged in the system of integral-differential equations for each separate electron of the molecule:

$$\hat{\mathbf{H}}\phi_i = E_i\phi_i \tag{2.4}$$

where E_i is the energies of the MO.

Usually the MO ϕ_i are represented in the form of the linear combinations of atomic orbitals (LCAO) η_i (v) which are the physical basis for the construction of the molecular wave function

$$\phi_j(v) = \sum C_{ji}\eta_i(v) \tag{2.5}$$

where C_{ji} is the coefficients which are determined with the use of the variational procedure of the minimization of the molecule total electron energy $\boldsymbol{E}_e$.

The coefficients C_{ji} corresponding to the minimal energy $\boldsymbol{E}_e$ are determined from the following set of equations

$$\sum_{\mu}(F_{i\mu} - E_{\kappa}G_{i\mu})C_{\kappa\mu} = 0 \tag{2.6}$$

where

$$F_{i\mu} = F'_{i\mu} + \sum_{k,l} P_{kl}(\langle i\mu|kl\rangle - \frac{1}{2}\langle ik|\mu l\rangle) \tag{2.7}$$

The condition of the normalization $C_{\kappa\mu}$ has the form

$$F_{i\mu} = F'_{i\mu} + \sum_{k,l} P_{kl}(\langle i\mu|kl\rangle - \frac{1}{2}\langle ik|\mu l\rangle) \tag{2.8}$$

where summing up is performed over the basic orbitals η_i, η_{μ}, η_{κ} and η_l respectively; $G_{i\mu}$ is the overlap interval AO η_i and η_{μ}; $F'_{i\mu}$ is the matrix element of the single-electron Hamiltonian taking into account the energy of the interaction of electrons and atomic nuclei and the kinetic energy of electrons; P_{kl} is the matrix of charges and the orders of bonds; $\langle i\mu|kl\rangle$ and $\langle ik|\mu l\rangle$ are the integrals of the Coulomb interaction of two electrons, which are determined according to the following relations:

$$\langle i\mu|kl\rangle = \iint \eta_i(v)\eta_{\mu}(v)\eta_k(\varsigma)\eta_l(\varsigma)r_{v\varsigma}^{-1}drd\tau_v \tag{2.9}$$

where $r_{v\varsigma}$ is the distance between the electrons v and ζ.

In eq 2.9, integration is performed over the entire space of the Cartesian coordinates.

The solving of nonlinear set of eq 2.6 is conducted by the self-consistent method. Arbitrary values of the $C^0_{\kappa\mu}$ coefficients are used as the zeroth approximation, and they are employed for the construction of the matrix $F^0_{i\mu}$. After that the $C^1_{\kappa\mu}$ coefficients are found from set (2.6) in the first approximation, and the new matrix $F^1_{i\mu}$ is constructed. The operation is repeated until the indicated quantities stop changing or the following condition will be fulfilled

$$\left|C^n_{\kappa\mu} - C^{n-1}_{\kappa\mu}\right| \le \delta_c; \left|F^n_{\kappa\mu} - F^{n-1}_{\kappa\mu}\right| \le \delta_F \tag{2.10}$$

where δ_c; δ_F are the small values determining the accuracy of the calculation of $C^n_{\kappa\mu}$ and $F^n_{\kappa\mu}$, respectively.

Nonempirical and semiempirical methods for the matrix calculation are distinguished, which describe the interaction of electrons with one another and of electrons and atomic nuclei in eq 2.6.

Nonempirical methods are employed for the analytical calculation of matrices. However, they are also approximate as they are used for the partial (small and average) bases of electrons. It introduces an error into the calculation results since such bases specify approximately the distribution of the electron density in molecules. The use of large bases providing the attainment of the Hartree–Fock limit, that is, when further increase in the number of basic orbitals does not influence the results obtained, requires much machine time. In addition, the calculation is carried out only for valence electrons, and the inner shell electrons are included into the molecular core. In this case, the minimal Slater basis is used[12] and part of the electron Coulomb interactions is not taken into account. This simplifies calculation significantly since the number of the matrix elements of the Coulomb interaction of electrons in eq 2.6 is proportional to the fourth power of the number of basic orbitals.

Semiempirical methods employ approximate empirical formulae and the parameters of atoms known from the experiment and neglect most Coulomb integrals (2.6) which have small values.

Note that in any case we have the solution of the Schrödinger equation in the Hartree–Fock approximation rather than the exact solution. For obtaining the exact solution of the Schrödinger equation, it is necessary to take into account the electron correlation which is a laborious task requiring a great amount of machine time. For this purpose, different variants of the configuration interaction method, the perturbation theory or the electron-pair method are more frequently used. In the configuration interaction method, the many-electron wave function Ψ is given in the form:

$$\Psi = \sum A_j \psi_j \tag{2.11}$$

where A_j is the expansion coefficients; ψ_j (where $j \geq 1$) is the determinants constructed from the single-electron wave functions of the ground state (ψ_js differ from the ground state wave function in that one or more electrons have passed from the occupied MO to the unoccupied ones).

It is often assumed that the electron correlation brings small changes into the ground state wave function. In this case, the correlation corrections are taken into account according to the Möller–Pleset perturbation theory or its Brillouin–Wigner variant. In the electron-pair method, the correlation energy is calculated as the sum of correlation effects produced by each electron pair. The simplest variant of this method is the approximation of independent electron pairs, in which the interaction of electron pairs is not taken into consideration, the equation for each electron pair is solved separately and the total correlation energy is calculated as the sum of correlation energy of all the electron pairs. If the interaction of electron pairs is taken into account, the problem becomes nonlinear and can be solved only by iteration methods. In this case, the calculation time increases significantly.

2.1.3 THE FUNCTIONAL OF ELECTRON DENSITY

Let us consider one of the approaches based on the theory of the functional of electron density (TFED). It is assumed in this method that all main properties of a quantum-mechanical system depend on the electron density in space $\rho(r)$. The electron density is similar to the wave function ψ, but in contrast to the latter, it depends only on three spatial coordinates and the wave function ψ depends on 3 N^{el} of the coordinates of all electrons.

Sholl and Stecker[138] proved the theorem according to which the total energy of a quantum-mechanical system $E(\rho)$ is the functional of the electron density $\rho(r)$:

$$E(\rho(\vec{\mathbf{r}})) = T(\rho(\vec{\mathbf{r}})) + \int V_{eff}(\vec{\mathbf{r}},\rho(\vec{\mathbf{r}}))\rho(\vec{\mathbf{r}})d\vec{\mathbf{r}} \tag{2.12}$$

In the system ground state, this functional is minimal, which allows us to express the exchange interaction via the electron density and to obtain single-electron equations in the simple form:

$$\left[-\Delta + V_{eff}(\vec{\mathbf{r}})\right]\psi_i = \varepsilon_i\psi_i \tag{2.13}$$

Here, the system of units is taken, in which $\dfrac{\hbar^2}{2m} = 1, \quad e = 1$; $V_{eff}(\vec{r})$ is the effective potential equal to

$$V_{eff}(\vec{\mathbf{r}}) = -\sum_I \frac{2Z_I}{\left|\vec{\mathbf{r}} - \vec{\mathbf{R}}_I\right|} + 2\int \frac{\rho(\vec{\mathbf{r}}')}{\left|\vec{\mathbf{r}} - \vec{\mathbf{r}}'\right|} d\vec{\mathbf{r}}' + V_{xc}(\vec{\mathbf{r}}) \tag{2.14}$$

The exchange-correlation potential $V_{xc}(\vec{\mathbf{r}})$ is often taken in the approximation of the local electron density:

$$V_{xc}(\vec{\mathbf{r}}) = \beta(\vec{\mathbf{r}}_s)V_{GKS}(\vec{\mathbf{r}}) \tag{2.15}$$

where $V_{GKS}(\vec{r})$ is the Gaspar–Kohn–Sham exchange potential:

$$V_{GKS}(\vec{\mathbf{r}}) = -\left[\frac{48}{\pi}\rho(\vec{\mathbf{r}})\right]^{1/3} \tag{2.16}$$

The exchange-correlation potential is also used in the approximation of generalized gradients

$$V_{xc}(\vec{\mathbf{r}}) = \int \rho(\vec{\mathbf{r}}) f\left[\rho(\vec{\mathbf{r}}), \vec{\nabla}\rho(\vec{\mathbf{r}})\right] d\vec{\mathbf{r}} \cdot \tag{2.17}$$

The above method eliminates the necessity to calculate exchange integrals and significantly decreases the calculation time in comparison with the direct solution of the set of Schrödinger equations.

2.1.4 CONCLUDING REMARKS

Note that within the framework of this book it is impossible to cover in detail all the approximate methods for the integration of the Schrödinger equations and to deal with the questions related to both the decrease of the configuration space dimension and the accuracy of various methods and to give a detailed description all the existing algorithms of the solution of quantum-chemical problems related to nanoscale systems. For those who are interested we can recommend special literature.

The above material is sufficient for the solution of the first problem of the formation of nanoparticles in the gas phase—the calculation of the spatial structure and the atomic composition of molecules which are formed at the initial stage of the gas mixture formation. In this case, the principle advantage of the calculation quantum-chemical methods ("*ab initio*") is used, which is that one does not need to know any empirical parameters to conduct the calculations such as, for example, the force and length of bonds, the values of angles, etc. It is sufficient to point out the chemical formula of the object under study as initial data and if the system elements are bound, to indicate the bond order of the elements and the number of bonds.

2.2 NUMERICAL SCHEMES OF THE SOLUTION OF THE MOLECULAR DYNAMICS PROBLEMS

In the first chapter of the book it is shown that the molecular dynamics method involves the solution of differential equations (1.14) with initial data (1.23). The equations in set (1.14) are independent; however, they require a combined solution since there are various forces present in the system, which reflect the result of the interactions between the atoms and molecules of a nanosystem. The forces are the nonlinear functions of the nanosystem state, which does not allow us to find an analytical solution for (1.14). Therefore, the solving of the above set of equations is conducted numerically. Let us consider the principle numerical schemes with regard to the problem of the calculation of a nano-aerosol gas mixture. For this purpose, we present eq 1.14 and initial data (1.23) in a more compact form, which is convenient for the construction of the integration numerical schemes:

$$\begin{cases} \dfrac{d\vec{\mathbf{x}}(t)}{dt} = \vec{\mathbf{V}}(t), \\ m\dfrac{d\vec{\mathbf{V}}(t)}{dt} = \vec{\mathbf{F}}\left(t, \vec{\mathbf{x}}(t), \vec{\mathbf{V}}(t)\right), \end{cases} \tag{2.18}$$

$$\vec{\mathbf{x}} = \vec{\mathbf{x}}_0, \vec{\mathbf{V}} = \vec{\mathbf{V}}_0, \ t = 0, \vec{\mathbf{x}} \subset \Omega \tag{2.19}$$

where the index i is omitted for brevity:

$$\vec{\mathbf{F}}\left(t, \vec{\mathbf{x}}(t), \vec{\mathbf{V}}(t)\right) = \sum_{j=1}^{N^a} \vec{\mathbf{F}}_{ij} + \vec{\mathbf{F}}_i(t) - \alpha_i m_i \vec{\mathbf{V}}_i(t), \quad i = 1, 2, .., N^a \tag{2.20}$$

The set of eq 2.18 is a Cauchy problem and can be integrated with the use of various numerical methods. Usually, for solving first-order differential equations, the Euler, Adams, and Runge–Kutta methods and predictor–corrector algorithms are employed.[71]

Based on the fact that the equations in set (2.18) are similar and have the form $du(t)/dt = f(t)$, we consider only the second equation of the set for the construction of numerical algorithms.

Let us assume that the solution of $\vec{\mathbf{V}}(t)$ for the time interval t is known. The task is to find the solution of $\vec{\mathbf{V}}(t_1)$ at a new time step $t_1 = t + \Delta t$, where Δt is the time step magnitude.

Owing to the simplicity of its realization, the Euler method occupies an important position in solving first-order differential equations. Consider the expansion of the $\vec{\mathbf{V}}(t + \Delta t)$ solution in a Taylor series of the kth order at the point t over powers $t_1 - t + \Delta t$

$$\vec{\mathbf{V}}(t+\Delta t)=\vec{\mathbf{V}}(t)+\Delta t\frac{d\vec{\mathbf{V}}(t)}{dt}+\frac{1}{2!}\Delta t^2\frac{d^2\vec{\mathbf{V}}(t)}{dt^2}+\ldots+$$
$$+\frac{1}{k!}\Delta t^k\frac{d^k\vec{\mathbf{V}}(t)}{dt^k}+\vec{O}\left(\Delta t^{k+1}\right). \quad (2.21)$$

The $\vec{\mathbf{V}}(t_1)=\vec{\mathbf{V}}(t+\Delta t)$ velocity is a desired solution and the summand $\vec{O}\left(\Delta t^{k+1}\right)$ indicates that at Taylor expansion of the solution the error will be the magnitude of the $(k + 1)$ th order with respect to the time step Δt.

If in the right-hand side of expansion eq 2.21, we use only two first summands, we obtain the ordinary Euler method:

$$\vec{\mathbf{V}}(t+\Delta t)=\vec{\mathbf{V}}(t)+\Delta t\frac{d\vec{\mathbf{V}}(t)}{dt}+\vec{O}\left(\Delta t^2\right) \quad (2.22)$$

The second equation in set (2.18) gives the relation

$$\frac{d\vec{\mathbf{V}}(t)}{dt}=\frac{\vec{\mathbf{F}}\left(t,\vec{\mathbf{x}}(t),\vec{\mathbf{V}}(t)\right)}{m} \quad (2.23)$$

and allows us to pass to the Euler method in the form:

$$\vec{\mathbf{V}}(t+\Delta t)=\vec{\mathbf{V}}(t)+\Delta t\frac{\vec{\mathbf{F}}\left(t,\vec{\mathbf{x}}(t),\vec{\mathbf{V}}(t)\right)}{m}+\vec{O}\left(\Delta t^2\right) \quad (2.24)$$

The error of the Euler method is characterized by the remainder term $\vec{O}\left(\Delta t^2\right)$. However, when the number of the integration steps is large, which is characteristic of molecular dynamics, the error will accumulate, and the order of the global error will be smaller by unity than that of a local error. Thus, the Euler method is a first-order method.

There exist different modifications of the Euler method allowing us to decrease the integration error. One of them can be obtained if in expansion (2.19) we consider three first summands rather than two:

$$\vec{\mathbf{V}}(t+\Delta t)=\vec{\mathbf{V}}(t)+\Delta t\frac{d\vec{\mathbf{V}}(t)}{dt}+$$
$$+\frac{1}{2!}\Delta t^2\frac{d^2\vec{\mathbf{V}}(t)}{dt^2}+\vec{O}\left(\Delta t^3\right) \quad (2.25)$$

The value of the first derivative is determined by relation (2.23). Upon differentiating $\bar{\mathbf{V}}(t)$ according to the formula of the total derivative, we obtain the value of the second derivative

$$\frac{d^2\vec{\mathbf{V}}(t)}{dt^2}=\frac{1}{m}\left[\frac{\partial\vec{\mathbf{F}}\left(t,\vec{\mathbf{x}}(t),\vec{\mathbf{V}}(t)\right)}{\partial t}+\frac{\partial\vec{\mathbf{F}}\left(t,\vec{\mathbf{x}}(t),\vec{\mathbf{V}}(t)\right)}{\partial\bar{\mathbf{x}}}\frac{d\vec{\mathbf{x}}(t)}{dt}\right]+$$

$$+\frac{1}{m}\frac{\partial\vec{\mathbf{F}}\left(t,\vec{\mathbf{x}}(t),\vec{\mathbf{V}}(t)\right)}{\partial\bar{\mathbf{V}}}\frac{d\vec{\mathbf{V}}(t)}{dt} \tag{2.26}$$

Taking into account the first equation in set (2.18) and formula (2.23) allows us to reduce the second derivative to the form:

$$\frac{d^2\vec{\mathbf{V}}(t)}{dt^2}=\frac{1}{m}\left[\frac{\partial\vec{\mathbf{F}}\left(t,\vec{\mathbf{x}}(t),\vec{\mathbf{V}}(t)\right)}{\partial t}+\frac{\partial\vec{\mathbf{F}}\left(t,\vec{\mathbf{x}}(t),\vec{\mathbf{V}}(t)\right)}{\partial\vec{\mathbf{x}}}\vec{\mathbf{V}}(t)\right]+$$

$$+\frac{1}{m^2}\frac{\partial\vec{\mathbf{F}}\left(t,\vec{\mathbf{x}}(t),\vec{\mathbf{V}}(t)\right)}{\partial\bar{\mathbf{V}}}\vec{\mathbf{F}}\left(t,\vec{\mathbf{x}}(t),\vec{\mathbf{V}}(t)\right) \tag{2.27}$$

The substitution of the expressions for finding the first and second derivatives from expansion (2.23) allows us to obtain the relation for the calculation of the velocity at the time step $t+\Delta t$:

$$\vec{\mathbf{V}}(t+\Delta t)=\vec{\mathbf{V}}(t)+\Delta t\frac{\vec{\mathbf{F}}\left(t,\vec{\mathbf{r}}(t),\vec{\mathbf{V}}(t)\right)}{m}+$$

$$+\frac{1}{2!}\Delta t^2\frac{1}{m}\frac{\partial\vec{\mathbf{F}}\left(t,\vec{\mathbf{r}}(t),\vec{\mathbf{V}}(t)\right)}{\partial t}+$$

$$+\frac{1}{2!}\frac{\Delta t^2}{m}\left[\begin{array}{l}\dfrac{\partial\vec{\mathbf{F}}\left(t,\vec{\mathbf{x}}(t),\vec{\mathbf{V}}(t)\right)}{\partial\bar{\mathbf{x}}}\vec{\mathbf{V}}(t)+\dfrac{\partial\vec{\mathbf{F}}\left(t,\vec{\mathbf{x}}(t),\vec{\mathbf{V}}(t)\right)}{\partial\vec{\mathbf{V}}}\times\\ \times\dfrac{\vec{\mathbf{F}}\left(t,\vec{\mathbf{x}}(t),\vec{\mathbf{V}}(t)\right)}{m}\end{array}\right]+ \tag{2.28}$$

$$+\vec{O}\left(\Delta t^3\right)$$

The method specified by formula (2.28) is called the corrected Euler method. The step error of this method is determined by the magnitude $\vec{O}\left(\Delta t^3\right)$, and, thus, the method has the second order of accuracy. The disadvantage of the method is the necessity to calculate the partial derivatives of the $\vec{\mathbf{F}}\left(t,\vec{\mathbf{x}},\vec{\mathbf{V}}\right)$ function.

If in formula (2.23) the value of the $\overline{\mathbf{F}}$ function is used at a new time step $t+\Delta t$, we obtain the implicit Euler method:

$$\vec{\mathbf{V}}\left(t+\Delta t\right)=\vec{\mathbf{V}}\left(t\right)+\Delta t\frac{\vec{\mathbf{F}}\left(t+\Delta t,\vec{\mathbf{x}}(t+\Delta t),\vec{\mathbf{V}}(t+\Delta t)\right)}{m} \\ ++\vec{O}\left(\Delta t^2\right) \tag{2.29}$$

The implementation of the implicit Euler method is difficult due to the fact that at each time step it is necessary to solve the nonlinear equation for the calculation of the value of $\vec{\mathbf{V}}\left(t+\Delta t\right)$. The method has the first order of accuracy.

The combination of the explicit and implicit Euler methods has led to the appearance of the trapezium method which refers to the methods of the second order of accuracy. It can be obtained with the help of the quadrature formula of trapezia.

$$\vec{\mathbf{V}}\left(t+\Delta t\right)=\vec{\mathbf{V}}\left(t\right)+\frac{\Delta t}{2}\frac{\vec{\mathbf{F}}\left(t,\vec{\mathbf{x}}(t),\vec{\mathbf{V}}(t)\right)}{m}+$$

$$+\frac{\Delta t}{2}\frac{\vec{\mathbf{F}}\left(t+\Delta t,\vec{\mathbf{x}}(t+\Delta t),\vec{\mathbf{V}}(t+\Delta t)\right)}{m}+\vec{O}\left(\Delta t^3\right) \tag{2.30}$$

The modification of the trapezium method with the help of the explicit Euler method has become a basis for the creation of a hybrid method, which is called the Heun method:

$$\vec{\mathbf{V}}\left(t+\Delta t\right)=\vec{\mathbf{V}}\left(t\right)+\frac{\Delta t}{2}\frac{\vec{\mathbf{F}}\left(t,\vec{\mathbf{x}}(t),\vec{\mathbf{V}}(t)\right)}{m}+$$

$$+\frac{\Delta t}{2}\frac{\vec{\mathbf{F}}\left(t+\Delta t,\vec{\mathbf{x}}(t+\Delta t),\vec{\mathbf{V}}(t)+\Delta t\vec{\mathbf{F}}\left(t,\vec{\mathbf{x}}(t),\vec{\mathbf{V}}(t)\right)\right)}{m}+ \\ +\vec{O}\left(\Delta t^3\right) \tag{2.31}$$

Better accuracy can be obtained with the use of several iterations at one time step according to the trapezium method. Such variant is called the improved Euler–Cauchy method with iterative processing. The essence of the method is presented by the following formulae:

$$\vec{\mathbf{V}}^{0}(t+\Delta t)=\vec{\mathbf{V}}(t)+\Delta t\frac{\vec{\mathbf{F}}\left(t,\vec{\mathbf{x}}(t),\vec{\mathbf{V}}(t)\right)}{m}+\vec{O}\left(\Delta t^{2}\right) \tag{2.32}$$

$$\vec{\mathbf{V}}^{p}(t+\Delta t)=\vec{\mathbf{V}}(t)+\frac{\Delta t}{2}\frac{\vec{\mathbf{F}}\left(t,\vec{\mathbf{x}}(t),\vec{\mathbf{V}}(t)\right)}{m}+$$

$$+\frac{\Delta t}{2}\frac{\vec{\mathbf{F}}\left(t+\Delta t,\vec{\mathbf{x}}(t+\Delta t),\vec{\mathbf{V}}^{p-1}(t+\Delta t)\right)}{m}+\vec{O}\left(\Delta t^{3}\right) \tag{2.33}$$

The accuracy of the Euler–Cauchy method with iterative processing is reached by the coincidence of a certain number of positions $\vec{\mathbf{V}}^{p-1}(t+\Delta t)$ and $\vec{\mathbf{V}}^{p}(t+\Delta t)$ at each time step and by the accuracy order of the method.

The integration of eq 2.18 in the time interval $[t-\Delta t,\ t+\Delta t]$ and the use of the quadrature formula of average rectangles for the calculation of the function $\vec{\mathbf{F}}$ integral allows us to obtain the more accurate Euler method:

$$\vec{\mathbf{V}}(t+\Delta t)=\vec{\mathbf{V}}(t-\Delta t)+2\Delta t\frac{\vec{\mathbf{F}}\left(t,\vec{\mathbf{x}}(t),\vec{\mathbf{V}}(t)\right)}{m}+\vec{O}\left(\Delta t^{3}\right) \tag{2.34}$$

The more accurate Euler method is a method of the second order. The specific feature of the method is that for finding the solution at the step $t+\Delta t$, it is necessary to know the solution in the t and $t-\Delta t$ time intervals; therefore it is a two-step method.

The essence of the Runge–Kutta method of the kth accuracy order is reflected in the use of the formula

$$\vec{\mathbf{V}}(t+\Delta t)=\vec{\mathbf{V}}(t)+\frac{\Delta t}{m}\vec{\varphi}\left(t,\vec{\mathbf{x}}(t),\vec{\mathbf{V}}(t),\Delta t\right)+\vec{O}\left(\Delta t^{k+1}\right) \tag{2.35}$$

where $\vec{\boldsymbol{\sigma}}\left(t,\vec{\mathbf{x}},\vec{\mathbf{V}},\Delta t\right)$ is the function approximating the Taylor series of the k th order (2.32), which does not comprise the $\vec{\mathbf{F}}$ function partial derivatives. For the Runge–Kutta methods, the multiparametric $\vec{\boldsymbol{\sigma}}\left(t,\vec{\mathbf{x}},\vec{\mathbf{V}},\Delta t\right)$ is chosen. Its parameters are determined based on the comparison with the expansion of the solution in Taylor series.

In the general case, the family of the Runge–Kutta methods of the k th order of accuracy is presented by the system of the relations:

$$\begin{cases} \vec{K}_1(t) = \dfrac{\vec{\mathbf{F}}\left(t, \vec{\mathbf{x}}(t), \vec{\mathbf{V}}(t)\right)}{m}, \\ \vec{K}_i(t) = \dfrac{1}{m}\vec{\mathbf{F}}\left(t + a_i\Delta t, \vec{\mathbf{x}}(t + a_i\Delta t), \vec{\mathbf{V}}(t) + \Delta t \sum_{j=1}^{i-1} b_{ij}\vec{K}_j(t)\right), \\ \vec{\mathbf{V}}(t+\Delta t) = \vec{\mathbf{V}}(t) + \Delta t \sum_{i=1}^{k} c_i \vec{K}_i(t) + \vec{O}\left(\Delta t^{k+1}\right), \end{cases} \tag{2.36}$$

where $i = 2, 3, \ldots, k$. The parameters a_i, b_{ij}, c_i of the system are found from the expansion of the solution in Taylor series of the required order of accuracy.

The Runge–Kutta methods of the second order of accuracy and better accuracy have a significant advantage in comparison with the Euler methods. When these methods are constructed, it is not necessary to calculate the $\vec{\mathbf{F}}$ function partial derivatives. In addition, the Runge–Kutta methods are one-step methods.

The Runge–Kutta method of the fourth order is a particular case and the most frequently used method from this family. The relations describing the method have the form:

$$\begin{cases} \vec{K}_1(t) = \dfrac{\vec{\mathbf{F}}\left(t, \vec{\mathbf{x}}(t), \vec{\mathbf{V}}(t)\right)}{m}, \\ \vec{K}_2(t) = \dfrac{1}{m}\vec{\mathbf{F}}\left(t + \dfrac{\Delta t}{2}, \vec{\mathbf{x}}(t + \dfrac{\Delta t}{2}), \vec{\mathbf{V}}(t) + \dfrac{\Delta t}{2}\vec{K}_1(t)\right), \\ \vec{K}_3(t) = \dfrac{1}{m}\vec{\mathbf{F}}\left(t + \dfrac{\Delta t}{2}, \vec{\mathbf{x}}(t + \dfrac{\Delta t}{2}), \vec{\mathbf{V}}(t) + \dfrac{\Delta t}{2}\vec{K}_2(t)\right), \\ \vec{K}_4(t) = \dfrac{1}{m}\vec{\mathbf{F}}\left(t + \Delta t, \vec{\mathbf{x}}(t + \Delta t), \vec{\mathbf{V}}(t) + \Delta t\vec{K}_3(t)\right), \\ \Delta\vec{\mathbf{V}}(t) = \dfrac{\Delta t}{6}\left(\vec{K}_1(t) + 2\vec{K}_2(t) + 2\vec{K}_3(t) + \vec{K}_4(t)\right), \\ \vec{\mathbf{V}}(t+\Delta t) = \vec{\mathbf{V}}(t) + \Delta\vec{\mathbf{V}}(t) + \vec{O}\left(\Delta t^5\right). \end{cases} \tag{2.37}$$

To derive the Adams numerical methods, the integral-interpolation approach is used. Upon integrating eq 2.21 at the time interval $[t, t + \Delta t]$, we arrive at the expression:

$$\vec{\mathbf{V}}(t+\Delta t)=\vec{\mathbf{V}}(t)+\frac{1}{m}\int\limits_{t}^{t+\Delta t}\vec{\mathbf{F}}\left(t,\vec{\mathbf{x}}(t),\vec{\mathbf{V}}(t)\right)dt \tag{2.38}$$

The $\vec{\mathbf{F}}\left(t,\vec{\mathbf{x}}(t),\vec{\mathbf{V}}(t)\right)$ integrand is replaced by an interpolating polynomial. The Newton interpolation formula allows us to obtain two types of polynomial for the $\vec{\mathbf{F}}$ function, namely, at the interpolation relative to the point t and the interpolation relative to the point $t + \Delta t$. Both variants of the interpolating polynomial comprise the $\vec{\mathbf{F}}$ function finite differences and are the basis of the multistep Adams methods.

When for the $\vec{\mathbf{F}}$ function the interpolation polynomial is used relative to the point t, we obtain the family of the Adams–Bashforth extrapolation methods. The order of the accuracy of the Adams–Bashforth methods varies depending on the number of steps in them. The methods are referred to the explicit methods for solving differential equations.

The first-order one-step Adams–Bashforth method known also as the Euler method is presented by the formula

$$\vec{\mathbf{V}}(t+\Delta t)=\vec{\mathbf{V}}(t)+\frac{\Delta t}{m}\vec{\mathbf{F}}\left(t,\vec{\mathbf{x}}(t),\vec{\mathbf{V}}(t)\right)+\vec{O}\left(\Delta t^{2}\right) \tag{2.39}$$

the second-order two-step method—

$$\vec{\mathbf{V}}(t+\Delta t)=\vec{\mathbf{V}}(t)+\frac{\Delta t}{2m}\Big[3\vec{\mathbf{F}}\left(t,\vec{\mathbf{x}}(t),\vec{\mathbf{V}}(t)\right)-$$

$$-\vec{\mathbf{F}}\left(t-\Delta t,\vec{\mathbf{x}}(t-\Delta t),\vec{\mathbf{V}}(t-\Delta t)\right)\Big]+\vec{O}\left(\Delta t^{3}\right) \tag{2.40}$$

the third-order three-step method—

$$\vec{\mathbf{V}}(t+\Delta t)=\vec{\mathbf{V}}(t)+\frac{\Delta t}{12m}\Big[23\vec{\mathbf{F}}\left(t,\vec{\mathbf{x}}(t),\vec{\mathbf{V}}(t)\right)-16\vec{\mathbf{F}}\left(t-\Delta t,\vec{\mathbf{x}}(t-\Delta t),\vec{\mathbf{V}}(t-\Delta t)\right)+$$

$$+5\vec{\mathbf{F}}\left(t-2\Delta t,\vec{\mathbf{x}}(t-2\Delta t),\vec{\mathbf{V}}(t-2\Delta t)\right)\Big]+\vec{O}\left(\Delta t^{4}\right) \tag{2.41}$$

the fourth-order four-step method—

$$\vec{\mathbf{V}}(t+\Delta t)=\vec{\mathbf{V}}(t)+\frac{\Delta t}{24m}\Big[55\vec{\mathbf{F}}\left(t,\vec{\mathbf{x}}(t),\vec{\mathbf{V}}(t)\right)-$$

$$-59\vec{\mathbf{F}}\left(t-\Delta t,\vec{\mathbf{x}}(t-\Delta t),\vec{\mathbf{V}}(t-\Delta t)\right)+37\vec{\mathbf{F}}\left(t-2\Delta t,\vec{\mathbf{x}}(t-2\Delta t),\vec{\mathbf{V}}(t-2\Delta t)\right)-$$

$$-9\vec{\mathbf{F}}\left(t-3\Delta t,\vec{\mathbf{x}}(t-3\Delta t),\vec{\mathbf{V}}(t-3\Delta t)\right)\Big]+\vec{O}\left(\Delta t^{5}\right) \tag{2.42}$$

The Adams–Moulton interpolation methods are based on the use of the interpolation polynomial relative to the $t+\Delta t$ point for the function $\vec{\mathbf{F}}$ in eq 2.27. The Adams–Moulton methods are implicit methods.

The one-step Adams–Moulton method of the first order, which is also called the implicit Euler method, is written in the form

$$\begin{aligned}\vec{\mathbf{V}}(t+\Delta t)=\vec{\mathbf{V}}(t)+\frac{\Delta t}{m}\vec{\mathbf{F}}\left(t+\Delta t,\vec{\mathbf{x}}(t+\Delta t),\vec{\mathbf{V}}(t+\Delta t)\right)+\\+\vec{O}\left(\Delta t^{2}\right)\end{aligned} \tag{2.43}$$

the one-step method of the second order known as the trapezium method

$$\begin{aligned}\vec{\mathbf{V}}(t+\Delta t)=\vec{\mathbf{V}}(t)+\frac{\Delta t}{2m}\vec{\mathbf{F}}\left(t,\vec{\mathbf{x}}(t),\vec{\mathbf{V}}(t)\right)+\\+\frac{\Delta t}{2m}\vec{\mathbf{F}}\left(t+\Delta t,\vec{\mathbf{x}}(t+\Delta t),\vec{\mathbf{V}}(t+\Delta t)\right)+\vec{O}\left(\Delta t^{3}\right)\end{aligned} \tag{2.44}$$

the two-step method of the third order—

$$\begin{aligned}\vec{\mathbf{V}}(t+\Delta t)=\vec{\mathbf{V}}(t)+\frac{\Delta t}{12m}\Big[5\vec{\mathbf{F}}\left(t+\Delta t,\vec{\mathbf{x}}(t+\Delta t),\vec{\mathbf{V}}(t+\Delta t)\right)+8\vec{\mathbf{F}}\left(t,\vec{\mathbf{x}}(t),\vec{\mathbf{V}}(t)\right)-\\-\vec{\mathbf{F}}\left(t-\Delta t,\vec{\mathbf{x}}(t-\Delta t),\vec{\mathbf{V}}(t-\Delta t)\right)\Big]+\vec{O}\left(\Delta t^{4}\right)\end{aligned} \tag{2.45}$$

the three-step method of the fourth order—

$$\begin{aligned}\vec{\mathbf{V}}(t+\Delta t)=\vec{\mathbf{V}}(t)+\frac{\Delta t}{24m}\Big[9\vec{\mathbf{F}}\left(t+\Delta t,\vec{\mathbf{x}}(t+\Delta t),\vec{\mathbf{V}}(t+\Delta t)\right)+19\vec{\mathbf{F}}\left(t,\vec{\mathbf{x}}(t),\vec{\mathbf{V}}(t)\right)-\\\left.\begin{aligned}-5\vec{\mathbf{F}}\left(t-\Delta t,\vec{\mathbf{x}}(t-\Delta t),\vec{\mathbf{V}}(t-\Delta t)\right)+\\+\vec{\mathbf{F}}\left(t-2\Delta t,\vec{\mathbf{x}}(t-2\Delta t),\vec{\mathbf{V}}(t-2\Delta t)\right)\end{aligned}\right]+\vec{O}\left(\Delta t^{5}\right)\end{aligned} \tag{2.46}$$

The methods of prediction and correction or the predictor–corrector methods are two-step methods and they combine the use of explicit and implicit methods. At the first step, the putative solution of $\vec{\mathbf{V}}^{E}(t+\Delta t)$ is the predictor. Then the predictor is corrected and the obtained value of $\vec{\mathbf{V}}(t+\Delta t)$ is the corrector, which is the desired solution.

The algorithms of prediction and correction based on the Adams methods are widely used. A predictor is calculated with the use of the explicit Adams–Bashforth methods, and a corrector—with the use of the Adams–Moulton methods. Based on the fact that the Adams–Bashforth and Adams–Moulton methods can have a different accuracy order, the following predictor–corrector Adams methods are possible:

the first-order method (the explicit–implicit Euler method)

$$\begin{cases} \vec{\mathbf{V}}^{E}\left(t+\Delta t\right)=\vec{\mathbf{V}}\left(t\right)+\dfrac{\Delta t}{m}\vec{\mathbf{F}}\left(t,\vec{\mathbf{x}}(t),\vec{\mathbf{V}}(t)\right)+\vec{O}\left(\Delta t^{2}\right), \\ \vec{\mathbf{V}}\left(t+\Delta t\right)=\vec{\mathbf{V}}\left(t\right)+\dfrac{\Delta t}{m}\vec{\mathbf{F}}\left(t+\Delta t,\vec{\mathbf{x}}(t+\Delta t),\vec{\mathbf{V}}^{E}(t+\Delta t)\right)+\vec{O}\left(\Delta t^{2}\right), \end{cases} \tag{2.47}$$

the second-order method

$$\begin{cases} \vec{\mathbf{V}}^{E}\left(t+\Delta t\right)=\vec{\mathbf{V}}\left(t\right)+\dfrac{\Delta t}{2m}\Big[3\vec{\mathbf{F}}\left(t,\vec{\mathbf{x}}(t),\vec{\mathbf{V}}(t)\right)- \\ \qquad -\vec{\mathbf{F}}\left(t-\Delta t,\vec{\mathbf{x}}(t-\Delta t),\vec{\mathbf{V}}(t-\Delta t)\right)\Big]+\vec{O}\left(\Delta t^{3}\right), \\ \vec{\mathbf{V}}\left(t+\Delta t\right)=\vec{\mathbf{V}}\left(t\right)+\dfrac{\Delta t}{2m}\Big[\vec{\mathbf{F}}\left(t+\Delta t,\vec{\mathbf{x}}(t+\Delta t),\vec{\mathbf{V}}^{E}(t+\Delta t)\right)+ \\ \qquad +\vec{\mathbf{F}}\left(t,\vec{\mathbf{x}}(t),\vec{\mathbf{V}}(t)\right)\Big]+\vec{O}\left(\Delta t^{3}\right), \end{cases} \tag{2.48}$$

the third-order method

$$\begin{cases} \vec{\mathbf{V}}^{E}\left(t+\Delta t\right)=\vec{\mathbf{V}}\left(t\right)+\dfrac{\Delta t}{12m}\Big[23\vec{\mathbf{F}}\left(t,\vec{\mathbf{x}}(t),\vec{\mathbf{V}}(t)\right)-16\vec{\mathbf{F}}\left(t-\Delta t,\vec{\mathbf{x}}(t-\Delta t),\vec{\mathbf{V}}(t-\Delta t)\right)+ \\ \qquad +5\vec{\mathbf{F}}\left(t-2\Delta t,\vec{\mathbf{x}}(t-2\Delta t),\vec{\mathbf{V}}(t-2\Delta t)\right)\Big]+\vec{O}\left(\Delta t^{4}\right), \\ \vec{\mathbf{V}}\left(t+\Delta t\right)=\vec{\mathbf{V}}\left(t\right)+\dfrac{\Delta t}{12m}\Big[5\vec{\mathbf{F}}\left(t+\Delta t,\vec{\mathbf{x}}(t+\Delta t),\vec{\mathbf{V}}^{E}(t+\Delta t)\right)+8\vec{\mathbf{F}}\left(t,\vec{\mathbf{x}}(t),\vec{\mathbf{V}}(t)\right)- \\ \qquad -\vec{\mathbf{F}}\left(t-\Delta t,\vec{\mathbf{x}}(t-\Delta t),\vec{\mathbf{V}}(t-\Delta t)\right)\Big]+\vec{O}\left(\Delta t^{4}\right) \end{cases} \tag{2.49}$$

The predictor–corrector method of the fourth order can be constructed similarly to (2.47)–(2.49) according to Adams–Bashforth and Adams–Mouton formulae (2.42) and (2.46), respectively. By comparing the solutions of $\vec{\mathbf{V}}^{E}(t+\Delta t)$ and $\vec{\mathbf{V}}\left(t+\Delta t\right)$ in the predictor–corrector methods, it is possible to monitor the value of the error, which is a significant advantage of these methods. There are other numerical algorithms for solving the molecular dynamics problems as well, for example, the Nordsick predictor–corrector.

In the calculation practice, Verlet numerical algorithms and predictor–corrector having numerous modifications and variations are most frequently used; therefore, let us consider these methods in detail.

The predictor–corrector algorithms are the most frequently used algorithms of the integration of equations of motion, which are the alternatives of the Verlet algorithms. Such algorithms consist of three parts:

1. Predictor: when the positions $\overline{x}(t)$ of atoms and their time derivatives are known, the positions of atoms, velocities, and accelerations are predicted at the next time step.
2. The determination of the acceleration error: the forces are calculated with the use of the potential gradient at the predicted positions. The obtained accelerations will differ from the acceleration values obtained at the predictor step. The difference between the two values is the signal of an error.
3. Corrector: the error signal is used for the correction of the positions $\overline{x}(t)$ of atoms and their derivatives. All the corrections are proportional to the error signal. The corrections are necessary for the stabilization of the algorithms.

Consider the Nordsic predictor–corrector algorithm. Let us determine the following variables:

$$\vec{\mathbf{R}}_1(t) = \Delta t \frac{d\vec{\mathbf{x}}(t)}{dt}, \tag{2.50}$$

$$\vec{\mathbf{R}}_2(t) = \frac{\Delta t^2}{2} \frac{d^2\vec{\mathbf{x}}(t)}{dt^2}, \tag{2.51}$$

$$\vec{\mathbf{R}}_3(t) = \frac{\Delta t^3}{6} \frac{d^3\vec{\mathbf{x}}(t)}{dt^3}. \tag{2.52}$$

The predictor step:

$$\vec{\mathbf{x}}(t+\Delta t) = \vec{\mathbf{x}}(t) + \vec{\mathbf{R}}_1(t) + \vec{\mathbf{R}}_3(t), \tag{2.53}$$

$$\vec{\mathbf{R}}_1(t+\Delta t) = \vec{\mathbf{R}}_1(t) + 2\vec{\mathbf{R}}_2(t) + 3\vec{\mathbf{R}}_3(t), \tag{2.54}$$

$$\vec{\mathbf{R}}_2(t+\Delta t)=\vec{\mathbf{R}}_2(t)+3\vec{\mathbf{R}}_3(t). \tag{2.55}$$

The step of the acceleration error determination:

$$\delta\vec{\mathbf{R}}_2(t+\Delta t)=\frac{\vec{\mathbf{F}}(t+\Delta t)}{2m}\Delta t^2-\vec{\mathbf{R}}_2(t+\Delta t), \tag{2.56}$$

The corrector step:

$$\vec{\mathbf{x}}^c(t+\Delta t)=\vec{\mathbf{x}}(t+\Delta t)+\frac{1}{6}\delta\vec{\mathbf{R}}_2(t+\Delta t), \tag{2.57}$$

$$\vec{\mathbf{R}}_1^c(t+\Delta t)=\vec{\mathbf{R}}_1(t+\Delta t)+\frac{5}{6}\delta\vec{\mathbf{R}}_2(t+\Delta t), \tag{2.58}$$

$$\vec{\mathbf{R}}_2^c(t+\Delta t)=\vec{\mathbf{R}}_2(t+\Delta t)+\delta\vec{\mathbf{R}}_2(t+\Delta t), \tag{2.59}$$

$$\vec{\mathbf{R}}_3^c(t+\Delta t)=\vec{\mathbf{R}}_3(t+\Delta t)+\frac{1}{3}\delta\vec{\mathbf{R}}_2(t+\Delta t) \tag{2.60}$$

Here $\vec{\mathbf{x}}^c(t+\Delta t)$ is the corrected position of an atom.

The Verlet algorithms are based on the solution expansion in Taylor series (2.36) (only several modifications of the Verlet method are known, which differ in the order of accuracy and the construction of numerical algorithms) and in the essence, they are the adaptations of the explicit and implicit Euler methods. The Verlet algorithm is an implicit symmetrical difference scheme.[174,175] The equation for the determination of the position of an atom i at the step $t+\Delta t$ taking into account the ith atom previous position $\overline{\mathbf{x}}(t)$ and the value of the force acting upon it has the form:

$$\vec{\mathbf{x}}(t+\Delta t)=2\vec{\mathbf{x}}(t)-\vec{\mathbf{x}}(t-\Delta t)+\Delta t^2\frac{\vec{\mathbf{F}}(\vec{\mathbf{x}})}{m_i}+\vec{O}(\Delta t^4) \tag{2.61}$$

Knowing the previous and the next positions of atoms, we calculate the velocities of the atoms:

$$\vec{\mathbf{V}}(t)=\frac{1}{2\Delta t}\left[\vec{\mathbf{x}}(t+\Delta t)-\vec{\mathbf{x}}(t-\Delta t)\right]+\vec{O}(\Delta t^4) \tag{2.62}$$

It is obvious that for obtaining the values of the velocities it is necessary to hold the values of the atoms positions at three steps in the computer memory. Note that knowing the position of atoms at the step $t+\Delta t$, the velocity value is determined only at the step t. There are modifications of the Verlet algorithm, which do not have such disadvantages, in particular, the Verlet velocity algorithm. When the Verlet velocity algorithm is used, the positions and velocities of atoms are calculated at the same moment. The basic equations have the form:

$$\vec{\mathbf{x}}(t+\Delta t)=\vec{\mathbf{x}}(t)+\vec{\mathbf{V}}(t)\Delta t+\frac{\vec{\mathbf{F}}(\vec{\mathbf{x}}(t))}{m_i}\Delta t^2+\vec{O}(\Delta t^3) \tag{2.63}$$

$$\vec{\mathbf{V}}(t+\Delta t)=\vec{\mathbf{V}}(t)+\frac{\Delta t}{2m}\left[\vec{\mathbf{F}}(\vec{\mathbf{x}}(t+\Delta t))+\vec{\mathbf{F}}(\vec{\mathbf{x}}(t))\right]+ +\vec{O}(\Delta t^3) \tag{2.64}$$

In practice the two above equations are broken into three:

$$\vec{\mathbf{V}}\left(t+\frac{1}{2}\Delta t\right)=\vec{\mathbf{V}}(t)+\Delta t\frac{\vec{\mathbf{F}}(\vec{\mathbf{x}}(t))}{2m} \tag{2.65}$$

$$\vec{\mathbf{x}}(t+\Delta t)=\vec{\mathbf{x}}(t)+\vec{\mathbf{V}}\left(t+\frac{1}{2}\Delta t\right)\Delta t \tag{2.66}$$

$$\vec{\mathbf{V}}(t+\Delta t)=\vec{\mathbf{V}}\left(t+\frac{1}{2}\Delta t\right)+\frac{\Delta t}{2m}\vec{\mathbf{F}}(\vec{\mathbf{x}}(t+1)) \tag{2.67}$$

Based on the first equation, the velocity at the $\left(t+\frac{1}{2}\Delta t\right)$ step is calculated with the use of the force and velocity at the step t. After that the displacement of the atom at the step $(t+\Delta t)$ is calculated. Finally, the velocity at the step $(t+\Delta t)$ is determined with taking into account the forces at a new position $\overline{\mathbf{x}}(t+\Delta t)$. This method occupies less computer memory, since it is necessary to hold only one set of coordinates, velocities of atoms and forces for any time step.

The economical version of the basic Verlet algorithm is a Leapfrog-algorithm. When the Leapfrog-algorithm is used, it is necessary to hold only one value of the position and velocities of atoms in the computer memory.

The Leapfrog-algorithm has the second order of accuracy and recalculates the coordinates of atoms and the velocity values for the next time step.

$$\vec{\mathbf{V}}\left(t+\frac{1}{2}\Delta t\right)=\vec{\mathbf{V}}\left(t-\frac{1}{2}\Delta t\right)+\Delta t\frac{\vec{\mathbf{F}}(\vec{\mathbf{x}}(t))}{m}+\vec{O}(\Delta t^3) \tag{2.68}$$

Knowing the velocity value at the $(t+\frac{1}{2}\Delta t)$ step we can find the next position of the atom:

$$\vec{\mathbf{x}}(t+\Delta t)=\vec{\mathbf{x}}(t)+\vec{\mathbf{V}}\left(t+\frac{1}{2}\Delta t\right)\Delta t+\vec{O}(\Delta t^4) \tag{2.69}$$

Then, the current velocities at the moment t are calculated:

$$\vec{\mathbf{V}}(t)=\frac{1}{2}\left[\vec{\mathbf{V}}\left(t+\frac{1}{2}\Delta t\right)+\vec{\mathbf{V}}\left(t-\frac{1}{2}\Delta t\right)\right]+\vec{O}(\Delta t^2) \tag{2.70}$$

The advantage of this method is the possibility to omit eq 2.64, which significantly speeds up the calculation. The technique does not influence the model adequacy, since during the calculation it is possible to use eq 2.64 at any moment for finding the velocity value at the entire step t. Note that the Leapfrog-algorithm is one of the quickest among all the above-described algorithms.

2.3 NUMERICAL SCHEMES FOR THE INTEGRATION OF THE MESODYNAMICS EQUATIONS

2.3.1 MOTION OF NANOPARTICLE WITHOUT ROTATION

Solving mesodynamics eq 1.62 at initial conditions (1.63) also requires the use of the integration numerical methods.[86] If the rotation of nanoelements is not taken into account, the set of eq 1.62 obtains the form

$$\begin{cases} M_k \dfrac{d^2X_1^k}{dt^2}=\sum\limits_{j=1}^{N_e}F_{X_1}^{kj}+F_{X_1}^{ke}, \\ M_k \dfrac{d^2X_2^k}{dt^2}=\sum\limits_{j=1}^{N_e}F_{X_2}^{kj}+F_{X_2}^{ke}, \\ M_k \dfrac{d^2X_3^k}{dt^2}=\sum\limits_{j=1}^{N_e}F_{X_3}^{kj}+F_{X_3}^{ke}. \end{cases} \tag{2.71}$$

In this case, all the numerical integration methods described in the previous section can be used for mesodynamics without changes.

2.3.2 MOTION OF NANOPARTICLES WITH ROTATION

For solving the complete set of eq 1.62 it is necessary to use numerical schemes taking into account both the linear displacement and the rotation of nanoelements. Here, it is practical to use the Runge–Kutta method, the scheme of which has the form with regard to mesodynamics eq 1.62:

$$(X_i^k)_{n+1} = (X_i^k)_n + (V_i^k)_n \Delta t + \frac{1}{6}(\mu_{1i}^k + \mu_{2i}^k + \mu_{3i}^k)\Delta t \tag{2.72}$$

$$(V_i^k)_{n+1} = (V_i^k)_n + \frac{1}{6}(\mu_{1i}^k + 2\mu_{2i}^k + 2\mu_{3i}^k + \mu_{4i}^k) \tag{2.73}$$

$$\mu_{1i}^k = \Phi_i^k(t_n;(X_i^k)_n,...;(V_i^k)_n...)\Delta t \tag{2.74}$$

$$\mu_{2i}^k = \Phi_i^k(t_n + \frac{\Delta t}{2};(X_i^k + V_i^k\frac{\Delta t}{2})_n,...;(V_i^k)_n + \frac{\mu_{1i}^k}{2},...)\Delta t \tag{2.75}$$

$$\mu_{3i}^k = \Phi_i^k(t_n + \frac{\Delta t}{2};(X_i^k + V_i^k\frac{\Delta t}{2} + \mu_{1i}^k\frac{\Delta t}{4})_n,...;(V_i^k)_n + \frac{\mu_{2i}^k}{2},...)\Delta t \tag{2.76}$$

$$\mu_{4i}^k = \Phi_i^k(t_n + \Delta t;(X_i^k + V_i^k\Delta t + \mu_{2i}^k\frac{\Delta t}{2})_n,...;(V_i^k)_n + \mu_{2i}^k,...)\Delta t \tag{2.77}$$

$$\Phi_i^k = \frac{1}{M_k}(\sum_{j=1}^{N_e} F_{X_3}^{kj} + F_{X_3}^{ke}) \tag{2.78}$$

$$(\Theta_i^k)_{n+1} = (\Theta_i^k)_n + (\frac{d\Theta_i^k}{dt})_n \Delta t + \frac{1}{6}(\lambda_{1i}^k + \lambda_{2i}^k + \lambda_{3i}^k)\Delta t \tag{2.79}$$

$$(\frac{d\Theta_i^k}{dt})_{n+1} = (\frac{d\Theta_i^k}{dt})_n + \frac{1}{6}(\lambda_{1i}^k + 2\lambda_{2i}^k + 2\lambda_{3i}^k + \lambda_{4i}^k) \tag{2.80}$$

$$\lambda_{1i}^k = \Psi_i^k(t_n;(\Theta_i^k)_n,...;(\frac{d\Theta_i^k}{dt})_n...)\Delta t \tag{2.81}$$

$$\lambda_{2i}^k = \Psi_i^k(t_n + \frac{\Delta t}{2};(\Theta_i^k + \frac{d\Theta_i^k}{dt}\frac{\Delta t}{2})_n,...;(\frac{d\Theta_i^k}{dt})_n + \frac{\lambda_{1i}^k}{2},...)\Delta t \tag{2.82}$$

$$\lambda_{4i}^k = \Psi_i^k(t_n + \Delta t; (\Theta_i^k + \frac{d\Theta_i^k}{dt}\Delta t + \lambda_{2i}^k \frac{\Delta t}{2})_n, ...; (\frac{d\Theta_i^k}{dt})_n + \lambda_{2i}^k, ...)\Delta t \quad (2.83)$$

$$\lambda_{4i}^k = \Psi_i^k(t_n + \Delta t; (\Theta_i^k + \frac{d\Theta_i^k}{dt}\Delta t + \lambda_{2i}^k \frac{\Delta t}{2})_n, ...; (\frac{d\Theta_i^k}{dt})_n + \lambda_{2i}^k, ...)\Delta t \quad (2.84)$$

$$\Psi_1^k = \frac{1}{J_{Z_1}^k}(-\frac{d\Theta_2^k}{dt} \cdot \frac{d\Theta_3^k}{dt}(J_{Z_3}^k - J_{Z_2}^k) + \sum_{j=1}^{N_e} M_{Z_1}^{kj} + M_{Z_1}^{ke}) \quad (2.85)$$

$$\Psi_2^k = \frac{1}{J_{Z_2}^k}(-\frac{d\Theta_1^k}{dt} \cdot \frac{d\Theta_3^k}{dt}(J_{Z_1}^k - J_{Z_3}^k) + \sum_{j=1}^{N_e} M_{Z_2}^{kj} + M_{Z_2}^{ke}) \quad (2.86)$$

$$\Psi_2^k = \frac{1}{J_{Z_2}^k}(-\frac{d\Theta_1^k}{dt} \cdot \frac{d\Theta_3^k}{dt}(J_{Z_1}^k - J_{Z_3}^k) + \sum_{j=1}^{N_e} M_{Z_2}^{kj} + M_{Z_2}^{ke}) \quad (2.87)$$

$$\Psi_3^k = \frac{1}{J_{Z_3}^k}(-\frac{d\Theta_1^k}{dt} \cdot \frac{d\Theta_2^k}{dt}(J_{Z_1}^k - J_{Z_2}^k) + \sum_{j=1}^{N_e} M_{Z_3}^{kj} + M_{Z_3}^{ke}) \quad (2.88)$$

2.3.3 MOTION OF NON-INTERACTING NANOPARTICLES

Note that for the problem of the motion of non-interacting nanoelements, when external forces and the rotation of nanoelements are absent, the set of the equations is broken into the equations of motion of individual nanoelements and we arrive at the form:

$$\begin{cases} M_k \frac{d^2 X_1^k}{dt^2} = 0, \\ M_k \frac{d^2 X_2^k}{dt^2} = 0, \\ M_k \frac{d^2 X_3^k}{dt^2} = 0. \end{cases} \quad (2.89)$$

This set of equations has an analytical solution, and all nanoelements move along the straight lines. However, if there are the processes of the coalescence of nanoelements leading to the formation of new nanoelements according to conditions (1.73), the analytical solution should be reconstructed every time the nanoelements coalesce.

2.3.4 ACCELERATION OF CALCULATIONS WITH MESODYNAMICS

The calculation of the motion of nanoelements with the use of the mesodynamics equations significantly saves the calculation time. Let us estimate the decrease of the time of the calculations conducted by the mesodynamics methods in comparison with the molecular dynamics method. For this purpose let us find the relation of the iteration number N_{md} of the molecular dynamics method to the iteration number N_{np} of the mesodynamics method at the calculation of the motion of the system consisting of N_p nanoparticles comprising N_{ai} atoms during the time t_c.

$$K_t = \frac{N_{md}}{N_{np}} \tag{2.90}$$

Here N_{md} is proportional to the number of the nanoparticles atoms $\sum_{i=1}^{N_p} N_{ai}$, to the number of the nearest neighbors of the atoms N_{ea}, and to the process time t_c and is in inverse proportion to the step of the integration of eq 1.14 in time Δt_{md}

$$N_{md} \to 6\sum_{i=1}^{N_p} N_{ai}\, N_{ea}\, t_c \,/\, \Delta t_{md} \tag{2.91}$$

N_{np} is proportional to the number of nanoparticles, the number of the nearest neighbors of nanoparticles N_{ep}, the process time t_c and is in inverse proportion to the integration step of eq 1.62 in time Δt_{md}

$$N_{np} \to 12 N_p N_{ep} t_c \,/\, \Delta t_{cp} \tag{2.92}$$

After the substitution of eqs 2.21 and 2.22 to eq 2.20 we arrive at

$$K_t = \frac{N_{md}}{N_{np}} = \frac{\sum_{i=1}^{N_p} N_{ai} N_{ea} \Delta t_{cp}}{2 N_p N_{ep} \Delta t_{md}} \tag{2.93}$$

For monodisperse particles consisting of the same number of atoms, formula (2.93) has the form

$$K_t = \frac{N_a N_{ea} \Delta t_{cp}}{2N_{ep} \Delta t_{md}} \tag{2.94}$$

Formula (2.94) allows us to estimate the total effectiveness of the mesodynamics method in comparison to that of the molecular dynamics method.

It can be seen that in this case the method effectiveness grows proportionally to the number of atoms N_a, the ratio of the numbers of the nearest neighbors N_{ea}/N_{ep}, and the ratio of the time integration steps $\Delta t_{cp}/\Delta t_{md}$.

For example, for $N_a = 1000$ and typical values $N_{ea} = 30, N_{ep} = 6, \Delta t_{cp} = 10^{-12}$, $\Delta t_{md} = 10^{-15}$ s, respectively, $K_t = 2{,}500{,}000$. It indicates the high efficiency of the method offered.

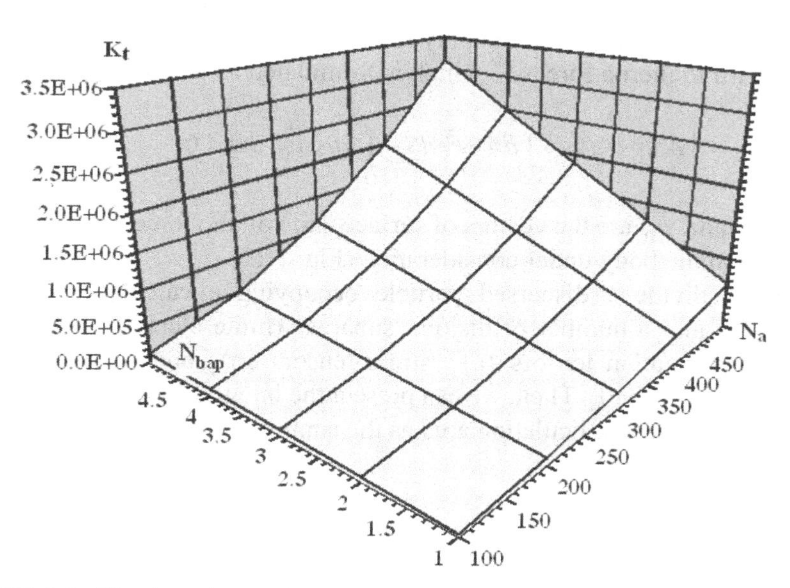

FIGURE 2.1 The dependence of the speedup of the calculations of K_t on the number of atoms N_a in a nanoparticle and the relation of the number of the nearest neighbors N_{ep}.

Figure 2.1 displays the plot of the K_t variation depending on the number of the parameters of a nanosystem: the number of atoms N_a in a nanoparticle and the relation of the number of the nearest neighbor N_{ep} of the calculation unit cell for the above methods. It can be seen that K_t increases proportionally to N_{ep} and N_a^2.

2.4 FINITE ELEMENT ALGORITHM OF THE DISPERSION OF NANOPARTICLES

In this section, the numerical algorithms are presented with the use of the finite-element method (FEM) for modeling isothermal and non-isothermal dispersion of powders with regard to inelastic deformations, initial defects, and phase transitions.

2.4.1 *FINITE ELEMENT FORMULATION OF THE PROBLEM OF ISOTHERMAL DISPERSION*

For the derivation of the finite-element set of equations of the problem of dynamic destruction we will employ the principle of possible displacements with regard to inertia forces,[196] which is formulated as follows

$$\int_{\Omega} \delta\, \hat{\varepsilon} \circ \hat{\sigma}\, d\Omega - \int_{S} \delta\vec{u} \circ \vec{F}_s\, dS - \int_{\Omega} \delta\vec{u} \circ \vec{F}_{\Omega}\, d\Omega = 0, \tag{2.95}$$

where $\vec{F}_s$ and $\vec{F}_{\Omega}$ are the vectors of surface and volume forces, respectively, acting upon the body under consideration (Fig. 1.12).

Let us divide a dispersed particle occupying a calculation area Ω (Fig. 1.12) into a number of disjoint subareas (finite elements (FEs)) and specify deformation tensors $\{\varepsilon\}_k$, stress tensors $\{\sigma\}_k$, and a displacement vector $\{u\}_k$ in each FE. Then, we can present the integrals over the volume Ω and surface S of the calculation area as the sum of integrals over all the FEs occupying this area

$$\sum_{k=1}^{N} \int_{\Omega_k} \delta\{\varepsilon\}_k^{\mathrm{T}} \{\sigma\}_k\, d\Omega_k - \sum_{k=1}^{N} \int_{S_k} \delta\{u\}_k^{T} \{F\}_k^{S}\, dS_k - \\ -\sum_{k=1}^{N} \int_{\Omega_k} \delta\{u\}_k^{T} \{F\}_k^{\Omega}\, d\Omega_k = 0 \tag{2.96}$$

where Ω_k and S_k are the FE volume and surface; $\{F\}_k^{\Omega}$ and $\{F\}_k^{S}$ are the vectors-matrices of the surface and volume forces acting within the volume and on the surface of the FEs, respectively; T is the symbol of transposition; δ is the symbol of variation.

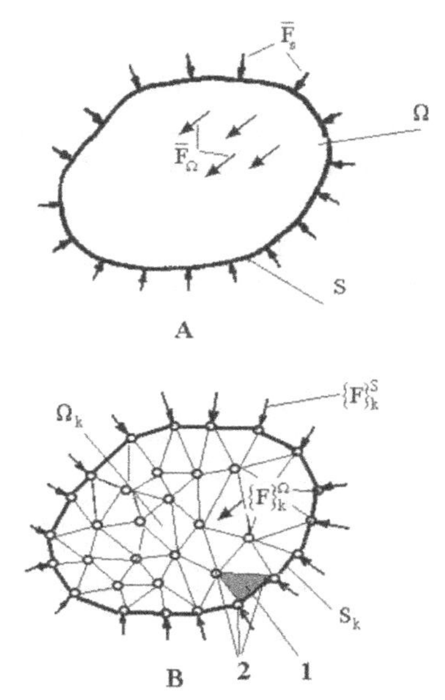

FIGURE 2.2 A dispersed particle (A) and its finite-element model (B): 1—a FE, 2—the FE nodes.

Let us specify the approximation of the displacement vector $\{u\}_k$ in the form

$$\{u\}_k = [N]\{\delta\} \tag{2.97}$$

where [N] is the shape functions; {δ} is the displacement vector for the nodes of the finite-element model.

The deformation within a FE can be found from the relation

$$\{\varepsilon\}_k = [B]\{\delta\} \tag{2.98}$$

where

$$[B] = [N] \otimes \bar{\nabla} + \bar{\nabla} \otimes [N] \tag{2.99}$$

is the matrix of the connection of the displacements of the FE nodes {δ}and the deformations $\{\varepsilon\}_k$ within a FE, which is calculated by eq 1.84.

Let us present the tensor of total deformations $\{\varepsilon\}_k$ in the FE as the sum of the tensors $\{\varepsilon\}_k^e$, $\{\varepsilon\}_k^p$, and $\{\varepsilon\}_k^c$ of elastic, plastic, and rheonomic deformations, respectively:

$$\{\varepsilon\}_k = \{\varepsilon\}_k^e + \{\varepsilon\}_k^p + \{\varepsilon\}_k^c \tag{2.100}$$

Using the connection between the components of the total deformation and the stresses in a linear form for the elastic part and in a quasi-linear form for the plastic and rheonomic deformations

$$\{\varepsilon\}_k^e = [D]^{-1}\{\sigma\}_k \tag{2.101}$$

$$\{\varepsilon\}_k^p = [R]^p * \{\sigma\}_k \tag{2.102}$$

$$\{\varepsilon\}_k^c = [R]^{\tilde{n}} * \{\sigma\}_k \tag{2.103}$$

we will find stresses $\{\sigma\}_k$ from eq 2.101 with regard to eqs 2.98 and 2.101–2.103

$$\{\sigma\}_k = [D][B]\{\delta\} - [D][R]^c * \{\sigma\}_k - [D][R]^p * \{\sigma\}_k \tag{2.104}$$

where [D] is the matrix of the material elastic constants; $[R]^p$ and $[R]^c$ are the matrices determining plastic and rheonomic properties of the material. The volume forces $\{F\}_k^\Omega$ are the inertia forces that depend on the system acceleration

$$\{F\}_k^\Omega = -\rho\left\{\frac{\partial^2 u}{\partial t^2}\right\} = -\rho[N]^T \frac{\partial^2}{\partial t^2}\{\delta\} \tag{2.105}$$

Upon substituting eqs 2.98, 2.100, 2.104, and 2.105 into eq 2.96, we arrive to the set of algebraic equations

$$[\mathbf{M}]\ \frac{\partial^2}{\partial t^2}\{\delta\} + [\mathbf{K}]\{\delta\} = \{\mathbf{f}\}_S + \{\mathbf{f}\}_P + \{\mathbf{f}\}_c \tag{2.106}$$

where

$$[\mathbf{M}] = \sum_{k=1}^{N} \int_{\Omega_k} \rho[\mathbf{N}]^T[\mathbf{N}]\, d\Omega_k \tag{2.107}$$

is the matrix of the "masses" of the FE system:

$$[\mathbf{K}] = \sum_{k=1}^{N} \int_{\Omega_k} [\mathbf{B}]^T [\mathbf{D}][\mathbf{B}] d\Omega_k \tag{2.108}$$

is the matrix of the system "rigidity:"

$$\{\mathbf{f}\}_S = \sum_{k=1}^{N} \int_{S_k} [N]^T \{F\}_k^S \, dS_k \tag{2.109}$$

is the vector of surface forces, reduced to the nodes of the FEs:

$$\{\mathbf{f}\}_P = \sum_{k=1}^{N} \int_{\Omega_k} [\mathbf{B}]^T [\mathbf{D}][R]^P * \{\sigma\} d\Omega_k \tag{2.110}$$

is the "fictitious volume forces" vector associated with plastic deformations:

$$\{\mathbf{f}\}_c = \sum_{k=1}^{N} \int_{\Omega_k} [\mathbf{B}]^T [\mathbf{D}][R]^c * \{\sigma\} d\Omega_k \tag{2.111}$$

is the "fictitious volume forces" vector associated with rheonomic deformations.

The set of eq 2.106 is non-linear since its right side and the rigidity matrix [K] depend on time due to the presence of rheonomic and inelastic deformations, which can appear due to the development of the stress and deformation fields within the particle.

Let us integrate the set of eq 2.106 using Newmark's method. The method relates to step-type methods. For the integration of set of eq 2.106 with respect to the time, the vectors of displacements, velocities, accelerations, and acceleration derivatives are used in the form:

$$\{\delta(t_i + \Delta t)\} = \{\delta_i\} + \Delta t \{\dot{\delta}_i\} + \frac{\Delta t^2}{2} \{\ddot{\delta}_i\} + \beta_1 \Delta t^3 \{\dddot{\delta}_i\} \tag{2.112}$$

$$\{\dot{\delta}(t_i + \Delta t)\} = \{\dot{\delta}_i\} + \Delta t \{\ddot{\delta}_i\} + \beta_2 \Delta t^2 \{\dddot{\delta}_i\} \tag{2.113}$$

where β_1, β_2 are the coefficients providing the stability of the time integration process.

In view of the relation

$$\{\dddot{\delta}_i\} = \left[(\{\ddot{\delta}_{i+1}\} - \{\ddot{\delta}_i\}) / \Delta t \right] \tag{2.114}$$

Newmark's method equations can be written in the following form:

$$\{\dot{\delta}_{i+1}\} = \frac{\beta_2}{\beta_1}\Delta t\left[\{\delta_{i+1}\} - \{\delta_i\}\right] - \left(1 - \frac{\beta_2}{\beta_1}\right)\{\dot{\delta}_i\} + \frac{\Delta t}{2}\left(2 - \frac{\beta_2}{\beta_1}\right)\{\ddot{\delta}_i\} \quad (2.115)$$

$$\{\ddot{\delta}_{i+1}\} = \frac{1}{\beta_1}\Delta t^2\left[\{\delta_{i+1}\} - \{\delta_i\}\right] + \frac{1}{\beta_1}\Delta t\{\dot{\delta}_i\} - \left(1 - \frac{1}{2\beta_1}\right)\{\ddot{\delta}_i\}, \quad (2.116)$$

Let us write eq 2.106 at the moment t_{i+1}

$$[M]\{\ddot{\delta}_{i+1}\} + [K]\{\delta_{i+1}\} = \{\mathbf{f}\}_{i+1} \quad (2.117)$$

where

$$\{\mathbf{f}\}_{i+1} = \{\mathbf{f}\}_S^{i+1} + \{\mathbf{f}\}_T^{i+1} + \{\mathbf{f}\}_p^{i+1} + \{\mathbf{f}\}_c^{i+1}. \quad (2.118)$$

Upon substituting eqs 2.115 and 2.116 in relations to eq 2.117, we obtain equations

$$[A]\{x_{i+1}\} = \{\mathbf{f}\}_{i+1}^{*} \quad (2.119)$$

where

$$[A] = [K] + \frac{1}{\beta_1\Delta t^2}[M] \quad (2.120)$$

$$\{\mathbf{f}\}_{i+1}^{*} = \frac{1}{\beta_1\Delta t^2}\{\delta_i\} + \frac{1}{\beta_1\Delta t}\{\dot{\delta}_i\} + \left(\frac{1}{2\beta_1} - 1\right)\{\ddot{\delta}_i\} + \{\mathbf{f}\}_{i+1} \quad (2.121)$$

Thus, the algorithm of the solution can be presented in the following form:

1. The initial values of vectors are specified.
2. With the use of eqs 2.118 and 2.119, the matrices [A] and $\{\mathbf{f}\}_{i+1}^{*}$ are calculated.
3. By solving the set of eq 2.117, the displacements of nodes by the end of the first step with respect to the time are determined.
4. With the use of relations (2.115) and (2.116), the velocity and acceleration of the nodes of the FEs by the end of the first step with respect to the time are calculated, which allows further integration of eq 2.121. In this case, at point 1 in the algorithm, the values are used which correspond to the end of the period t_{i+1}.

Next, using relations (2.114), we find stresses in each FE. When the equivalent stress exceeds the yield stress in an element, irreversible plastic deformations that are constituents of the vector of forces (2.111) are taken into account. If the equivalent stress is less than the yield stress according to condition (2.112), the plastic deformations are zero and, consequently, $\{\mathbf{f}\}_P = 0$ as well. However, if in the element the plastic deformations were observed at the previous time step, they remain constant during the element unloading until the yield stress is exceeded again. When destruction condition (2.109) is fulfilled in an FE, the rigidity matrix of this element is rearranged depending on the type of the element material destruction. In the case of "non-directional" destruction, the elastic characteristics of the element are equated to zero.[138] The appearance of a crack leads to the formation of the anisotropy of mechanical properties of the material in the element, and its matrix is constructed in the system associated with a crack.[12] It should also be noted, that the mass matrix of the "destructed" element, which is determined by relation (2.107), depends on the material density. For example, if the element is within a dispersed powder particle, the mass matrix does not change since the element continues its movement together with the particle. However, when the FE is destructed on the particle surface, the mass matrix is taken equal to zero, since the element separates from the particle, simulating breaking off.

2.4.2 FINITE ELEMENT FORMULATION OF THE NON-ISOTHERMAL DISPERSION PROBLEM

To solve the problem of nonisothermal dispersion, the simultaneous calculation of the parameters of the stressedly-deformed state and temperature fields is required, because the parameters of the problem of the stressedly-deformed state calculation depend on temperature. Therefore, the finite-element definition of the problem of isothermal dispersion given in the previous section of the present part should be supplemented with regard to temperature. First, it is necessary to add the temperature deformation $\{\varepsilon\}_k^T$

$$\{\varepsilon\}_k = \{\varepsilon\}_k^e + \{\varepsilon\}_k^p + \{\varepsilon\}_k^c + \{\varepsilon\}_k^T \quad (2.122)$$

where

$$\{\varepsilon\}_k^T = \alpha_T (T - T_o)\{I\} \quad (2.123)$$

α_T is the coefficient of the material temperature expansion; T is temperature; $\{I\}$ is the unity matrix.

In view of eqs 2.122 and 2.123, eq 2.102 takes the form:

$$\begin{gathered}\{\sigma\}=[D][B]\{\delta\}-[D][R]^{p}*\{\sigma\}_{k}-\\-[D][R]^{c}*\{\sigma\}_{k}-[D]\alpha_{T}(T-T_{o})\{I\}\end{gathered} \tag{2.124}$$

In view of eqs 2.122–2.123, the set of eq 2.106 for the calculation of the displacements of FE nodes takes the following form:

$$[\mathbf{M}]\frac{\partial^2}{\partial t^2}\{\delta\}+[\mathbf{K}]\{\delta\}=\{\mathbf{f}\}_S+\{\mathbf{f}\}_T+\{\mathbf{f}\}_P+\{\mathbf{f}\}_c \tag{2.125}$$

where

$$\{\mathbf{f}\}_T=\sum_{k=1}^{N}\int_{\Omega_k}[\mathbf{B}]^T[\mathbf{D}]\alpha_T(T-T_0)\{I\}\,d\Omega_k \tag{2.126}$$

is the "fictitious forces" vector conditioned by the temperature deformation.

In eq 2.125, a number of parameters depend on the temperature, which is calculated from the solution of the thermal conductivity problem. To solve this problem, we also use the method of FEs. For deriving the finite-element set of equations for the thermal conductivity problem, we use the condition of the functional minimum, which determines the solution of thermal conductivity eq 2.106,[196]

$$\begin{gathered}\Psi=\int_{\Omega}\frac{1}{2}\left(\lambda\begin{pmatrix}\left(\left(\frac{\partial T}{\partial x}\right)^2+\left(\frac{\partial T}{\partial y}\right)^2+\left(\frac{\partial T}{\partial z}\right)^2\right)-\\-2\left(q_p-\frac{\lambda}{a_T}\frac{\partial T}{\partial t}\right)T\end{pmatrix}\right)d\Omega+\\+\int_{S}\frac{\alpha}{2}\left(T^2-2TT_S-T_S^2)\right)dS=0.\end{gathered} \tag{2.127}$$

Similar to the stress calculation problem, let us divide the calculation area Ω (Fig. 2.2) into FEs and specify the temperature $\{T\}_k$ for each element. Then, the functional Ψ transforms into $\breve{\Psi}$, in which the integrals over the volume Ω and surface S in eq 2.127 are calculated as the sum of the integrals over the FEs

$$\breve{\Psi} = \sum_{k=1}^{N} \int_{\Omega_k} \frac{1}{2} \left(\lambda \left(\left(\frac{\partial \{T\}_k}{\partial x} \right)^2 + \left(\frac{\partial \{T\}_k}{\partial y} \right)^2 + \left(\frac{\partial \{T\}_k}{\partial z} \right)^2 \right) - \\ -2 \left(q_p - \frac{\lambda}{a_T} \frac{\partial \{T\}_k}{\partial t} \right) \{T\}_k \right) d\Omega_k + \quad (2.128)$$

$$+\sum_{k=1}^{N} \int_{S_k} \frac{\alpha}{2} \left(\{T\}_k^{\ 2} - 2\{T\}_k T_S - T_S^2 \right) dS_k = 0.$$

We take the order of the approximation of the temperature fields within the element similar to that in the stress calculation problem

$$\{T\}_k = [N]\{\mathbf{T}\} \quad (2.129)$$

where $\{T\}_k$ is the temperature within the FE; $\{\mathbf{T}\}$ is the temperature in the finite-element model nodes.

Upon substituting eq 2.129 in the functional eq 2.128, we obtain a linear algebraic set of equations relative to the temperature $\{\mathbf{T}\}$ in the FE nodes based on the condition of functional stationarity

$$[\mathbf{L}]_1 \{\mathbf{T}\} + [\mathbf{L}]_2 \frac{\partial}{\partial t} \{\mathbf{T}\} = \{\mathbf{f}\} \quad (2.130)$$

where

$$[\mathbf{L}]_1 = \sum_{k=1}^{N} \int_{\Omega_k} \lambda \left(\frac{\partial}{\partial x} [\mathbf{N}]^T + \frac{\partial}{\partial y} [\mathbf{N}]^T + \frac{\partial}{\partial z} [\mathbf{N}]^T \right) \left(\frac{\partial}{\partial x} [\mathbf{N}] + \frac{\partial}{\partial y} [\mathbf{N}] + \frac{\partial}{\partial z} [\mathbf{N}] \right) d\Omega_k + \\ + \sum_{k=1}^{N} \int_{S_k} \alpha [\mathbf{N}]^T [\mathbf{N}] dS_k \quad (2.131)$$

$$[\mathbf{L}]_2 = \sum_{k=1}^{N} \int_{\Omega_k} \frac{\lambda}{a_T} [\mathbf{N}]^T [\mathbf{N}] d\Omega_k \quad (2.132)$$

$$\{\mathbf{f}\} = \sum_{k=1}^{N} \int_{\Omega_k} [\mathbf{N}]^T q_p d\Omega_k + \sum_{k=1}^{N} \int_{S_k} [\mathbf{N}]^T \alpha T_S \, dS_k \quad (2.133)$$

where T_s is the temperature of the particle surface; α is the heat transfer coefficient; λ is the thermal conductivity coefficient; a_T is the temperature conductivity coefficient.

Thus, the finite-element formulation of the nonisothermal dispersion problem leads to the necessity of the simultaneous time integration of the interconnected sets of eqs 2.125 and 2.129. This requires special approaches, which would provide the simultaneous stability of the solutions of the heat problem and the problem of the calculation of the particle stressed-deformed state. Note that in the general case, the conditions of the numerical integration stability of the above systems do not coincide. To solve eq 2.125, we use an implicit integration scheme, which provides the count stability at an arbitrary time step and its value is determined from the condition of the stability of the integration of eq 2.130. For that purpose, we use the Euler explicit iterative scheme and the integration formula based on the method of trapezoids

$$\{T\}_t - \{T\}_0 = \frac{\Delta t}{2}(\{\frac{\partial \mathbf{T}}{\partial t}\}_t - \{\frac{\partial \mathbf{T}}{\partial t}\}_0) \quad (2.134)$$

where $\{T\}_t$ is the values of the temperature in the FE nodes at the end of the time step Δt; $\{T\}_0$ is the initial temperature distribution in the FE nodes.

Upon substituting eq 2.134 in eq 2.130, we arrive at

$$([\mathbf{L}]_1 + \frac{2}{\Delta t}[L]_2)\{\mathbf{T}\}_t = \{\mathbf{f}\}_t + [\mathbf{L}]_2\{\frac{\partial \mathbf{T}}{\partial t}\}_0 + \frac{2}{\Delta t}[L]_2)\{\mathbf{T}\}_0 \quad (2.135)$$

Using eq 2.130 at the initial moment, we obtain a recurrent relation for the step-by-step calculation of temperatures in the FE nodes.

$$[Q_1]\{T\}_{t_{i+1}} = [Q_2]\{T\}_{t_i} + \{\mathbf{f}\}_{t_{i+1}} + \{\mathbf{f}\}_{t_i} \quad (2.136)$$

where

$$[Q_1] = ([L_1] + 2/\Delta t[L_2]) \quad (2.137)$$

$$[Q_2] = (2/\Delta t[L_2] - [L_1]) \quad (2.138)$$

t_{i+1} is the moment considered; t_i is the previous moment.

The stability of the integration with respect to the time of eq 2.130 is determined from the condition

$$\Delta t \leq \frac{1}{2}\frac{\Delta x^2}{\lambda} \quad (2.139)$$

where Δx is the specific size of the smallest FE.

2.4.3 FINITE ELEMENT MODELING OF THE DISPERSION PROBLEM WITH REGARD TO THE PHASE TRANSITIONS

As it was mentioned above, the intensive heating of a particle can cause the transition of its solid phase into the liquid or even gaseous state. In this case, additional stresses appear in the particle due to the change of the specific volumes and the pressure of gases formed within the particle. The set of equations describing the stressed state of the solid phase is presented in the previous section and remains unchanged.

When a solid material passes into the liquid state, its constitutive equations take a different form. The material becomes low compressible and its inner viscosity changes. This can be taken into account if the elastic matrix constants are specially selected. In addition, the solid-liquid phase transition is characterized by the change in the volume deformation $\{\varepsilon\}_{\phi}$. This requires that additional forces should be introduced in eq 2.106

$$\{\mathbf{p}\}_{\varphi} = \int_{\Omega}[\mathbf{B}]^{T}[\mathbf{D}]\{\varepsilon\}_{\varphi}d\Omega_{\varphi} \tag{2.140}$$

where Ω_{ϕ} is the volume of the elements, which have passed from the solid into the liquid state.

It is natural that the plastic deformation in such elements "disappears," and the rheonomic deformation is described by the relations, which are characteristic of viscous liquid with a small viscosity parameter.

The separation of a "liquid" final element from the particle is a special matter. If the element is within a solid particle, it does not separate. If it is on the surface, then it is necessary to evaluate the relation between the surface forces, which press the element against the surface, and the forces, which detach it from the surface. In this case, the external pressure and the forces of the surface tension hold the element and the forces of inertia (mass forces) detach it. The relation between the above forces is determined from the condition of the inseparability of the element:

$$-\int_{\Omega_m}[\mathbf{M}]\frac{\partial^2}{\partial t^2}\{\delta\}d\Omega_m + \int_{S_m}[\mathbf{N}]^T\{\mathbf{f}\}_{S_m}dS_m + \{\mathbf{p}\}_{SF} \geq 0 \tag{2.141}$$

where Ω_m is the volume of the considered FE numbered m that is on the surface of the dispersed particle; S_m is the free surface of the element; $\{\mathbf{p}\}_{SF}$ is the sum vector of the forces of the surface tension.

Upon the "evaporation" of the element that is on the surface, it is sufficient to zero its relative weight and to take the modulus of elasticity of the material within the element equal to zero in order to take into account its "disappearance." If the element is within some other solid or liquid elements, the pressure of the gas, which has formed upon its evaporation, should be taken into consideration.

In this case, instead of the whole set of the acting "forces," the force appearing due to the gas pressure should be introduced into eq 2.106

$$\{\mathbf{p}\}_g = \int_{S_e} [\mathbf{N}]^T P_g \{I\}\, dS_e \tag{2.142}$$

where P_g is the gas pressure; S_e is the surface of the element numbered e that has passed into the "gaseous" state.

The gas pressure P_g can be found based on the law of conservation of mass. The mass of a "non-evaporated" element is

$$m = \int_{\Omega_e} \rho_e d\Omega_e \tag{2.143}$$

Then

$$P_g = \frac{mRT}{\Omega} = \frac{\int_{\Omega_e} \rho d\Omega_e RT}{\Omega_e} \tag{2.144}$$

If ρ is constant,

$$P_g = \rho_e RT. \tag{2.145}$$

The element elastic constants should be taken equal to zero.

At the phase transitions, the heat problem solution should take into account the heat required for the evaporation or melting during the phase transition and the variation of the thermal and physical constants of the material such as thermal conductivity, density, temperature conductivity, and specific heat.

The heat required for the phase transitions is taken into consideration in the thermal conductivity equation by the introduction of the specific heat absorption, which leads to the appearance of an additional summand in eq 2.134.

2.5 MODELING OF NANOCOMPOSITE COMPACTION BY THE METHOD OF FINITE ELEMENTS

In this section, a calculation of the succession of technological problems of pressing nanocomposites using the method of FEs is presented; this method allows taking into account non-linearity and evolution in time of the processes under study, real boundary conditions and other specific features of the problems that are being solved.

At each operating step (Fig. 1.21), both scleronomic and rheonomic properties of a powder material become apparent. Depending on the technological process (step duration and the change of physical-mechanical properties composites) at this step, the influence of the above properties is different. At active pressing and unloading (short-time steps), the material scleronomic properties are the most essential; the influence of the composite rheonomic properties is more noticeable during technological "wait" and storage of a finished product. Naturally, a complete model of the process should take into account the whole range of material properties. In practice, the use of such a model is very difficult, since complicated algorithms and programs are required and calculations are time-consuming. Therefore, let us consider the development of finite-element schemes for the calculation of separate steps of producing products from powder metal composites (PMC) based on the mathematical models, which were described in Chapter 1, having regard to the scleronomic and rheonomic material properties. At the same time, let us show the way to realize the connection of the steps into a single technological process.

2.5.1 *FINITE ELEMENT DEFINITION OF THE PROBLEM OF PRESSING*

The generation of residual stresses,[151] which hold powder particles, takes place at the step of forming (pressing) a powder mixture. The value and distribution of these stresses in the composite at the end of pressing are initial information for the analysis of the redistribution of these stresses with time and their influence on the future product properties.

Like most other technological problems, the problem of elasto-plastic deformation of a powder body in a limited volume is non-linear.[142] Non-linearity is due to the character of the relation between stresses and strains and possible geometric changes as well, which occur during plastic processing. It is shown in Ref. 14 that the solution of such problems is based on known variational principles. Pozdeyev et al.[120] developed the method to analyze the processes of cold strain by means of the principle of virtual displacements. This method in combination with the FEM allows us to reflect a real pattern of deformation rather accurately with regard to true properties of a material and real boundary conditions. In this case, the solution of a nonlinear problem is reduced to the solution of the sequence of "elastic" linear problems with the help of a certain iterative process.

To make a mathematical model of the pressing process, we use the finite-element approach; its detailed description is given in Ref.14 By using a variational definition of the FEM in the form of displacements,[196] we obtain a set of equations for the FEM of a powder body compacted (Fig. 2.3) relative to the increment of the nodes of FEs

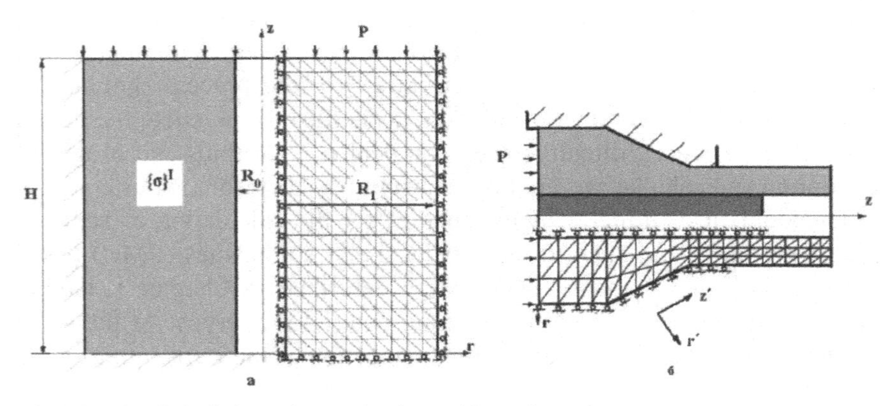

FIGURE 2.3 Calculation schemes for the problem of pressing.

$$[K]^{ep}\{\Delta\delta\} = \{\Delta\mathbf{f}\}_S \tag{2.146}$$

where

$$\{\Delta\mathbf{f}\}_S = \sum_{k=1}^{N} \int_{S_k} [N]^T \{\Delta F\}_k^S \, dS_k \tag{2.147}$$

is the vector of the increments of the node forces corresponding to the increments of the surface loads, that is, pressures of pressing and frictional forces;

$$[\mathbf{K}]^{ep} = \sum_{k=1}^{N} \int_{\Omega_k} [\mathbf{B}]^T [\mathbf{D}]^* [\mathbf{B}] d\Omega_k \tag{2.148}$$

is the elasto-plastic matrix of rigidity of the system of elements; $[D]^*$ is the matrix which establishes the connection between the increments of the components of stress $\{\Delta\sigma\}$ and strains $\{\Delta\varepsilon\}$ in an element:

$$\{\Delta\sigma\} = [D]^* \{\Delta\varepsilon\}. \quad (2.149)$$

With regard to the theory of plasticity of an isotropically strengthened powder-porous medium, the matrix $[D]^*$ components in expression (2.4) in the matrix form at axially symmetric pressing are as follows[55]

$$[D]^* = \frac{E^*(1-c^*)}{(1+c^*)(1-2c^*)} \times$$

$$\times \begin{vmatrix} 1 & \frac{c^*}{1-c^*} & \frac{c^*}{1-c^*} & 0 \\ & 1 & \frac{c^*}{1-c^*} & 0 \\ & & 1 & 0 \\ \textit{симм.} & & & \frac{(1+c^*)(1-2c^*)(1+\alpha_\theta)}{3(1-c^*)} \end{vmatrix}, \quad (2.150)$$

where

$$E^* = \frac{\beta_\theta^{3n_\theta} \sigma_{eq}}{d\varepsilon_{eq}^p}, \quad (2.151)$$

$$c^* = \frac{\frac{1}{2} - \alpha_\theta}{1+\alpha_\theta}, \quad (2.152)$$

is the material constant taking into account the configuration of pores; α_θ and β_θ are the functions of porosity θ described in detail in Ref. 55

The porosity increment caused by the increment of external loads is found as

$$d\theta = \frac{9\alpha_\theta(1-\theta)d\varepsilon_{eq}^p \sigma_0}{\beta^{3n_\theta} \sigma_{eq}}. \quad (2.153)$$

The increments of equivalent plastic strains

$$d\varepsilon^{p}{}_{eq} = \beta_{\theta}{}^{2n_{\theta}-0,5}\left(\frac{2}{3}d\varepsilon_{i}{}^{2} + \frac{d\varepsilon_{0}^{2}}{\alpha_{\theta}}\right)^{1/2}, \tag{2.154}$$

are related to equivalent stresses

$$\sigma_{eq} = \frac{1}{\beta_{eq}{}^{n_{\theta}+0,5}}\left(\frac{2}{3}\sigma_{i}{}^{2} + 9\alpha_{\theta}\sigma_{0}{}^{2}\right)^{1/2} \tag{2.155}$$

by the dependence, which corresponds to the strain diagram of the material compacted

$$\sigma_{eq} = \sigma^{p}{}_{eq} + b_{1}\cdot\left(\int d\varepsilon^{p}_{eq}\right)^{b_{2}}, \tag{2.156}$$

where $d\varepsilon_i$ is the intensity increment for strains; $d\varepsilon_0$ is the increment of a mean strain.

At the axial symmetry of the process, we consider only half of the meridian section of the pressing in the calculation scheme (Fig. 2.3). On the pivot pin, we specify kinematic boundary conditions of the pivot. On the contact surface of the pressing and the die, we introduce kinematic restriction normal to the surface:

at pressing in a (blanking die) blind die (Fig. 2.3)

$$\{\Delta\delta_{r}\}_{r=R} = 0,\ \{\Delta\delta_{z}\}_{z=0} = 0; \tag{2.157}$$

at extrusion (Fig. 2.3b), it is necessary to add the following condition on the cone surface

$$\{\Delta\delta_{r}\}' = \{\Delta\delta_{r}\}\cos\gamma, \tag{2.158}$$

where $\{\Delta\delta_z\}$ and $\{\Delta\delta_r\}$ are axial and radial displacements of nodes; γ is the angle of pressing.

On the contact surface, we simultaneously consider concentrated nodal loads that are due to the presence of frictional forces:

$$\{\Delta F\}_{S} = -\operatorname{sign}\{\Delta\delta\}_{S}\, f\{\sigma_{n}\}, \tag{2.159}$$

where $\{\Delta\delta\}_s$ is the displacement of the nodes of FEs on the friction surface (here and further, the friction between the pressing and the die is meant);

f is the friction coefficient; $\{\sigma_n\}$ are stresses normal to the contact surface.

In accordance with the incremental definition of problem (2.146), the loading process can be presented as the sequence of small increments of external pressures $\sum_{i=1}^{m}\{\Delta F\}_i$; within the limits of each of them, we determine current properties (E_i^*, c_i^*, and) of the PC compacted using the method of variable parameters. The first step in the solution of the problem of pressing is carried out at the parameters of elasticity E_0^* and c_0^* of non-compacted powder, which correspond to the specified initial density θ_0 of a semi-finished product and $\{\Delta F\}_1$. Refinement of the solution of eq 2.146 at each $\{\Delta F\}_i$ is carried out according to the scheme

$$\{\Delta\delta\}_{i(n)} = \left([K]^{ep}\right)^{-1}_{i(n-1)} \cdot \{\Delta\mathbf{f}\}_i \qquad (2.160)$$

until the point within the coordinates $\sigma_{eq} - \int d\varepsilon_{eq}^{p}$, which expresses a stressed-strained state of a model element at the end of the increment $\{\Delta F\}_i$ of the load, is close enough to the strain curve of the PMC. When a maximal pressure is reached on the puncheon, stresses, and strains in the PMC compacted are found as follows

$$\{\sigma\}^{\mathrm{I}} = \{\sigma\}_0 + \sum_{i=1}^{m}[D]_i^*[B]\{\Delta\delta\}_i, \qquad (2.161)$$

$$\{\varepsilon\}^{\mathrm{I}} = \{\varepsilon\}_0 + \sum_{i=1}^{m}[B]\{\Delta\delta\}_i, \qquad (2.162)$$

where $\{\sigma\}_0$ and $\{\varepsilon\}_0$ are initial stresses and initial strains in the product being pressed.

2.5.2 FINITE ELEMENT DEFINITION OF THE UNLOADING PROCESS AFTER PRESSING

After the puncheon is removed, elastic "expansion" of the pressing is observed in the die. Naturally, in this case, the regions of tensile stresses appear in the material, which can lead to the appearance of cracks on the side surface of the product or the split of the "end plate." In investigations of the unloading process, a discrete model of the pressing is used, which appeared at the end of its pressing (Fig. 2.4). Here, every FE of the region under study presents the information on the properties of the material and its stressed state at the end of the compaction with regard to the position of the element in the entire system.

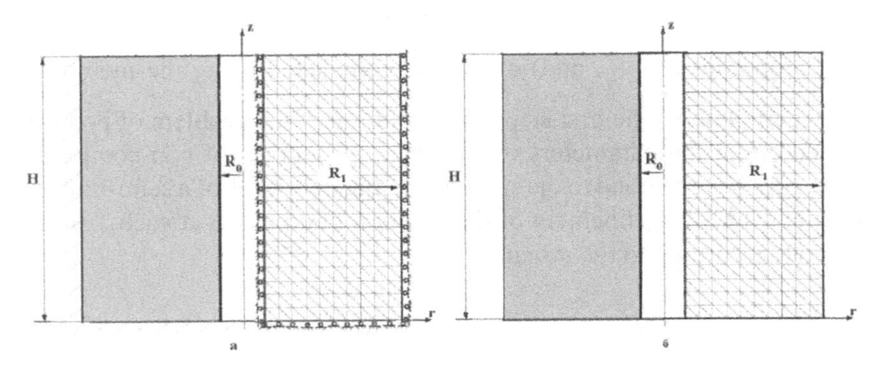

FIGURE 2.4 The calculation schemes for the problems of unloading and relaxation.

After finite-element transformations, the set of equations for the problem of unloading is reduced to the following form

$$[K]_\theta\{\delta\} = \{\mathbf{f}\}_{\sigma^{\mathrm{I}}} + \{\mathbf{f}\}_S \tag{2.163}$$

where $[\boldsymbol{K}]_\theta$ is a general matrix of rigidity that is determined by the equation

$$[\boldsymbol{K}]_\theta = \sum_{k=1}^{N} \int_{\Omega_k} [\mathbf{B}]^T [\mathbf{D}]_\theta [\mathbf{B}] d\Omega_k \tag{2.164}$$

$$\{\mathbf{f}\}_{\sigma^{\mathrm{I}}} = \sum_{k=1}^{N} \int_{\Omega_k} [B]^T \{\sigma\}^{\mathrm{I}} d\Omega_k \tag{2.165}$$

is the vector of nodal forces due to the presence of stresses $\{\sigma\}^{\mathrm{I}}$ in the pressing,

$$\{\mathbf{f}\}_S = \sum_{k=1}^{N} \int_{S_k} [N]^T \{F\}_k^S dS_k \tag{2.166}$$

is the vector of nodal forces resulting from the action of frictional forces,

$[\mathbf{D}]_\theta$ is the matrix of elastic constants dependent, in the general case, on the porosity of the material of the pressing at the maximal pressing pressure.

In this case, the boundary conditions includes force conditions in the form of the condition similar to that of eq 2.159, which are written in the form of finite displacements

$$\{F\}_S = -\text{sign}\{\delta\}_S f\{\sigma_n\} \tag{2.167}$$

and kinematic (conjugation conditions), under which some necessary limiting (one-sided) conditions are understood; they are taken into account in solving the problem. In the case of the rigid matrix, they can be written in the form of the conditions of "non-penetration"

$$\{\delta_r\}_{r=R} = 0 \text{ at } \{\sigma_r\}_{r=R} < 0; \ \{\delta_z\}_{z=0} = 0 \text{ at } \{\sigma_z\}_{z=0} < 0 \tag{2.168}$$

and simple-connectedness

$$\{\delta_r\}_{r=R} < 0 \text{ at } \{\sigma_r\}_{r=R} > 0; \ \{\delta_z\}_{z=0} > 0 \text{ at } \{\sigma_z\}_{z=0} > 0. \tag{2.169}$$

Thus, the problem defined is reduced to the solution of the set of linear algebraic equations (2.163) relative to the displacements of the FE nodes. In general, however, the problem is non-linear due to uncertainty of the vector $\{\mathbf{f}\}_s$. Therefore, the calculations are carried out with the use of the step-by-step method based on the sequential solution of an ordinary problem of the theory of elasticity for a heterogeneous body, which is corrected for the frictional force and in which at a certain step the conditions are verified (2.168 and 2.169).

At the first step, we apply the vector $\{\mathbf{f}\}_{\sigma^I}$ and frictional forces $\{\mathbf{f}\}_s$ to the element nodes, for the determination of which we use the epure of normal stresses $\{\sigma\}^I$ at the end of pressing. By solving eq 2.163, we find the components $\{\delta\}_1$, $\{\sigma\}_1$, and $\{\varepsilon\}_1$ of the stressed-strained state of the pressing, which correspond to the release of one of the ends from pressing pressures under the conditions of the initially acting frictional forces on the briquette surface. By correcting the frictional force vector for "new" normal stresses $\{\sigma\}_1$ of the system, we repeat the solution procedure until the required accuracy of the result is reached. After that, as the numerical experiment shows, it is practical to verify the conditions of simple-connectedness (2.169) (the conditions of "non-penetration" (2.169) are satisfied in solving the problem). As a result, we obtain the distribution of residual stresses and strains in the pressing after the puncheon has been removed, which is the only one possible at the specified initial stress state $\{\sigma\}^I$.

After the press-tool is parted (or the pressing is extruded from the die), the pressing undergoes an additional expansion which is accompanied by the redistribution of the previous (after the puncheon removal) residual stresses. The new stress state is determined from equations

$$[K]_\theta\{\delta\} = \{\mathbf{f}\}_\sigma \tag{2.170}$$

where $\{\mathbf{f}\}_\sigma$ is the vector of equivalent loads caused by the stresses $\{\sigma\}^{\mathrm{I}}$ at the end of the pressing process or the stresses after partial unloading which is, for example, due to the puncheon removal.

By solving eq 2.25 relative to the displacements of the elements' nodes $\{\delta\}$ and using physical equations for elastic body with the initial stresses $\{\sigma\}^{\mathrm{I}}$, we find the stresses after unloading from the expression

$$\{\sigma\}^{\mathrm{II}} = [D][B]\{\delta\} + \{\sigma\}^{\mathrm{I}}. \tag{2.171}$$

2.5.3 FINITE ELEMENT MODEL OF TECHNOLOGICAL "WAIT" PROCESSES OR STORAGE OF A PRESSING

The change of residual stresses with time (relaxation) occurs due to the rheological properties of a PC. Since strains are small in this case, we believe that total deformation is the sum of elastic and rheonomic components. Under the action of self-balanced residual stresses throughout the bulk of the product, the total deformation $\hat{\varepsilon} = \hat{\varepsilon}^c + \hat{\varepsilon}^e$ is redistributed between the increasing creep strain $\hat{\varepsilon}^c$ and decreasing elastic component $\hat{\varepsilon}^e$. The latter is the reason of the change of the stress state of the product.

The analysis of the relaxation of residual stresses is done based on the information on the product stressed-strained state after unloading and the law of creeping of a particular powder pressed under definite conditions. This information is introduced into the "memory" of the discrete model of the object under study that was made previously (e.g., at the calculation of unloading) (Fig. 2.4). The time interval considered is divided into rather small periods Δt_i, within which the stresses can be considered constant. Let us suppose that in the beginning of each time increment Δt_i, the total pattern of the stress state $\{\sigma\}_{i-1}$ is known (the stresses at the end of the previous step Δt_{i-1}). Then the increments of creep strains $\{\Delta\varepsilon\}_i^c$ related to stresses by a specific dependence are determined by the value of this step with respect to the time and $\{\sigma\}_{i-1}$ at this Δt_i

$$\{\Delta\varepsilon\}_i^c = [R]^c * \{\sigma\}_{i-1} \cdot \Delta t_i, i = 1, 2, 3... \tag{2.172}$$

where $[R]^c$ is the operator characterizing the reological properties of the PMC.

At the first step of the increments of the creep strains $\{\Delta\varepsilon\}_1^c$, caused by the elastic stresses $\{\sigma\}^{\text{II}}$ after the unloading of the pressing, in their turn, provide the vector increments reduced to the loads of the elements' nodes $\{\Delta\mathbf{f}\}_1^c$, which take place at each Δt_i

$$\{\Delta\mathbf{f}\}_i^c = \int\limits_{\Omega}[B]^T[D]\{\Delta\varepsilon\}_i^c d\Omega \,. \tag{2.173}$$

We think that the elastic properties of the compacted material are unchangeable during the redistribution of the residual stresses. To determine the changes in the product stressed-strained state for Δt_i taking place due to its strain under the loads $\{\Delta F\}_i^c$, let us again use the principle of possible displacements in the system of FEs, which imitates the pressing after unloading:

$$\int\limits_{\Omega}\delta\{\varepsilon\}^T\{\Delta\sigma\}_i d\Omega = \delta\{\delta\}_i^T\{F\}_i^c \,. \tag{2.174}$$

By assuming that within the step with respect to the time, the relationship between $\{\Delta\delta\}$, $\{\Delta\varepsilon\}$, and $\{\Delta\sigma\}$ increments remains in the form

$$\{\Delta\sigma\} = [D]_\theta\{\Delta\varepsilon\}, \tag{2.175}$$

$$\{\Delta\varepsilon\} = [B]\{\Delta\delta\}, \tag{2.176}$$

and summing over all the FEs, we obtain the set of equations relative to the displacement increments for the nodes of the FEs

$$[K]_\theta\{\Delta\delta\}_i = \{\Delta\mathbf{f}\}_i^c, \tag{2.177}$$

where $[K]_\theta$ has been determined previously.

$$\{\Delta\mathbf{f}\}_c = \sum_{k=1}^{N}\int\limits_{\Omega_k}[\mathbf{B}]^T[\mathbf{D}][R]^c * \{\sigma\}\,\Delta t_i\, d\Omega_k$$

The stress increments obtained for Δt_i

$$\{\Delta\sigma\}_i = [D][B]\{\Delta\delta\}_i, \tag{2.178}$$

added to the stresses existing at the beginning of the time increment Δt_{i-1} determine the stress state at its end

$$\{\sigma\}_i = \{\sigma\}_{i-1} + \{\Delta\sigma\}_i. \tag{2.179}$$

This calculation can be repeated n times until a stressed-strained state in each FE and the entire product has been obtained by the moment

$$\{\sigma\}_t = \{\sigma\}^{\text{II}} + \sum_{i=1}^{n} [D][B]\{\Delta\delta\}_i. \tag{2.180}$$

2.5.4 MODELING CONTACT CONDITIONS

The deformations of the press-tool essentially influence the stressed-strained state of the body compacted. It is believed, for example, that the die expansion, when the powder is compacted, and the die compression during unloading due to the reversed resistance of the body compacted (the pressing) are the reasons of transverse cracking of the finished product.[172] In some cases, this is also the reason of the press-tool failure. Therefore, in the works concerning with the investigation of the theoretical basis of the pressing processes or the solution of particular problems in this sphere, the above interaction has always been given much attention to. Since obtaining exhaustive information on the regularities of combined deformations of the pressing and the press-tool requires overcoming the same experimental and theoretical difficulties as those with which one confronts while solving the problem of pressing itself, simplifications are often used. As a rule, they are used for the calculation of the matrices of empirical laws of the distribution of contact stresses at the compacted-mass and press-tool (or puncheon) interface. Moreover, when the parameters of compacting the powders are calculated, the press-tool is often considered hard (non-deformable). In this case, the use of numerical methods allows expanding the boundaries of the phenomenon under study.

To model the combined deformation of the system consisting of the bodies "puncheon" and "the pressing-die" that are not interconnected (Fig. 2.5), let us introduce a thin contact layer (CL) on the boundary between the above bodies. The CL model should provide continuous transfer of the stresses that are normal to the surface of the interaction of the bodies, and allow their free movement along the above surface within the limits of acting frictional forces. This can be attained, for example, by imparting the properties of transversally isotropic medium to the layer elements. Let us differentiate the properties of the bottom and lateral CLs (Fig. 2.5). In the case of axially symmetric loading, the strain of the CL elements is described by the corresponding equations of Hooke's generalized law with regard to transversal isotropy.

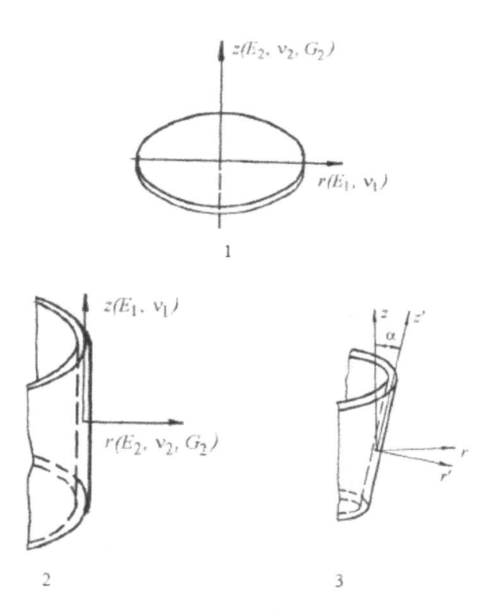

FIGURE 2.5 Contact layers modeling the interaction of a pressed powder body and an elastic press-tool: (1) a bottom (upper and lower ends of a pressing); (2) a lateral CLs; and (3) a CL on the oblique surface.

For the bottom layer (Fig. 2.5-1),

$$\varepsilon_{\phi} = -\nu_2 \frac{\sigma_z}{E_2} - \nu_1 \frac{\sigma_r}{E_1} + \frac{\sigma_{\phi}}{E_1}$$

$$\varepsilon_r = -\nu_2 \frac{\sigma_z}{E_2} + \frac{\sigma_r}{E_1} - \nu_1 \frac{\sigma_{\phi}}{E_1}, \tag{2.181}$$

$$\varepsilon_z = \frac{\sigma_z}{E_2} - \nu_2 \frac{\sigma_r}{E_2} - \nu_2 \frac{\sigma_{\phi}}{E_2}$$

$$\tau_{rz} = G_2 \gamma_{rz}$$

for the lateral layer (Fig. 2.5-2)

$$\varepsilon_\phi = -\nu_1 \frac{\sigma_z}{E_1} - \nu_2 \frac{\sigma_r}{E_2} + \frac{\sigma_\phi}{E_1}$$

$$\varepsilon_r = -\nu_2 \frac{\sigma_z}{E_2} + \frac{\sigma_r}{E_2} - \nu_2 \frac{\sigma_\phi}{E_2}, \tag{2.182}$$

$$\varepsilon_z = \frac{\sigma_z}{E_1} - \nu_2 \frac{\sigma_r}{E_2} - \nu_1 \frac{\sigma_\phi}{E_1}$$

$$\tau_{rz} = G_2 \gamma_{rz}$$

E_1, ν_1, E_2, ν_2, and G_2 are the elasticity constants of the transversally isotropic layer.

Or in the matrix form

$$\{\sigma\} = [D]_{cl} \{\varepsilon\}, \tag{2.183}$$

where $[D]_{cl}$ is a matrix of connection, which determines the stressed-strained state of the CL elements. The rest of designations are the same as above. For the bottom CL (the upper and the lower ends of the pressing) and the lateral CL, the matrix $[D]_{cl}$ has the following forms, respectively:

$$[D]_{cl}^{d} = \frac{E_2}{(1+\nu_1)(1-\nu_1-2n\nu_2^2)} \times$$

$$\times \begin{bmatrix} n(1-n\nu_2^2) & n(\nu_1+n\nu_2^2) & n\nu_2(1+\nu_1) & 0 \\ & n(1-n\nu_2^2) & n\nu_2(1+\nu_1) & 0 \\ & & 1-\nu_1^2 & 0 \\ \text{Симм.} & & & m(1+\nu_1)\times \\ & & & \times(1-\nu_1-2n\nu_2^2) \end{bmatrix}, \tag{2.184}$$

$$[D]_{cl}^{b} = \frac{E_2}{(1+\nu_1)(1-\nu_1-2n\nu_2^2)} \times$$

$$\times \begin{bmatrix} n(1-n\nu_2^2) & n\nu_2(1+\nu_1) & n(\nu_1+n\nu_2^2) & 0 \\ & 1-\nu_1^2 & n\nu_2(1+\nu_1) & 0 \\ & & n(1-n\nu_2^2) & 0 \\ \text{Симм} & & & m(1+\nu_1)\times \\ & & & \times(1-\nu_1-2n\nu_2^2) \end{bmatrix}, \quad (2.185)$$

$$n = \frac{E_1}{E_2}, \quad m = \frac{G_2}{E_2}. \quad (2.186)$$

For the CL on the sloping surface of the press-tool (Fig. 2.5-3), it is necessary to determine a new matrix $[D]'_{cl}$ in the rotated coordinate system $z'r'$ related to the normal to the surface

$$[D]'_{cl} = [T]^{-1}[D]_{cl}[T], \quad (2.187)$$

where the matrix of the rotation is

$$[T] = \begin{bmatrix} 1 & 0 & 0 & 0 \\ 0 & \cos\alpha & \sin\alpha & 0 \\ 0 & -\sin\alpha & \cos\alpha & 0 \\ 0 & 0 & 0 & 1 \end{bmatrix}. \quad (2.188)$$

The introduction of a jointing CL into the finite-element scheme of the bodies interacting in the technological process allows us to essentially simplify the calculation algorithms, since, in this case, the problem of the combined deformation of several bodies, which have different properties, becomes a problem of the deformation of a "single" multilayer system characterized by internal non-linearity along the boundary conditions merely. Depending on the technological stage studied (pressing, unloading, relaxation, etc.), the deformation of this system layers obeys corresponding constitutive equations. This approach allows us to reduce the time of calculation. The use of the CL enables modeling various conditions for the interaction of the elements of technological equipment and a product formed.

2.5.5 INVESTIGATION OF ACCURACY AND STABILITY OF NUMERICAL SOLUTION BASED ON THE FINITE ELEMENT METHOD

To provide convergence of calculation processes during solving the problem of pressing, it is required to follow the recommendations concerning with the selection of the value of the step at loading.[196] In this case, the number of successive additional loads directly depends on the specified load increment and the degree of strain required.

When solving the problem of relaxation, special attention should be given to the selection of the step with respect to the time. As is seen from numerical experiments, its arbitrary assignment can cause an unstable (divergent) count.

The analysis of the numerical investigation results shows that in solving the equations of creeping, the count stability with time is observed when the following condition is fulfilled

$$\left|\{\Delta\varepsilon\}_i^c\right|_q < \left|\{\varepsilon\}_{i-1}^e\right|_q, \tag{2.189}$$

where the index q corresponds to a FE with the largest increment $\{\Delta\varepsilon\}_i^c$ of the vector of the creep strain for the ith step Δt_i with respect to the time. The value of the increment should not exceed the value of the elastic strain $\left|\{\varepsilon\}_{i-1}^e\right|_q$ of the qth element at the considered moment t_i–1. Based on this condition, the value of the current step with respect to the time is assigned in accordance with the minimal $\left|\{\varepsilon\}_{i-1}^e\right|_q$ that is selected from the set of $\left|\{\varepsilon\}_{i-1}^e\right|$ the entire system of FEs constituting the area under study.

In the strain space, the terms of inequality (5.41) are determined by the relations

$$\left|\{\varepsilon\}_{i-1}^e\right| = \sqrt{3\left(\varepsilon_0^e\right)_{i-1}^2 + 3/2\left(\varepsilon_i^e\right)_{i-1}^2}, \tag{2.190}$$

$$\left|\{\Delta\varepsilon\}_i^c\right| = \sqrt{3\left(\Delta\varepsilon_0^c\right)_i^2 + 3/2\left(\Delta\varepsilon_i^c\right)_i^2}, \tag{2.191}$$

where $\left(\varepsilon_0^e\right)_{i-1}$ and $\left(\Delta\varepsilon_0^c\right)_i$ are the volume elastic strains of an element at the moment t_{i-1} and an increment of the volume creep strain of the same element during Δt_i;

$\left(\varepsilon_i^e\right)_{i-1}$ and $\left(\Delta\varepsilon_i^c\right)_i$ are the intensities of elastic strains and the increments of creep strains of an element during corresponding periods.

Taking into account that

$$\left(\varepsilon_0^e\right)_{i-1} = (\sigma_0)_{i-1}/K,\ \left(\varepsilon_i^e\right)_{i-1} = (\sigma_i)_{i-1}/3G, \qquad (2.192)$$

$$\left(\Delta\varepsilon_0^c\right)_i = [R_1]^c * \{\sigma\}_{i-1}\cdot\Delta t_i,\ \left(\Delta\varepsilon_i^c\right)_i = [R_2]^c * \{\sigma\}_{i-1}\cdot\Delta t_i, \qquad (2.193)$$

(K and G are the modules of the volume compression and shear of powder material, respectively; $[R_1]^c$ and $[R_2]^c$ are the operators determining the volume creep and creep in shear of the material), we can determine the norm for the time step Δt_i:

$$\Delta t_i \le \min\left(\frac{\sqrt{\left((\sigma_0)_{i-1}^2/3K^2\right)+\left((\sigma_i)_{i-1}^2/6G^2\right)}}{\sqrt{3\left([R_1]^c * \{\sigma\}_{i-1}\right)^2+\frac{3}{2}\left([R_2]^c * \{\sigma\}_{i-1}\right)^2}}\right)_q . \qquad (2.194)$$

The use of the norm (2.49) in the calculations of stress relaxation allows arranging automatic selection of an optimal value for the step with respect to the time ensuring stability of the count. Suitability of the stability criterion offered was verified in test calculations for all the basic output parameters. The numerical processes were notable for their stability in this case.[172]

2.6 FINITE ELEMENT ALGORITHM FOR THE CALCULATION OF THE LOADING OF NANOCOMPOSITES

In the present section, finite-element algorithms are formulated for the calculation of the stress and deformation fields in compacted powder composites at the complex action of external parameters, first of all, elevated temperatures and high humidity and some others associated with swelling and force loads.

The interaction of a nanocomposite with the environment leads to both swelling and shrinkage, that is, to the appearance of volume deformations determined by eq 1.122. In their turn, these deformations provide the appearance of additional stresses, when some limitations are applied to the displacement of a certain part of the item surface. To describe the process, it is necessary to solve simultaneously the problems of diffusion of substances from the environment into the item and the variations of the volume deformations of a nanocomposite, and to calculate the appearing stress fields and their relaxations due to the appearance of the rheonomic properties of a material.

The analytical solution of such problem is possible only for a number of restricted calculation cases. Therefore, for solving the problem stated we also use a numerical approach based on the FEM.

Upon deriving the FE set of equations for the problem of diffusion we use the condition of the functional minimum similar to the functional determining the solution of the thermal conductivity equation.

The utilization of the finite-element approximation for the above problem in accordance with the method described in the previous chapters leads to the construction of two sets of algebraic equations:

the set of equations for the calculation of the substance diffusion from the surface inside the material

$$[K]_w\{\mathbf{C}\}+[\Phi]_w\frac{\partial}{\partial t}\{\mathbf{C}\}=0 \tag{2.195}$$

and the set of equations determining the stress-deformed state of the composite

$$[K]\{\delta\}=\{\mathbf{f}\}_w+\{\mathbf{f}\}_c+\{\mathbf{f}\}_S \tag{2.196}$$

where

$$[K]_w=\sum_{k=1}^{N}\int_{\Omega_k}\chi[\sum_{i=1}^{3}\frac{\partial}{\partial x_i}\mathbf{N}]^T[\sum_{i=1}^{3}\frac{\partial}{\partial x_i}\mathbf{N}]d\Omega_k \tag{2.197}$$

$$[\Phi]_w=\sum_{k=1}^{N}\int_{\Omega_k}[\mathbf{N}]^T[\mathbf{N}]d\Omega_k \tag{2.198}$$

$$[\mathbf{K}]=\sum_{k=1}^{N}\int_{\Omega_k}[\mathbf{B}]^T[\mathbf{D}][\mathbf{B}]d\Omega_k \tag{2.199}$$

the matrix of the system "rigidity;"

$$\{\mathbf{f}\}_S=\sum_{k=1}^{N}\int_{S_k}[N]^T\{F\}_k^S\,dS_k \tag{2.200}$$

the vector of the surface forces reduced to the nodes of the FEs;

$$\{\mathbf{f}\}_c = \sum_{k=1}^{N} \int_{\Omega_k} [\mathbf{B}]^T [\mathbf{D}][R]^c * \{\sigma\} d\Omega_k \tag{2.201}$$

the vector of "the fictitious volume forces" specified by rheonomic deformations.

$$\{\mathbf{f}\}_w = \sum_{k=1}^{N} \int_{\Omega_k} [D][B] \frac{\{\varepsilon\}^w}{3} d\Omega_k \tag{2.202}$$

the vector of "the fictitious volume forces" specified by volume deformations.

$$\{C\}_k = [N]\{\mathbf{C}\} \tag{2.203}$$

where $\{C\}_k$ and $\{\mathbf{C}\}$ are the concentrations of the substance penetrated into the composite from the environment, in the FE and in the nodes of the FE model, respectively.

The time integration of the interconnected sets of eqs 2.195 and 2.196 requires special approaches providing simultaneous stability both of the diffusion problem and the calculation of the stress-deformed state. It should be noted that the stability conditions for the numerical integration of the indicated systems do not coincide in the general case. Therefore, for solving eq 2.195 we employ an implicit integration scheme providing the count stability at the arbitrary time step, and we determine the value of the latter from the condition of the integration stability of eq 2.196.

The integration with the use of the difference scheme of the method of trapezoids[88] reduces eq 2.1 to the form:

$$[\Psi_1]\{C\}_{t_{i+1}} = [\Psi_2]\{C\}_{t_i} \tag{2.204}$$

where

$$[\Psi_1] = \left([K]_W + \frac{2}{\Delta t}[\Phi]_W \right) \tag{2.205}$$

$$[\Psi_2] = \left(\frac{2}{\Delta t}[\Phi]_w - [K]_w \right) \tag{2.206}$$

t_{i-1} is the considered moment; t_i is the previous moment; Δt is the time step.

Set (2.196) is integrated with the use of the explicit scheme

$$[K]\{\delta\}_{t_{i+1}} = \sum_{n=1}^{i} \Delta\{\mathbf{f}\}_{w}^{n} + \sum_{n=1}^{i} \Delta\{\mathbf{f}\}_{c}^{n} + \{\mathbf{f}\}_{S} \tag{2.207}$$

where

$$\Delta\{\mathbf{f}\}_{w}^{n} = \sum_{k=1}^{N} \int_{\Omega_k} [D][B] \frac{\{\varepsilon\}_{t_{i-1}}^{w} \Delta t}{3} d\Omega_k \tag{2.208}$$

$$\Delta\{\mathbf{f}\}_{c}^{n} = \sum_{k=1}^{N} \int_{\Omega_k} [\mathbf{B}]^T [\mathbf{D}][\frac{\partial}{\partial t} R]^c * \{\sigma\}_{t_{i-1}} d\Omega_k \tag{2.209}$$

The criterion of the stability is not obtained yet for explicit calculation schemes solved by the FEM; therefore, we use the criterion offered in the previous chapter, which has been verified by a large series of computer-aided calculations:

$$\Delta t \leq \min \left(\frac{\sqrt{\{\sigma_0\}_{t_{k-1}}^2 / (3K^2) + \{\sigma_i\}_{t_{k-1}}^2 / (6G^2)}}{\sqrt{3\left(\left[\frac{\partial R_1}{\partial t}\right] * \{\sigma\}_{t_{k-1}}\right)^2 + \frac{3}{2}\left(\left[\frac{\partial R_2}{\partial t}\right] * \{\sigma\}_{t_{k-1}}\right)^2}} \right) \tag{2.210}$$

where $\{\sigma_i\}$ and $\{\sigma_0\}$ are the matrices of the intensity of stresses and the mean stress; R_1 and R_2 are the matrices determining the volume and shear creep of the material; K and G are the modules of the three-dimensional compression and volume shift, respectively.

In the count process, the time step is chosen based on condition (2.210). In this case, the integration stability both of eqs 2.195 and 2.196 is provided and the machine time is reduced significantly due to the use of the explicit time integration scheme for eq 2.196.

2.7 STRUCTURE OF PROGRAM COMPLEX FOR NANOSTRUCTURE SYSTEM MULTISCALE MODELING

It should be pointed out that program complexes for multi-level modeling of nanosystems, as applied to different types of tasks, currently can be quite easily completed from standard software packages, both free for use and commercial ones. There are many program complexes, which contain the sets of rather well selected parameters for single-type molecules and combine them into special libraries—force fields. The force fields most frequently used include: Assisted Model Building with Energy Refinement (AMBER), Chemistry at HARvard Macromolecular mechanics (CHARMM), GROningen MOlecular Simulation (GROMOS), Optimized Potential for Liquid Simulation (OPLS), Consistent Valence Force Field (CVFF), and Merck Molecular Force Field (MMFF).[16,28,64,101,132]

The force field AMBER is used for proteins, nucleic acids, and a number of other classes of molecules. The filed CHARMM is intended for modeling different systems: from small molecules to solvated complexes of biological macromolecules.

There are force fields built on the calculations of molecular systems with quantum-mechanical methods. The example of such field is MMFF. Force field CVFF comprises specifying contributions of anharmonicity and interactions of force field components. The field is parameterized for the calculations peptides and proteins. The field GROMOS is intended to model aqueous or non-polar solutions of proteins, nucleic acids, and sugars.

HyperChem serves for calculating equilibrium configurations of molecules from *ab initio* principles and defining some constants required for the calculations with the methods of molecular dynamics.

A widely used program Nanoscale Molecular Dynamics (NAMD) is developed by the group of biophysicists from Illinois University together with Backman University. NAMD is intended for parallel modeling of biomolecular systems with the method of molecular dynamics. This program is supplemented with the visualization program of calculation results— Visual Molecular Dynamics (VMD). VMD is specially developed for the visualization and analysis of such biological systems as proteins, nucleic acids, molecular systems based on lipids (e.g., components of cell membranes). The program understands the format Protein Data Bank (PDB) and allows using various variants and methods of visualization and coloring of molecules. VMD can be applied for the animation and analysis of phase theory obtained in the result of molecular-dynamic modeling. An interesting feature of the program is the fact that it can be used as the graphic component of computer modeling system and operates on a distant computer.

Let us describe the typical structure of the program complex for nanosystem modeling. The program complex NANOMULTI represents a multistep structure, the interaction of blocks and modules of which is demonstrated in Figure 2.6. The program complex is arranged in such a way that the modules of the aforesaid programs could be easily integrated into it.

The formation and application of the nanoparticle systems are modeled stage by stage, from block to block, as shown in this figure. The program complex consists of the blocks written on Fortran, Borland C++ Builder 6, and Borland Delphi 7.

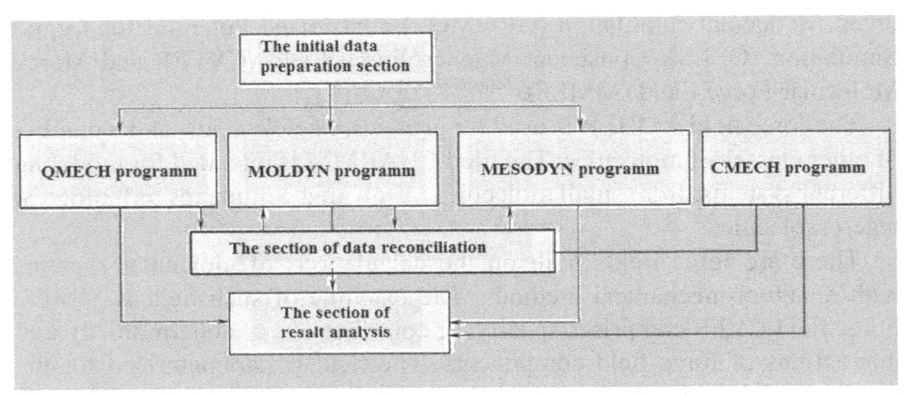

FIGURE 2.6 General structure of program complex NANOMULTI.

The initial conditions of computational experiment are determined and calculation files are prepared in the preparation block. For this the coordinates and velocities of atoms and molecules are set, the rated cell volume, temperature, and pressure of the system being modeled are calculated. The types of atoms in molecules are defined; the constants of interaction parameters are set. The structures of molecules and nanoparticles are determined. At the final stage of the initial data preparation the files of certain formats containing the aforesaid data are formed.

Further the files are used by successively computational modules: programs for quantum-chemical modeling (QMECH), molecular-dynamic modeling (MOLDYN), modeling with mesodynamics methods (MESODYN), and modeling of continue (MC) mechanics processes with FEM.

Function of the matching block consists in arranging the interaction between different computational modules according to the general scheme of solving the problem.

The processing, visualization, and computation of the investigated parameters of nanoaerosoles and analysis of the modeling results are carried out in the block of results processing.

The certain structure of each program block is defined with the computational task type. For example, let us consider the structure of separate blocks as applicable to the task of nanoparticle formation in gaseous phase.

The block of initial data preparation is intended for collecting and processing initial data for modeling the task of composite nanoparticle formation and has a linear structure, shown in Figure 2.7.

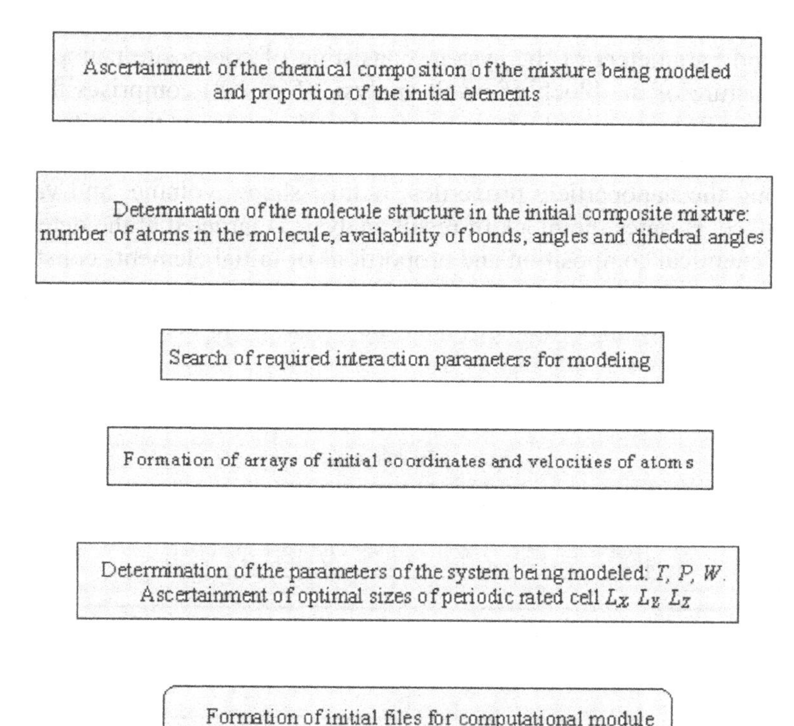

FIGURE 2.7 Structure of the block of initial data preparation.

The sequence of actions is indicated with arrows. To solve the task of composite nanoparticle formation it is necessary to determine the chemical composition of the mixture being modeled and proportions in which the initial elements are taken. Based on the chemical composition the structure of molecules in the composite mixture is defined: number of molecules in each molecule, availability and type of bonds, values of angles, and dihedral angles.

To use the MD methods the necessary interaction parameters are searched: constants of bond extension-compression, dihedral angles, Van der Waals interactions, and electrostatic interactions. At the next stage the arrays of initial coordinates and velocities of atoms are formed, the nano-system parameters, such as temperature T, pressure P, and volume of the rated cell W are set, optimal sizes of the periodic cell L_x, L_y, L_z are defined. At the final stage of initial data preparation the input files for computational module are formed.

The block of result analysis in the task of nanoparticle formation in gaseous phase is intended to solve a number of problems for defining properties and parameters of the system consisting of composite nanoparticles. The structure of the block of result analysis (Fig. 2.8) comprises the tasks on finding atoms grouped into nanoparticles, determining the uniformity of the formed composite mixture of atoms, molecules and nanoparticles, and searching the nanoparticle properties: radius, shape, volume, and value of 3D surface. Besides the block of result analysis is targeted at the determination of chemical composition and proportions of initial elements constituting nanoparticles, and the share of atoms and molecules condensed into nanoparticles. One of the capabilities of analysis block is the determination of the

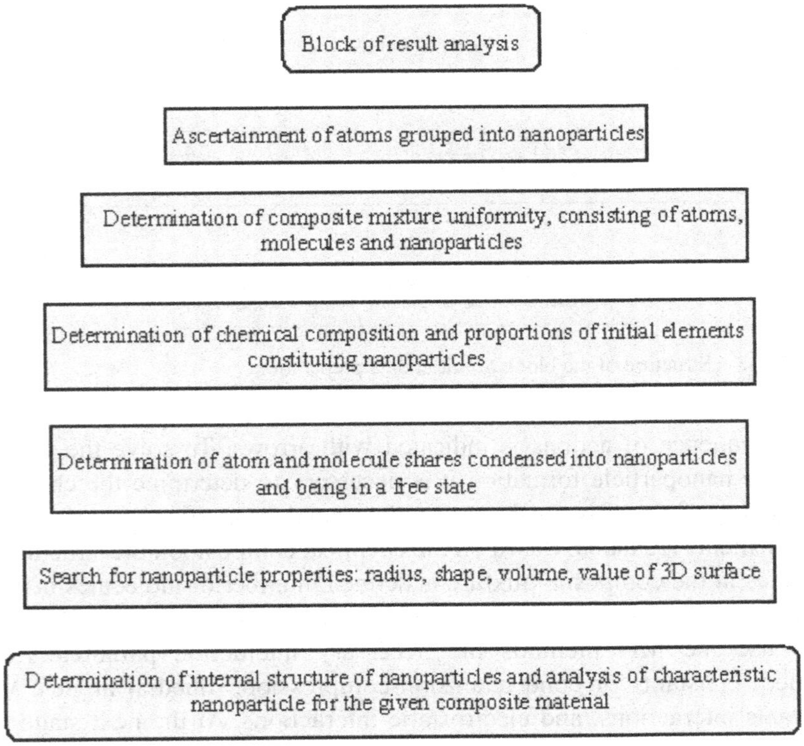

Block of result analysis

Ascertainment of atoms grouped into nanoparticles

Determination of composite mixture uniformity, consisting of atoms, molecules and nanoparticles

Determination of chemical composition and proportions of initial elements constituting nanoparticles

Determination of atom and molecule shares condensed into nanoparticles and being in a free state

Search for nanoparticle properties: radius, shape, volume, value of 3D surface

Determination of internal structure of nanoparticles and analysis of characteristic nanoparticle for the given composite material

FIGURE 2.8 Structure of the block of result analysis.

internal structure, physical properties, and values of nanoparticles, search and analysis of characteristic nanoparticle for the given composite material.

The important function of the block of result analysis is finding the atoms grouped into nanoparticles. Composite nanoparticles are searched through the enumeration of all system atoms. For the current *i*-atom (i = 1..*N*) the sphere with the radius *R* and center in x_i, y_i, z_i is considered, where x_i, y_i, z_i —coordinates of *i*-atom (Fig. 2.9).

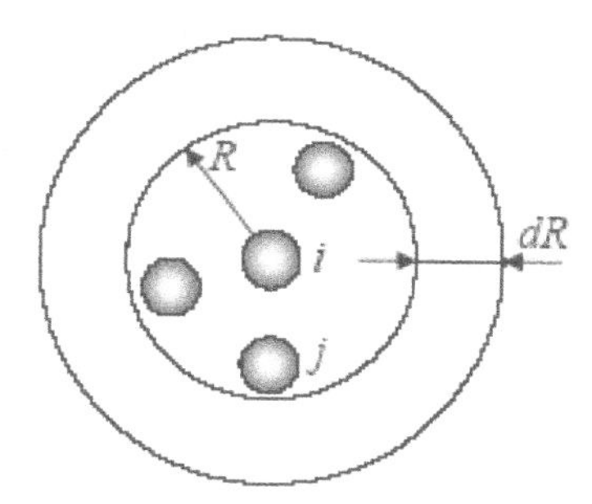

FIGURE 2.9 Scheme of searching for atoms constituting nanoparticles.

It is found what atoms get into the given sphere, these atoms are marked as constituting the nanoparticle. Then the sphere radius increases by the value *dR*, and it is checked whether the atoms not marked before got into the sphere with the new radius. If there are additional atoms in the sphere with the new radius, they are marked and the radius is increased one more time. The increase in the sphere radius and search for new atoms continues until new atoms in the sphere with the increased radius are present. If new atoms are not available, it is considered that the search for atoms grouped into the current nanoparticle is over and the transition to (*i* + 1)-atom takes place to search for new composite nanoparticles.

After analyzing all the system atoms for constituting nanoparticles, the number of composite nanoparticles is determined, the chemical composition and proportions of initial elements constituting the nanoparticles are found, the share of atoms and molecules condensed into nanoparticles is calculated. Based on the data obtained, the properties of nanoparticles are searched: radius, shape, volume, and value of 3D surface.

The uniformity of composite mixture formed of atoms, molecules, and nanoparticles is defined by the mass distribution in the volume rated. The rated cell representing a parallelepiped with the ribs L_x, L_y and L_z, is divided into N_v parts by one of the directions as shown in Figure 2.10. The number of parts N_v is selected in such a way that the volume of each part is much larger than the size of one molecule. In each volume part the values of absolute and relative mass are calculated. Insignificant mass fluctuations are conditioned by the transition of atoms, molecules, and nanoparticles, and confirm the uniformity of composite mixture. In case of the availability of significant mass fluctuations, we can speak of heterogeneities in the rated cell volume.

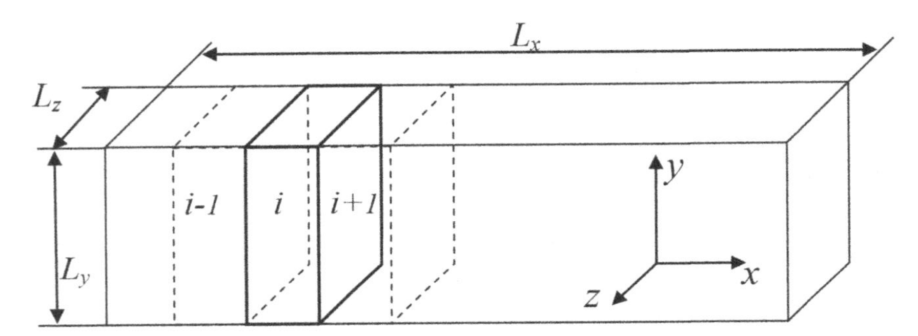

FIGURE 2.10 Scheme for checking the uniformity of composite gaseous mixture.

Let us consider the definition of internal structure of nanoparticles on the example of a separate particle consisting of N_p atoms, having the mass M_p, density ρ_p. For many pairs of atoms in the particle, the distances between them R_{ij} are calculated using the following correlation (Fig. 2.11).

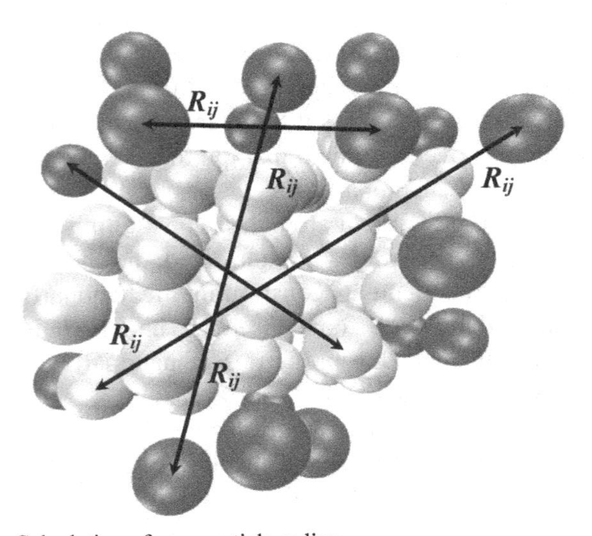

FIGURE 2.11 Calculation of nanoparticle radius.

$$R_{ij} = \sqrt{(x_i - x_j)^2 + (y_i - y_j)^2 + (z_i - z_j)^2}, \ i, j = 1..N_p, \tag{2.211}$$

where x_i, y_i, z_i—coordinates of i-atom, x_j, y_j, z_j—coordinates of j-atom.

The half of the largest of the distances R_{ij} is taken as the nanoparticle radius R_p:

$$R_p = \frac{1}{2}\max_{i,j} R_{ij}, \ i, j = 1..N_p \tag{2.212}$$

The location of nanoparticle center is determined as the arithmetic mean value of the coordinates of all atoms:

$$x_p = \frac{1}{N_p}\sum_{i=1}^{N_p} x_i, y_p = \frac{1}{N_p}\sum_{i=1}^{N_p} y_i, z_p = \frac{1}{N_p}\sum_{i=1}^{N_p} z_i. \tag{2.213}$$

The nanoparticle structure is analyzed by layers. The nanoparticle is split into N_{sl} layers (Fig. 2.12), the thickness of each layer is $dR_p = R_p / N_{sl}$. The central layer is the sphere with the center in (x_p, y_p, z_p) and radius $R_p^1 = dR_p$. Each next layer represents the external part of the sphere with the radius $R_p^k = k \cdot dR_p, k = 2...N_{sl}$, limited inside the previous layer of the nanoparticle. The volume of layers is calculated by the following correlation:

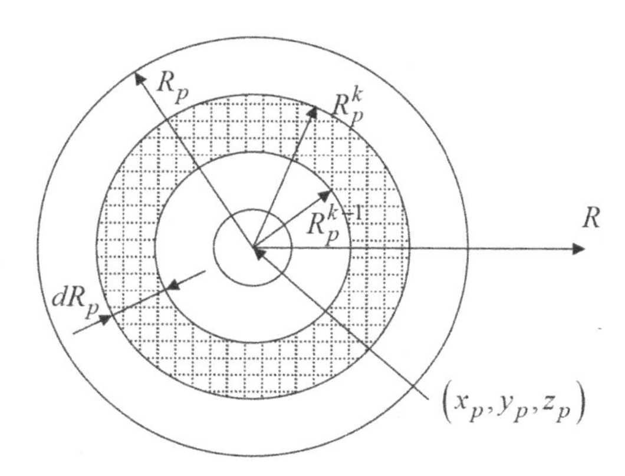

FIGURE 2.12 Scheme of nanoparticle structure analysis.

$$V^1 = \frac{4}{3} \cdot \left(R_p^1\right)^3, V^k = \frac{4}{3} \cdot \left(\left(R_p^k\right)^3 - \left(R_p^{k-1}\right)^3\right), k = 2...N_{sl}. \quad (2.214)$$

Checking the nanoparticle atoms for belonging to the given layer, the properties and characteristics of each layer are calculated: M_{El}^k —absolute mass of initial element *El* in *k*-layer, $Motn_{El}^k = M_{El}^k \cdot 100 / M_p$—relative mass of initial element *El* in *k*-layer in percentage, absolute density $\rho_{El}^k = M_{El}^k / V^k$ of element *El* and relative density $\rho otn_{El}^k = \rho_{El}^k \cdot 100 / \rho_p$ of element *El* in *k*-layer in percentage. The relative mass reflects the share of initial element contained in the given layer, and relative density characterizes the deviation of the layer density from the average density of the nanoparticle.

We can speak of the belonging of *i*-atom to the given layer if the following correlation is realized:

$$R_p^{k-1} \leq \sqrt{\left(x_i - x_p\right)^2 + \left(y_i - y_p\right)^2 + \left(z_i - z_p\right)^2} < R_p^k, \quad (2.215)$$
$$i = 1..N_p$$

where x_i, y_i, z_i coordinates of *i*-atom.

Having analyzed the structure of several nanoparticles and formulated the dependencies of layer parameters depending upon the dimensionless radius, we can speak of the structure of characteristic average statistical nanoparticle for the given composite material.

To compare the modeling results and experimental data, the molecular-dynamic modeling of the formation of copper nanoparticles was carried out with the method of vacuum evaporation and further condensation. The stage of atom condensation into nanoparticles was realized at normal temperature (300 K) within 20 ns.

The important moment preceding the numerical modeling is the investigation of the convergence of numerical schemes, stability of calculations with time and test computations giving the possibility to compare the numerical solution with the experimental or analytical solution. Let us give the example of test computation.

On the modeling results the dynamics of copper nanoparticle distribution by size is defined, and the change in the average size with time is calculated. Based on the data obtained in the time interval 0–20 ns, the trend line is arranged, and the changes in the average diameter till the time moment of 15 min, that is 12 units in logarithmic scale of time relative to nanoseconds, are prognosticated. The curve of dynamics of average diameter of nanoparticles and corresponding trend line are given in Figure 2.13. The line with diamond markers demonstrates the modeling results, the firm line corresponds to the prediction of modeling data in the form of trend line, the square marker shows the experimental data.

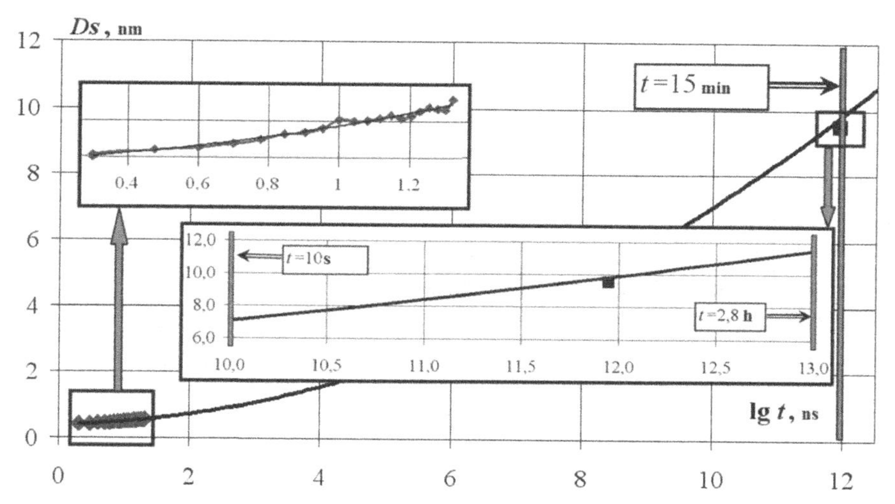

FIGURE 2.13 Comparison of average diameter of copper nanoparticles modeled and experimental data.

The type of approximating curve is selected from the point of the most reliability of approximation. For linear approximation the reliability value is $R^2 = 0963$, for logarithmic $R^2 = 0873$, power $R^2 = 0897$, and for exponential $R^2 = 0973$. Polynomial approximation gives the most reliability value $R^2 = 0978$, therefore it is used to prognosticate the modeling data.

The analysis of Figure 2.13 demonstrates a good compliance of experimental data[48] and modeling results. For the synthesis of copper nanoparticles during 15 min the average diameter is 9.5 nm according to the experimental data and 10 nm—according to modeling results. It should be pointed out that the decimal logarithm of time is the abscissa axis of the graph in Figure 2.13, therefore even the significant increase in the synthesis time up to 2.8 h results in the change in the average diameter of nanoparticles only up to 11 nm. This value corresponds to the deviation from the experimental data by 15.78%.

Good coincidence of the results of test calculations with experimental data allows speaking of the adequacy and correctness of the mathematical model used in the work, and verity of the modeling data obtained.

The application of the program complex NANOMULTI to various nanotechnological tasks is described in detail in the Chapters 3, 4, and 5 of this book.

KEYWORDS

- **numerical algorithms**
- **quantum-chemical problems**
- **molecular dynamics problems**
- **mesodynamics numerical schemes**
- **finite element modeling**

CHAPTER 3

NUMERICAL SIMULATION OF NANOPARTICLE FORMATION

CONTENTS

ABSTRACT

The examples of the numerical simulation of nanoparticle formation are described. In first part of chapter, the examples of the calculation of nanoparticle formation with the method of condensation from the gaseous phase are given. The regularities of composite nanoparticle formation are considered, their structure and shape formation processes are analyzed, and the probabilistic regularities of nanoparticle parameter distribution are discussed. The influence of various parameters on the processes of nanoparticle formation in gaseous phase is demonstrated. The second part of chapter is dedicated to the investigation of nanoparticle formation kinetics in "up down" processes of powder dispersion under different types of their dynamic loading: fast high-pressure release, particle impact against an obstacle, and particle destruction in the impact wave.

3.1 FORMATION OF NANOPARTICLES DURING THE CONDENSATION IN A GASEOUS PHASE

This section contains the calculation of processes of nanoparticle formation in a gaseous phase. The results of calculations of structural characteristics of nanoparticles and evolution of basic variables of nanosystem with time depending on the initial state of the nanosystem are given: correlations of masses of initial materials, initial temperature and pressure, parameters of the environment action upon the nanosystem (medium cooling or heating rate, changes in the external pressure, characteristics of different physical fields), interaction of nanoparticles, etc.

3.1.1 FORMATION OF METAL COMPOSITE NANOPARTICLES

The formation of metal nanoparticles was investigated under the vacuum evaporation of metals and further condensation. This method is implemented as follows: initial metals are placed in the required proportion in the closed volume and heated up to the evaporation temperature. The atoms mix in the gas mixture. Further cooling results in the condensation of atoms into composite metal nanoparticles.

Modeling was carried out in the cell (Fig. 3.1) with the dimensions: $L_x = 6.10^{-8}$, $L_y = 2.10^{-8}$ and $L_z = 2.10^{-8}$ m with periodic boundary conditions (Fig. 3.2).

Two-component mixtures Ag–Cu and Ag–Zn were investigated as initial metals. The data on the properties of the metals indicated and their parameters required for the calculation are given in Table 3.1.

TABLE 3.1 Parameters of Metals Used In the Calculations.

Characteristics of metals	Ag	Cu	Zn
Type of crystal lattice	Cubic face-centered	Cubic face-centered	Hexagonal
Parameter of crystal lattice, 10^{-10} m	4.09	3.61	2.66
Molecular mass, a.m.u.	107.868	63.546	65.39
Melting temperature, K	1235.1	1356.6	692.7
Boiling temperature, K	2485	2840	1180
ε, 10^{-21} J	55.276	65.626	1.739
$e^*/2$, 10^{-10} m	1.4839	1.3122	1.9422

Let us consider the processes of formation of composite metal nanoparticles from silver and copper. At the initial time interval, silver and copper atoms were located in the nodes of crystal lattice of the corresponding nanocrystals (Fig. 3.1). The initial correlation of metal atoms was: 770 atoms of Ag, 2390 atoms of Cu. At the same time, the silver share by the number of atoms was 24.6%, and by mass—35.4%.

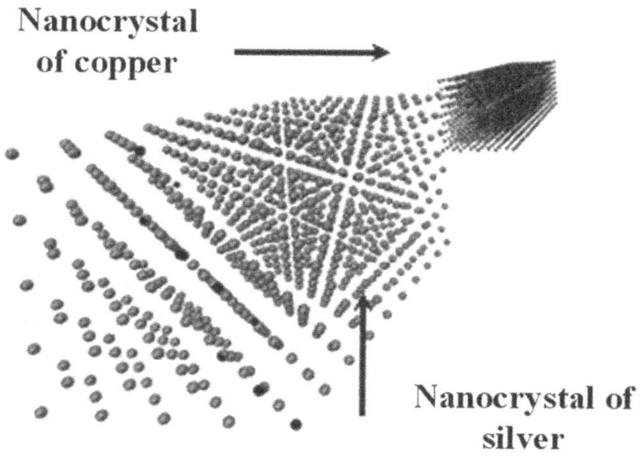

FIGURE 3.1 Nanocrystals of metals at the initial time interval.

Short-time high-temperature heating within 1 nanosecond (ns) up to 3000 K was carried out for metal evaporation. As a result, metal atoms formed a gaseous mixture. Further cooling to normal temperature (300 K) resulted in the condensation of metal atoms into nanoparticles containing different number of silver and copper atoms (Fig. 3.2).

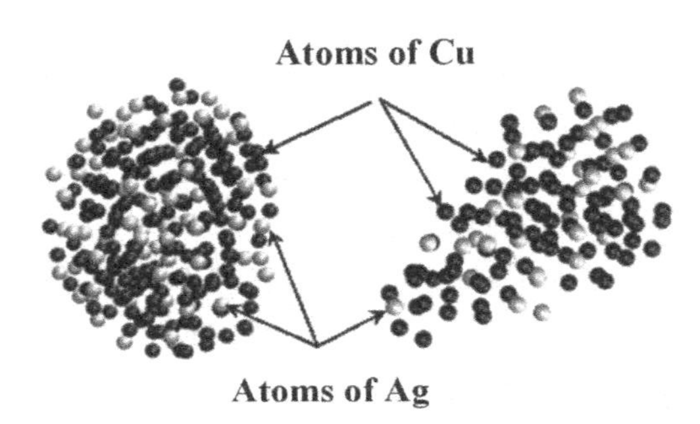

FIGURE 3.2 Composite nanoparticles formed.

It should be pointed out nanoparticles are solid, there are no cavities inside them observed.

In the process of condensation, the nanoparticle tends to spherical shape corresponding to the minimum value of its potential energy. Nanoparticles mainly have the mixed composition (although nanoparticles consisting of one atom are also observed). Atoms of different metals are uniformly distributed in the nanoparticle volume. This is explained by the fact that melting temperatures of initial materials differ insignificantly; therefore, the process of condensation of copper and silver atoms starts practically simultaneously.

The condensation process is split into two phases: first, the atoms combine, then, the condensation process progresses by the fusion of nanoparticles. Figure 3.3 illustrates five stages of the process of combination of two nanoparticles with total duration of ~0.01 ns. At the first stage, the nanoparticles approach each other under the attractive force. The repulsive forces increase, consequently, the nanoparticles are able not only "to oscillate" near each other for a long time, but also to start moving away. At the next stage, the contact between the nanoparticles is observed, which means that for these nanoparticles, the attractive forces prevail over the repulsive forces. Afterwards, the process of combining the nanoparticles into one nanoparticle continues. At first, this nanoparticle has an elongated shape, which "is optimized" with time becoming spherical.

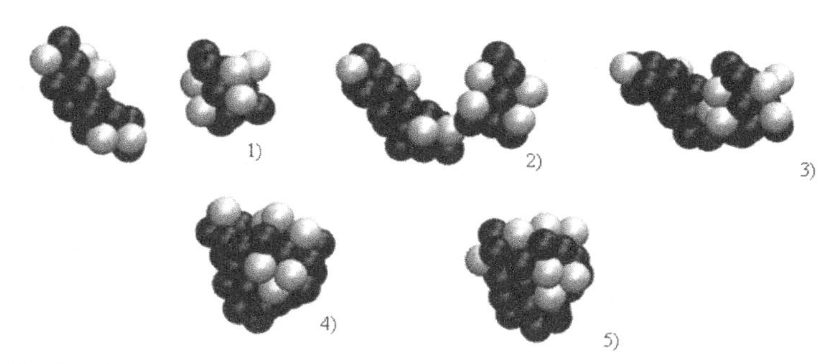

FIGURE 3.3 Stages of combining composite nanoparticles (Ag atoms are marked with gray, Cu atoms—with black).

In Figure 3.4, you can see the change by time in the mass share of atoms condensed into nanoparticles. In the period till 1 ns, all metal atoms leave the nodes of crystal lattice and transform into the gaseous state due to high-temperature impact. In the time interval of 1–3 ns, atoms actively condense into nanoparticles. After the first nanosecond, all atoms are located in nanoparticles. The condensation process continues further, but it is already directed not at the formation of new nanoparticles but their enlargement caused by fusion.

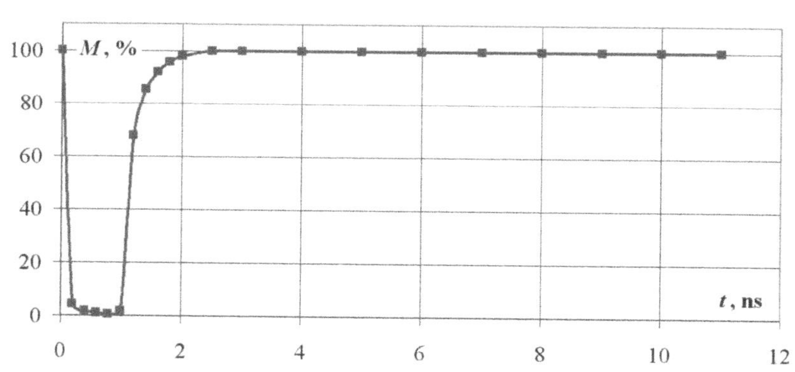

FIGURE 3.4 Change in the share of atoms condensed into nanoparticles.

Since the melting temperatures of initial metals are quite similar, the curves of mass share changes for atoms of copper and silver, and common curve for all condensed particles are close to each other. More detailed change in the mass share of condensed particles can be seen in Figures 3.5 and 3.6.

Figure 3.5 demonstrates the change in decimal logarithm of mass share of condensed atoms by time. Figure 3.6 characterizes the behavior of decimal logarithm of percentage of non-condensed atoms in time. The analysis of graphs shows that silver atoms leave the nodes of crystal lattice faster and stay in non-condensed state longer in comparison with copper atoms. Such effect can be explained by the fact that silver has a lower melting temperature than copper. After $t =$ 2.5 ns, the share of condensed atoms is constant. By this time, all metal atoms are grouped into nanoparticles and there are no "free" atoms left.

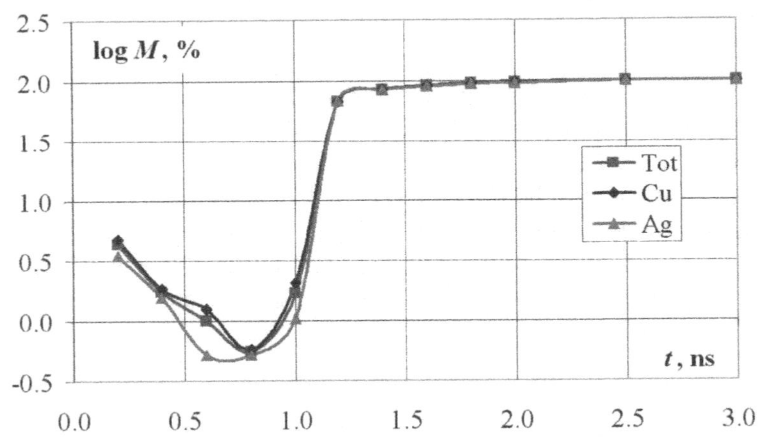

FIGURE 3.5 Change in the share of condensed atoms.

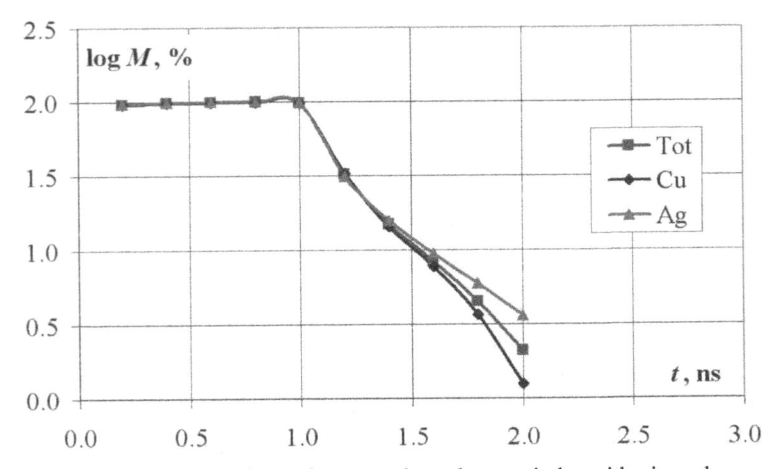

FIGURE 3.6 Change in the share of non-condensed atoms in logarithmic scale.

Let us discuss the change in the distribution of nanoparticles by size in different time intervals. The maximum distance between two atoms in nanoparticle is taken as the characteristic nanoparticle size. Figure 3.7 demonstrates the dependence of the number of nanoparticles upon the diameter at the time moment of 3, 6, and 11 ns. Since in the beginning of condensation process small nanoparticles are formed, by the time moment of 3 ns, the number of nanoparticles with the diameter of 4 angstroms (Å) prevails, and the number of bigger particles is small: 235 nanoparticles with the diameter of 4 Å and one nanoparticle with the diameter of 8 Å.

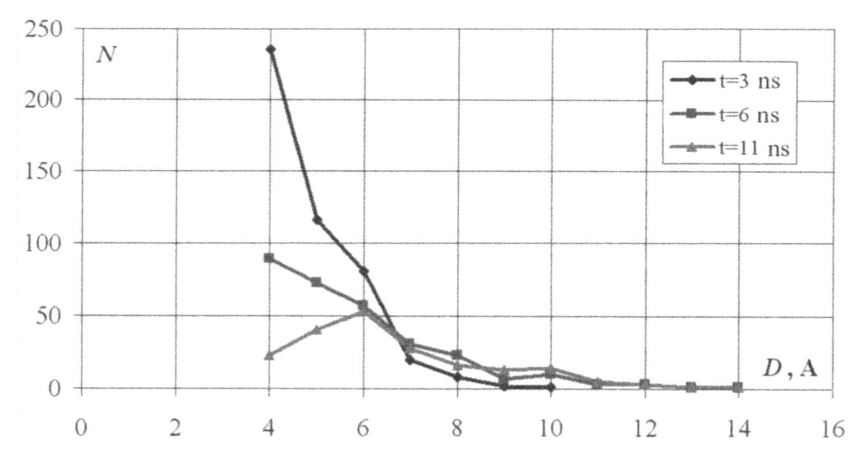

FIGURE 3.7 Distribution of nanoparticles of Ag–Cu by sizes.

In the course of time, nanoparticles are getting bigger; therefore, the number of particles of small diameter decreases considerably. At the time moment of 6 ns, the number of nanoparticles with the diameter of 4 Å decreases to 89; however, bigger nanostructures with the diameter of 14 Å, the formation of which is not observed at earlier stages of condensation, appeared. The peak of the distribution of nanoparticles by sizes to 11 ns in the graph shifts to the right. The number of nanoparticles with the size of 4 Å decreases to 23, and the number of nanoparticles with the size of 9 Å increases from 7 up to 13.

The enlargement of nanoparticles during the condensation process can be also traced in Figure 3.8, in which the change in the average size of nanoparticles increased from 4 Å at the time moment of 1 ns up to 6.6 Å by the moment of 11 ns. The dependence of the average size of nanoparticles is practically linear upon time. However, the gradual decrease in the intensity of the increase in the sizes of nanoparticles with time is observed. This effect is explained by the fact that nanoparticles enlarge with time, and due to the mass increase, lose their mobility. Besides, also the number of particles decreases. When the number and the mobility of nanoparticles decrease, the value of approach probability of two nanoparticles at the distance at which the attracting forces are enough for their "fusion" decreases as well.

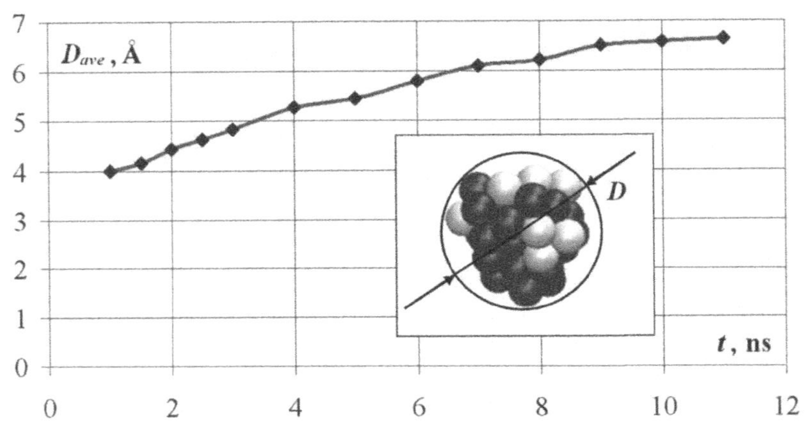

FIGURE 3.8 Change in the average diameter of Ag–Cu nanoparticles.

Dependence of the number of nanoparticles in the volume unit upon time is given in Figure 3.9.

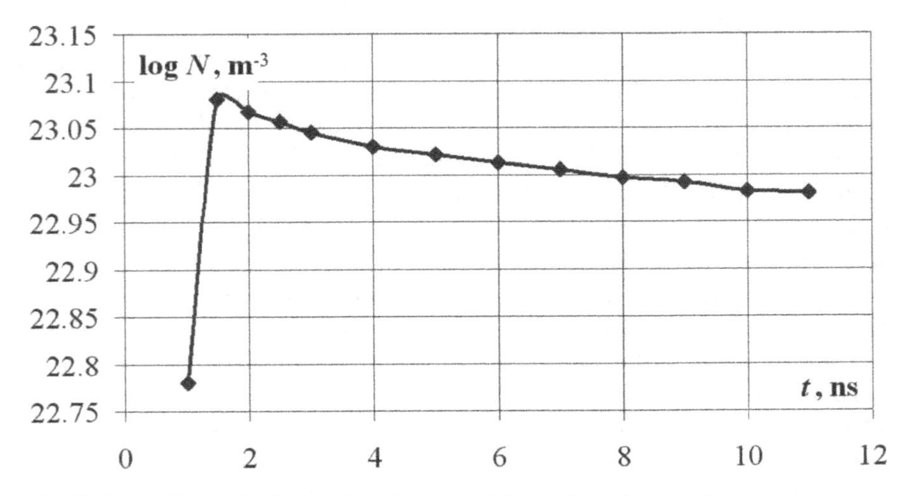

FIGURE 3.9 Change in the number of nanoparticles in the volume unit.

The beginning of condensation is characterized by the sharp increase in the number of nanoparticles. However, closer to $t = 2$ ns, the growth of the number of nanoparticles stops. Such behavior of the system is explained by the fact that by the given moment of time, all free metal atoms are already condensed into nanoparticles. Further decrease in the number of nanoparticles in the volume unit proves their enlargement.

Figure 3.10 demonstrates the change in the cumulative volume of nanoparticles with time. Rapid increase in the given volume of nanoparticles in the time interval from 1 up to 1.7 ns corresponds to the active condensation of metal-free atoms into nanoparticles. After the transition of all free atoms into the condensed state, the fracture on the curve of the change in the total volume is observed at 1.7 ns. The total volume of nanoparticles with time after 1.7 ns remains practically unchanged. Its insignificant increase with time is explained by the reconstruction of atomic structure of nanoparticles during their unification.

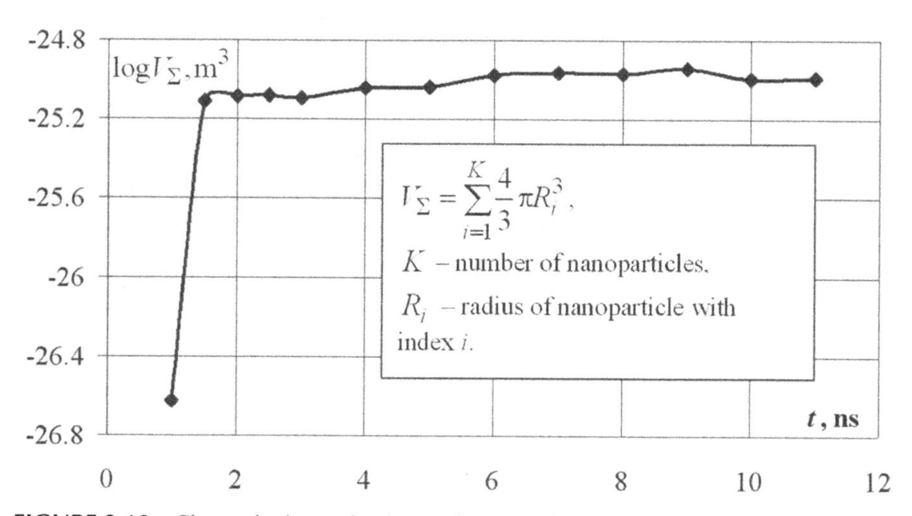

FIGURE 3.10 Change in the total volume of nanoparticles.

In Figure 3.11, you can see the graph of the dependence of total surface of nanoparticles upon time. It is observed that the transition of metal atoms from free state into the condensed (1–1.7 ns) results in the increase in the total surface of nanoparticles in the given time interval. After $t = 1.7$ ns, the curve of nanoparticle surface changes insignificantly.

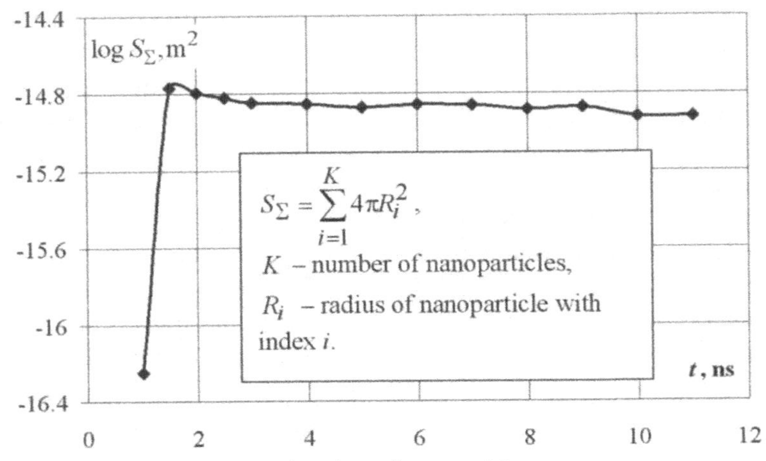

FIGURE 3.11 Change in the total surface of nanoparticles.

Figure 3.12 demonstrates the trajectory of the nanoparticle with the diameter of 6 Å. By the time moment of 11 ns, the particle consisted of two silver and five copper atoms (Fig. 3.13). The movement trajectory of a large nanoparticle comprising 18 silver and 29 copper atoms (Fig. 3.15) is given in Figure 3.14. Since the first nanoparticle weighs 533.466 atomic mass units (a.m.u.)—less than the second one—3784.458 a.m.u., the distance it passes (0.2159 mcm) in the time interval 9–11 ns is much longer than the distances passed by a bigger nanoparticle (0.0713 mcm). The trajectories of both nanoparticles are chaotic, but the bigger one moves "smoother."

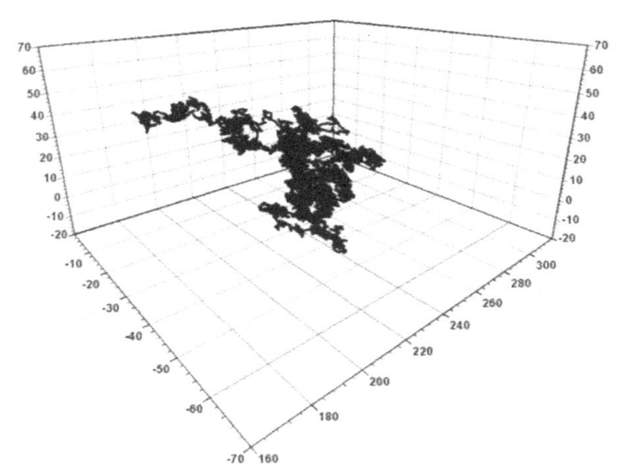

FIGURE 3.12 Movement trajectory of Ag_2Cu_5 nanoparticle.

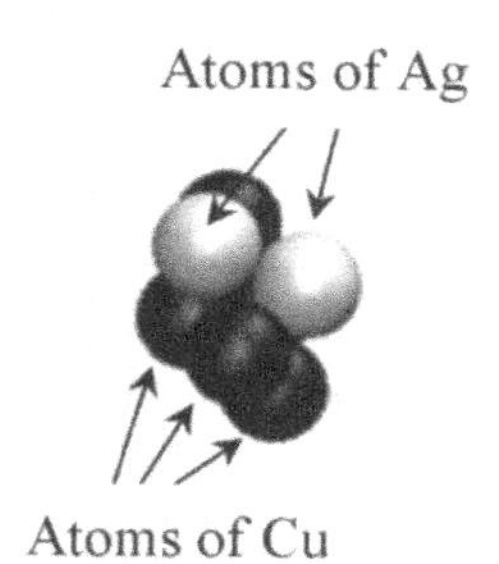

FIGURE 3.13 Type and structure of Ag_2Cu_5 nanoparticle.

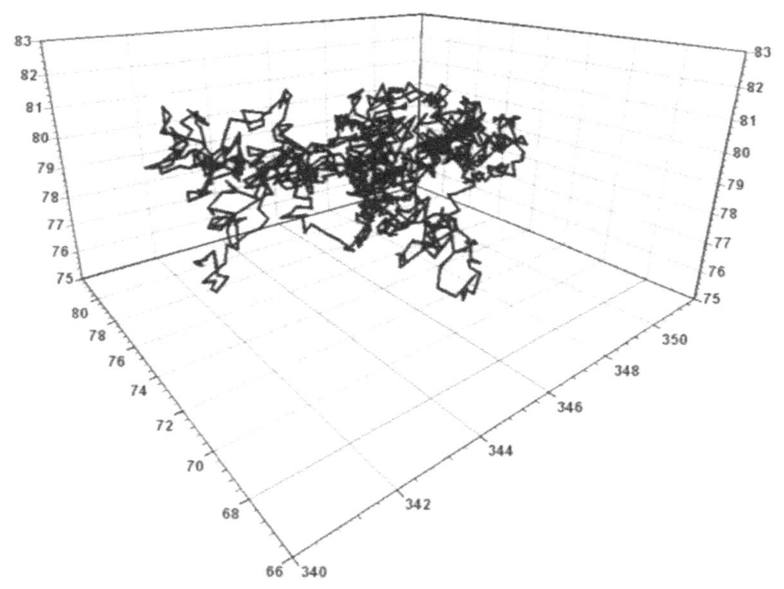

FIGURE 3.14 Movement trajectory of $Ag_{18}Cu_{29}$ nanoparticle.

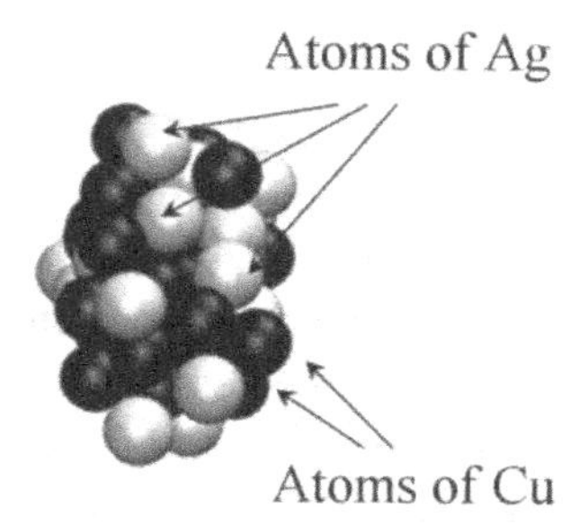

FIGURE 3.15 Type and structure of $Ag_{18}Cu_{29}$ nanoparticle.

The composition of composite nanoparticles formed is of great interest. Figure 3.16 demonstrates the standardized distribution by composition and mass built up for 198 nanoparticles at the time moment of 11 ns. The calculations revealed that the share of silver atoms in nanoparticles is within 0–60%. From the graph, it is seen that among the nanoparticles formed, 10 do not contain silver. Besides, there are nanoparticles with a large share of silver: 42% Ag atoms—9 nanoparticles; 48% Ag atoms—9; 54% Ag atoms—10; 60% Ag atoms—1 nanoparticle. However, the maximum of distribution by the share of silver atoms falls on the ratio: 25% Ag and 75% Cu atoms. This practically corresponds to the initial proportion of initial materials.

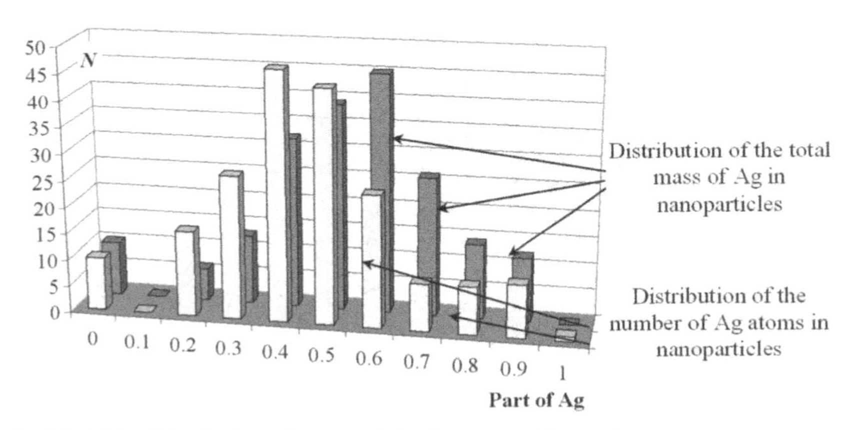

FIGURE 3.16 Distribution of nanoparticles by composition and mass.

The value of silver share in nanoparticles was from 0% (some nanoparticles did not contain silver) up to 71.8%. The distribution by mass is somewhat shifted to the right in comparison with the distribution by the number of atoms.

To form the homogeneous nanocomposite from the nanoparticles calculated, it is necessary to provide the uniform distribution of the material through the volume of nanoparticle mixture. The distribution of metal mass constituting the nanocomposite through the rated cell volume can serve as a criterion of homogeneity (Fig. 3.17). At the time moment of 0.02 ns (Fig. 3.17a), a considerable irregularity in the composition is observed: the majority of silver atoms are located in the left side of the volume rated, and copper atoms are mainly located in the right region of the cell. At the time moment of 1 ns, all atoms are in a gaseous state and there is no irregularity of the composition due to their mixing, that is, all the mass is homogeneously distributed through the region rated (Fig. 3.17b). After the nanoparticles are formed, the irregularity of the composition is observed again by 11 ns (Fig. 3.17c). However, the deviation from the homogeneous distribution does not exceed 5–10%, thus indicating the possibility to form a homogeneous nanocomposite.

The distribution of silver and copper atoms through the nanoparticles is characterized by the dependence demonstrated in Figure 3.18. Here, the change in the mass share of silver in nanoparticles is given along the abscissa axis, the value of total mass of nanoparticles containing the definite share of silver—along the ordinate axis. The half of nanoparticles (55%) corresponds to 40% of silver relative mass in nanoparticles. The value of the material mass share for which the silver relative mass in nanoparticles deviates considerably from the maximum (below 20% or over 65%) totals to only 10%.

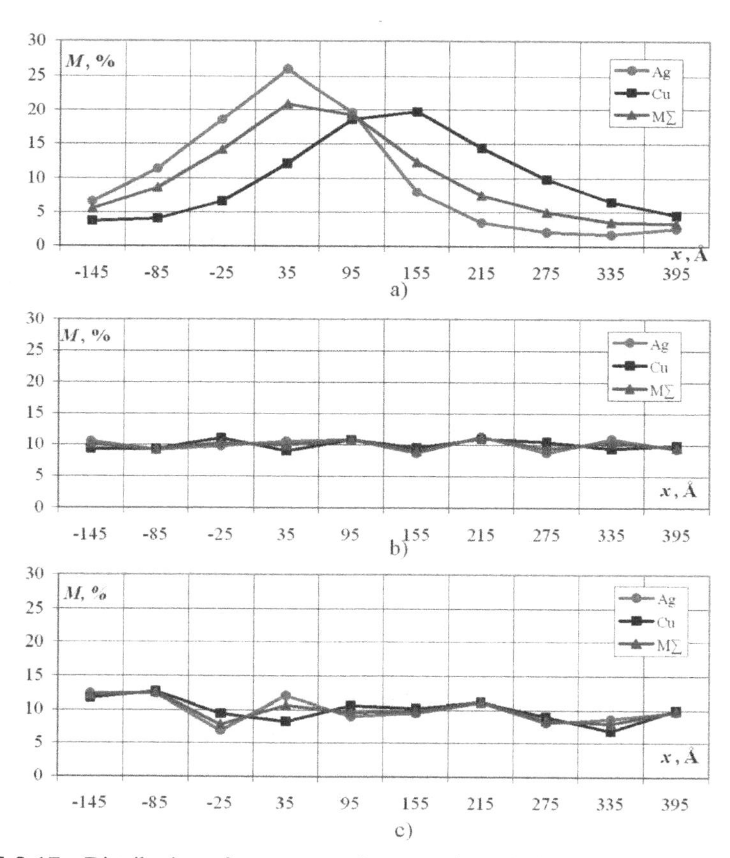

FIGURE 3.17 Distribution of nanocomposite mass through the volume rated at the time moments: (a) $t = 0.02$ ns, (b) $t = 1$ ns, and (c) $t = 11$ ns. The lines with circle, square, and triangular markers indicate the distribution of Ag, Cu atoms, and total mass distribution, respectively.

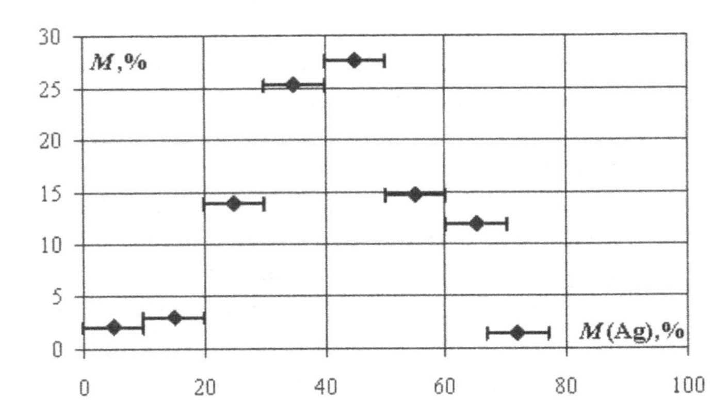

FIGURE 3.18 Dependence of the mass share of composite material upon the silver relative mass in nanoparticles.

The structure of nanoparticles plays an important role. To reveal the main structural properties, let us consider five nanoparticles: $Ag_{40}Cu_{60}$, $Ag_{26}Cu_{35}$, $Ag_{26}Cu_{18}$, $Ag_{19}Cu_{26}$, and $Ag_{16}Cu_{34}$. The diameter of nanoparticles varies in the range from 8.32 to 11.02 Å, and molecular mass—from 3701.69 to 8127.48 a.m.u. More detailed properties and parameters of nanoparticles investigated are given in Table 3.2. Average characteristics were determined based on the properties of these nanoparticles with subsequent preparation of the distributions of relative mass and absolute density through the nanoparticle spherical layers.

TABLE 3.2 Type and Properties of Nanoparticles Investigated.

Nanoparticle	**Diameter, Å**	**Molecular mass, a.m.u.**	**Mass share of Ag atoms, %**	**Mass share of Cu atoms, %**	**Density, a.m.u./Å**
$Ag_{40}Cu_{60}$	9.42	8127.48	53.09	46.91	18.55
$Ag_{26}Cu_{18}$	9.18	3948.40	71.03	28.97	9.78
$Ag_{19}Cu_{26}$	9.56	3701.69	55.37	44.63	8.10
$Ag_{16}Cu_{34}$	11.02	3886.45	44.41	55.59	5.55
$Ag_{26}Cu_{35}$	8.32	5028.68	55.77	44.23	16.74
Average characteristics of nanoparticles	9.50	4938.54	55.93	44.07	11.74

Figure 3.19 demonstrates the average relative mass distribution of nanoparticles consisting of silver and copper atoms, depending on their

relative radius. The mixed proportion of masses of Cu and Ag atoms that can be explained by similar melting and boiling temperatures of the initial materials is observed in the layers. Therefore, the processes of condensation of these materials from gaseous phase flow simultaneously, and Ag and Cu atoms are uniformly distributed through the nanoparticle volume. The smaller mass share in the nanoparticle central layers is explained by the increase in the volume of spherical layers with the radius growth.

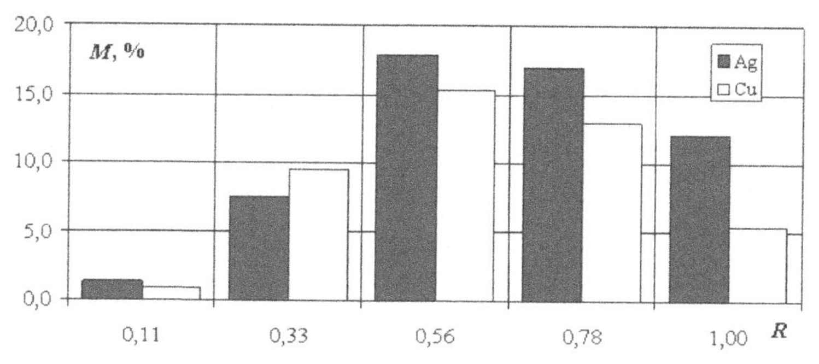

FIGURE 3.19 Average distribution of relative mass of Ag–Cu nanoparticles depending upon the relative radius in mass %.

Let us consider the distribution of substance density depending upon the nanoparticle relative radius (Fig. 3.20). From the graph, it is seen that the nanoparticle inner layers are denser than external layers.

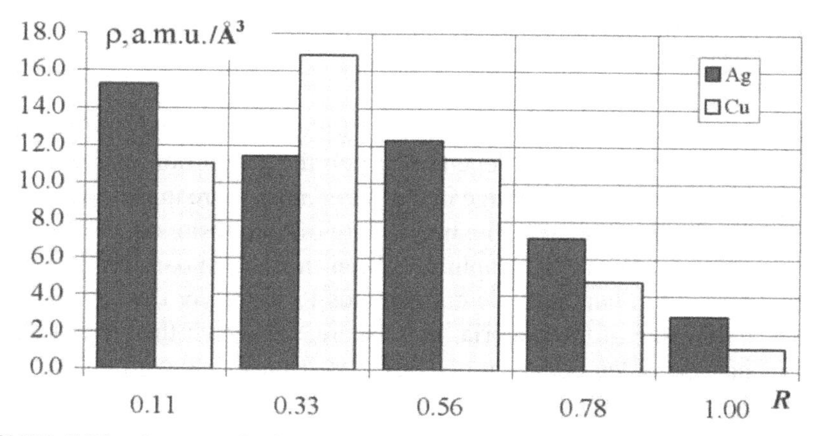

FIGURE 3.20 Average distribution of the substance density of Ag–Cu nanoparticles depending upon the relative radius.

The distribution of silver through the nanoparticles of silver and copper depending upon the cooling rate of the composite mixture is shown in Figure 3.21. Here, the change in the silver mass share in nanoparticles is given along the abscissa axis, and the mass share of composite material—along the ordinate axis. The calculations are carried out for three condensation periods: 1 fs; 0.5 ns, and 5 ns. The difference in the distribution of silver share is insignificant: curves are located quite close to each other. Thus, the condensation rate in the ranges investigated does not significantly influence the silver share in the composite nanoparticles from silver and copper.

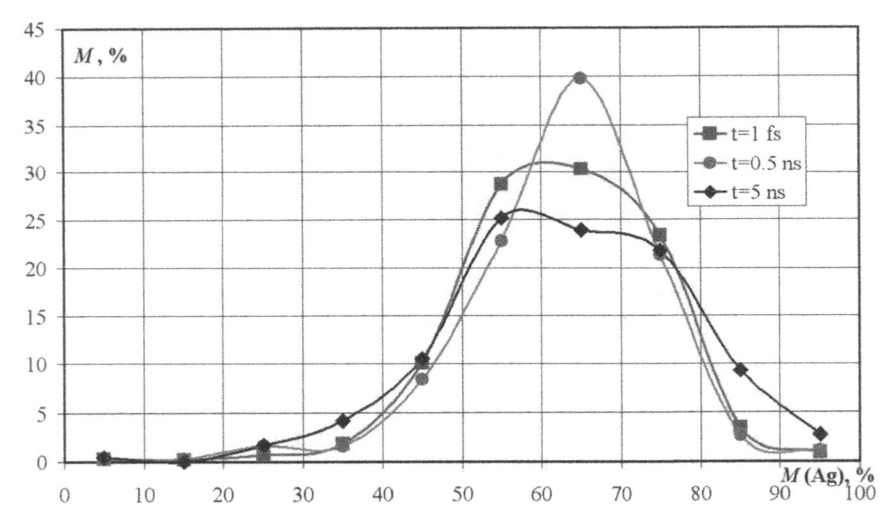

FIGURE 3.21 Dependence of the mass share of composite material upon the relative mass of Ag in nanoparticles for different cooling rates.

Let us consider the silver distribution (Fig. 3.22) through the nanoparticles from silver and copper depending upon the proportion of these materials in initial crystals. The change in the silver mass share in nanoparticles is given along the abscissa axis, the mass share of composite material—along the ordinate axis. From the graph, it is seen that the maximum of percent content of silver in nanoparticle corresponds to the silver content in initial crystals. The value of the material mass share, for which the silver relative mass in nanoparticles considerably deviates from the maximum value, is only 10–15%. This indicates the relative distribution uniformity of atoms of initial metals through nanoparticles.

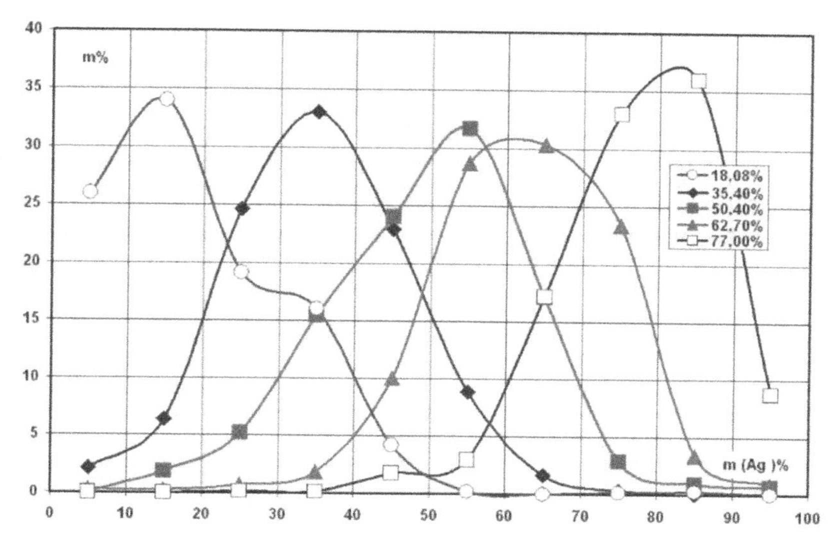

FIGURE 3.22 Dependence of silver mass share in nanoparticles upon the silver relative mass in initial crystals.

Let us consider the formation of composite metal nanoparticles from silver and zinc. The calculations are carried out for the nanosystem consisting of 1702 atoms, including 608 silver atoms and 1094 zinc atoms. This corresponds to 50% of silver and 50% of zinc by mass in initial nanocrystals.

The modeling of composite nanoparticles includes two stages: short-time heating and further cooling. The heating is carried out at $T = 3000$K within 1 ns. The condensation lasts for 170 ns at $T = 300$K, significantly exceeding the condensation period for nanoparticles consisting of silver and copper. The condensation period increase is explained by the smaller value of Van der Waals forces between zinc atoms (Table 3.1).

The change in the share of atoms condensed into nanoparticles is shown in Figure 3.23. From the graph, it is seen that silver atoms are being condensed much faster than zinc atoms. At 1 ns, that is the end of heating and evaporation stage, all atoms are in gaseous state. There is an active condensation within 10 ns, and by 11 ns, all silver atoms are in nanoparticles, but only 6.03% zinc atoms are grouped into nanoparticles. By the time moment of 171 ns, the number of such zinc atoms is 15.72%.

A more detailed change in the share of atoms condensed into nanoparticles with time is given in Figure 3.24, which shows the change in the decimal logarithm of the concentration of condensed atoms with time. Since the melting temperatures of zinc and silver differ considerably, the concentration curves for zinc and silver atoms, as well as the common curve for all condensed atoms also differ considerably. The smooth change in the share of grouped silver atoms (line with triangular markers) demonstrates the stable character of silver condensation. The variable (pulse) with time change in the percent of condensed zinc atoms (dependence marked with circle markers) shows the unstable state of zinc atoms in nanoparticles. Due to small value of attractive forces between zinc and silver atoms, zinc atoms can be separated from the nanoparticle surface and return to the gaseous medium.

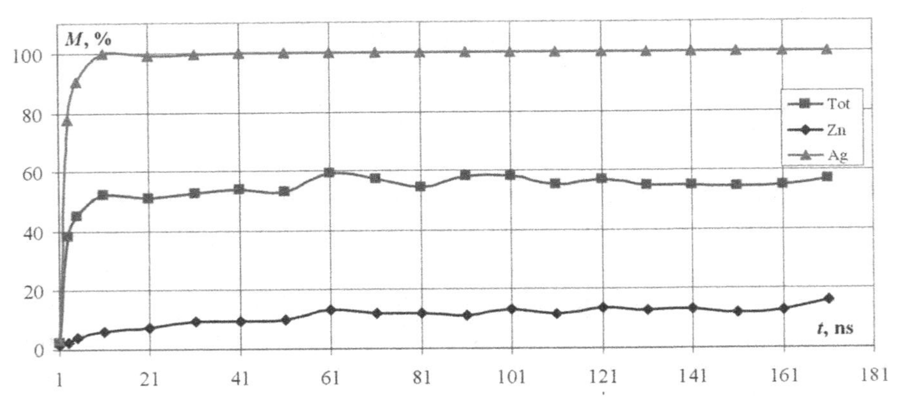

FIGURE 3.23 Change in the percent of silver and zinc atoms condensed into nanoparticles.

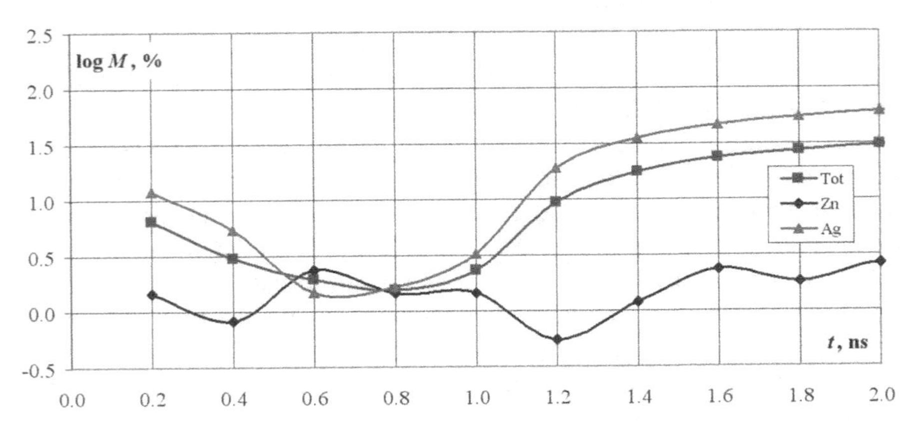

FIGURE 3.24 Change in the share of atoms condensed into nanoparticles in the logarithmic scale.

Let us consider the distribution of nanoparticles by size at different time moments. In Figure 3.25, the bar charts of the distribution of nanoparticles at different time moments are given: 11 ns (line with square markers), 71 ns (line with triangular markers), and 161 ns (line with circle markers).

In the beginning of the condensation process, small nanoparticles prevail. Thus for t = 11 ns, the number of nanoparticles with the diameter of 4 Å equals 60, and only one particle has the diameter of 11 Å—maximum for this moment.

With time, the number of small nanoparticles decreases, but the large ones increases. By 161 ns, nanoparticles with the diameter of 18 Å appear, and there are only eight small particles with the diameter of 4 Å.

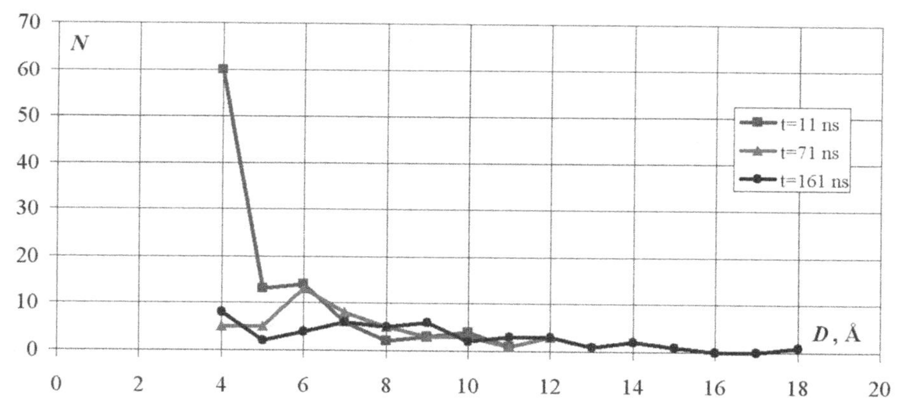

FIGURE 3.25 Bar chart of the distribution of Ag–Zn nanoparticles by size.

The tendency of the enlargement of nanoparticles is seen in Figure 3.26, demonstrating the change in the average size of nanoparticles with time. The average size of nanoparticles increases from 4 Å for the time moment of 1 ns up to 8.59 Å for t = 171 ns. When composite nanoparticles are formed from copper and silver (Fig. 3.7), a gradual decrease in the intensity of the increase in the nanoparticle size is observed due to their enlargement and attenuation of mobility. For Ag–Zn nanoparticles, the average size increases slowly but stably during the entire condensation process. Besides, the rate of the increase in the average diameter of composite nanoparticles consisting of silver and copper atoms is much higher than for nanoparticles from zinc and silver.

The change in the number of nanoparticles in the volume unit with time is given in Figure 3.27. The transition from heating to condensation corresponds to the considerable increase in the number of nanoparticles in the interval 1–11 ns. Further decrease in the number of nanoparticles for the time period exceeding 11 ns is characterized by their combining and enlarging: for t =11 ns, 103 nanoparticles are present in the volume rated, and by t =171 ns, the number of nanoparticles decreases to 32.

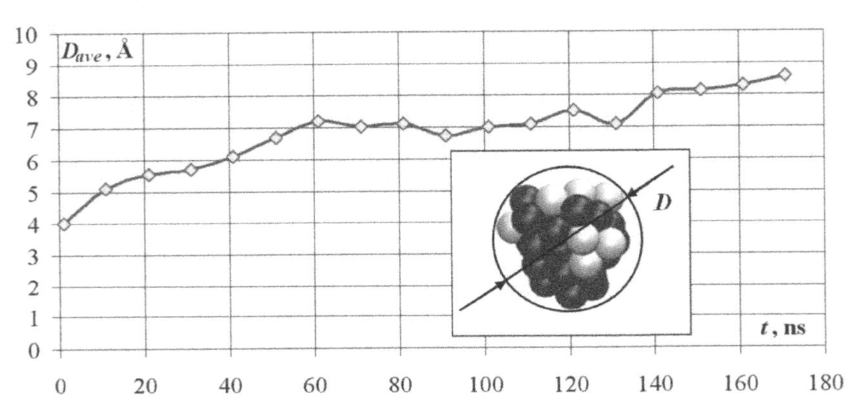

FIGURE 3.26 Change in the average diameter of Ag–Zn nanoparticles.

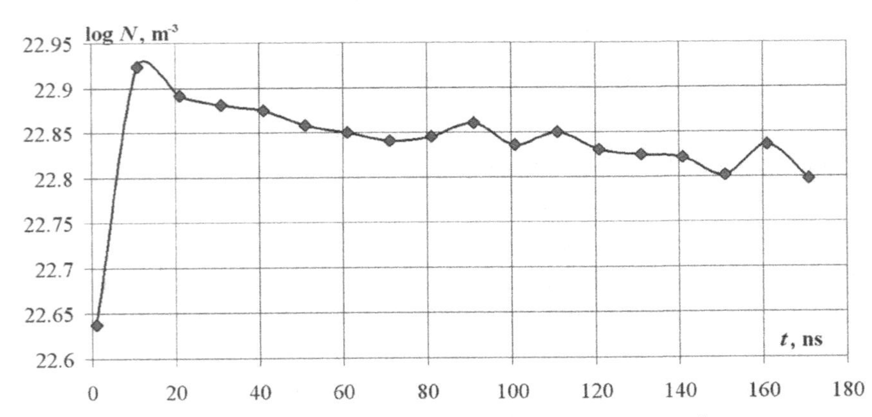

FIGURE 3.27 Change in the number of nanoparticles in the volume unit.

The dependence of the total volume and the area of nanoparticle surfaces in logarithmic scale upon time are shown in Figures 3.28 and 3.29. The behavior of these curves is mainly explained by the change in the volume rated and increase in the nanoparticle average diameter (Figs. 3.59 and 3.60). Nevertheless, the growth of the total volume at the condensation stage for t =11–171 ns does not result in the increase in total surface of nanoparticles.

Besides, the change in the volume and surface area of silver–zinc nanoparticles, on the contrary to silver–copper ones, is unstable. This is conditioned by the weak interaction potential of zinc atoms.

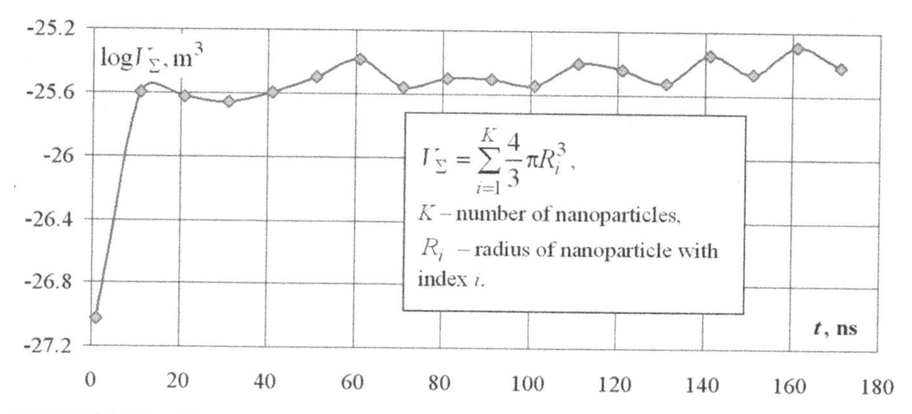

FIGURE 3.28 Change in the total volume of nanoparticles.

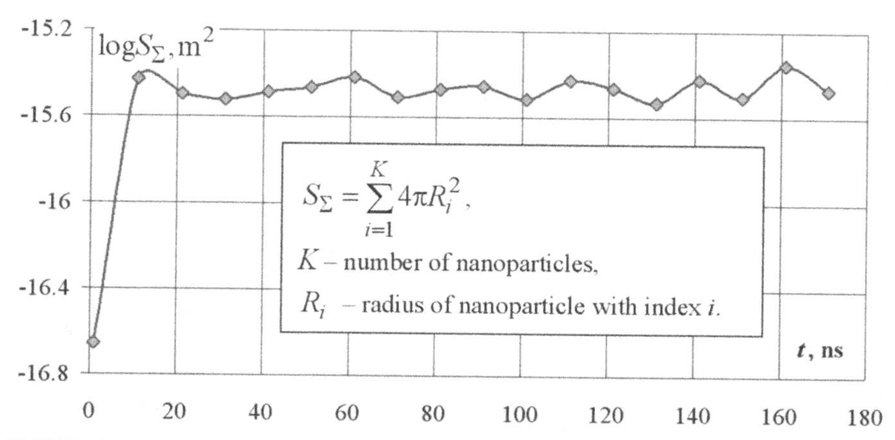

FIGURE 3.29 Change in the total surface of nanoparticles.

The difference in atomic structure of composite nanoparticles Ag–Cu and Ag–Zn should be pointed out. Silver–copper nanoparticles mainly have a mixed structure, though the nanoparticles consisting only of one metal are also observed. Nanoparticles from silver and zinc consist of "the nucleus" formed by silver atoms and "the shell" from zinc atoms (Fig. 3.30). Thus, when initial materials with considerably different melting temperatures are used, multi-layer nanoparticles can be obtained.

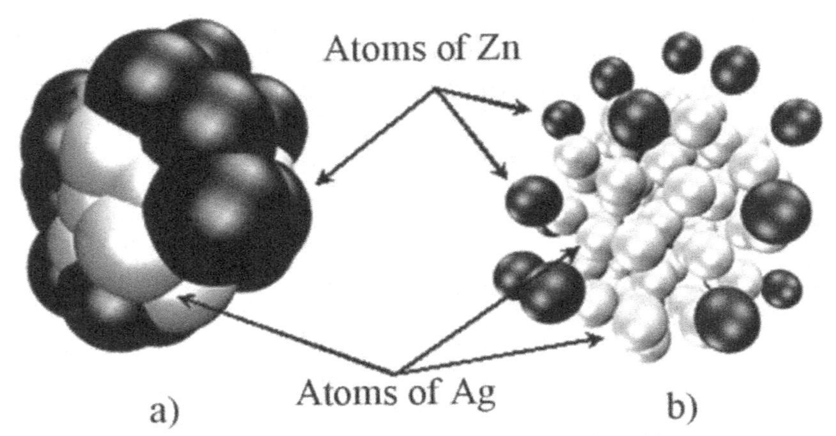

FIGURE 3.30 Composite nanoparticle $Ag_{55}Zn_{22}$: (a) general view, (b) structure.

The stages of the fusion of two nanoparticles consisting of silver and zinc atoms are given in Figure 3.31. The combining of nanoparticles starts with their attraction and contact. Due to different properties of initial metals, each nanoparticle represents "a nucleus" by its structure, formed by the silver atoms surrounded with zinc atoms. After the contact of nanoparticles, their "nuclei" start attracting to each other, and zinc atoms are ejected to the nanoparticle surface (Figs. 3.31–3.34). In the center of the nanoparticle obtained by the fusion of two nanoparticles, there are only silver atoms, and zinc atoms compose "the shell" of a new nanoparticle. This effect is explained by the fact that Van der Waals attractive forces between the silver atoms are much greater than similar forces arising between the silver–zinc

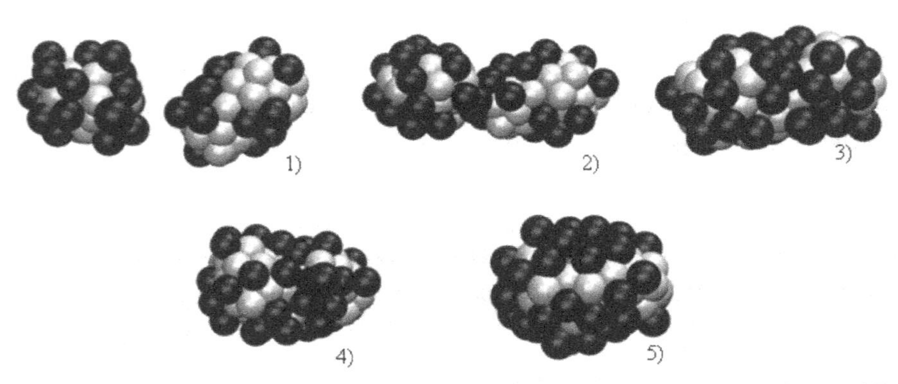

FIGURE 3.31 Stages of combining two composite nanoparticles consisting of Ag and Zn atoms (Ag atoms are marked with gray, Zn atoms with black).

and zinc–zinc atoms. Therefore, the silver "nucleus" tends to the most compact shape and takes the position in the nanoparticle center and zinc atoms stay on the surface. The shape of the nanoparticle combined resembles the sphere (Figs. 3.31–3.35), thus corresponding to the minimum of the nanoparticle potential energy.

We will consider the structure of Ag–Zn nanoparticles formed by analogy with silver–copper nanoparticles. For this, we will study five nanoparticles: $Ag_{33}Zn_5$, $Ag_{29}Zn_7$, $Ag_{31}Zn_{10}$, $Ag_{57}Zn_{17}$, and $Ag_{55}Zn_{22}$. The molecular mass of nanoparticles changes in the range from 3585.90 to 7371.32 a.m.u., the diameters vary within 7.70–9.60 Å. The detailed characteristics of nanoparticles investigated are given in Table 3.3. The average characteristics are determined based on the properties of these nanoparticles: average diameter—8.70 Å; average molecular mass—5220.35 a.m.u.; average mass share of Ag atoms—85.53%; average mass share of Zn atoms—14.47%; average density—14.72 a.m.u./$Å^3$. Besides, the distributions of relative mass and absolute density by the nanoparticle layers are calculated.

In Figure 3.32, you can see the average distribution of relative mass of silver–zinc nanoparticles depending upon their relative radius.

TABLE 3.3 Type and Properties of Nanoparticles Investigated.

Nanoparticle	Diameter, Å	Molecular mass, a.m.u.	Mass share of Ag atoms, %	Mass share of Zn atoms, %	Density, a.m.u./A^3
$Ag_{33}Zn_5$	7.82	3886.59	91.59	8.41	15.56
$Ag_{29}Zn_7$	7.70	3585.90	87.24	12.76	14.95
$Ag_{31}Zn_{10}$	8.88	3997.81	83.64	16.36	10.87
$Ag_{57}Zn_{17}$	9.60	7260.11	84.69	15.31	15.64
$Ag_{55}Zn_{22}$	9.48	7371.32	80.48	19.52	16.57
Average values	8.70	5220.35	85.53	14.47	14.72

Only silver atoms are present in the nanoparticle central region (Fig. 3.32), in surface layers—silver and zinc atoms. Melting and boiling temperatures of silver and zinc differ, therefore, first silver condense from gaseous phase, and nanoparticle "nucleus" is formed. Later, zinc atoms get involved into the condensation process and form the nanoparticle shell.

The obtained results on the nanoparticle structure fit the data in Ref. 133, where the copper nanoparticles with zinc shell were obtained experimentally. Owing to similar physical and chemical properties of silver and copper atoms (Ag and Cu are in one subgroup of elements in periodic system and possess similar melting temperatures), composite copper–zinc and silver–zinc nanoparticles have a shell structure.

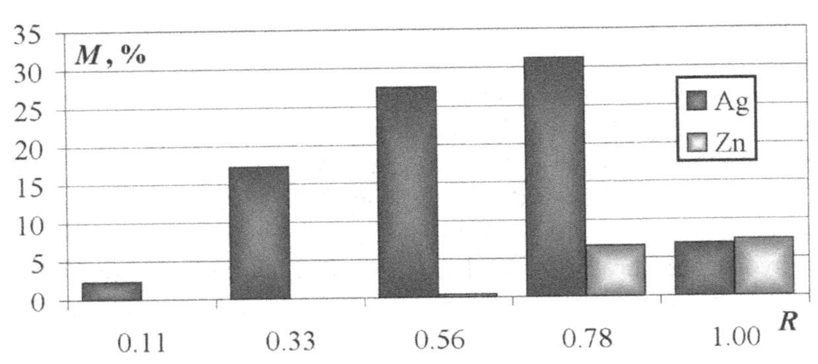

FIGURE 3.32 Average mass distribution of Ag and Zn atoms depending upon the nanoparticle relative radius.

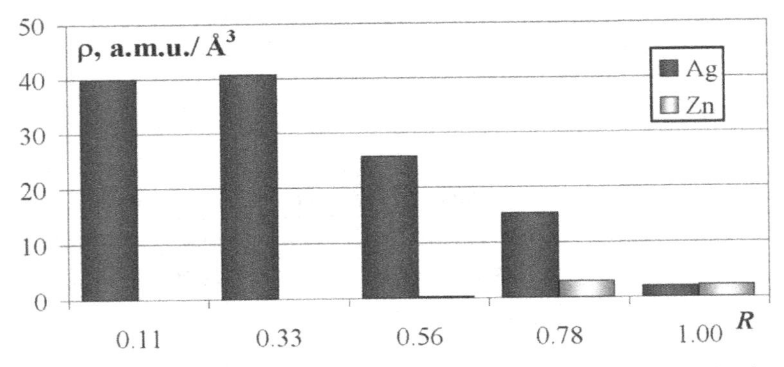

FIGURE 3.33 Average distribution of substance density in Ag–Zn nanoparticles depending upon the nanoparticle relative radius.

Average distribution of substance density depending upon the nanoparticle relative radius is shown in Figure 3.33. The density of silver atoms is marked with black and zinc atoms with white. It is seen that external layers of Ag–Zn nanoparticles are denser than the internal ones, in the same way as with silver–copper nanoparticles.

By analogy with Ag–Cu nanoparticles, let us consider the distribution of silver and zinc atoms through the cell rated for the time moment of 171 ns given in Figure 3.34. The mass distribution of Ag atoms is marked with the line with circle markers, Zn—with square markers, total mass distribution—with triangle markers.

In Figure 3.34, it is seen that the composition irregularity to 5% of mass is caused by silver atoms. By this time, all silver atoms are in nanoparticles. Mass fluctuations are caused by the movement of large particles in the volume rated. However, there are no considerable fluctuations in the mass of different atoms in the volume of the cell rated.

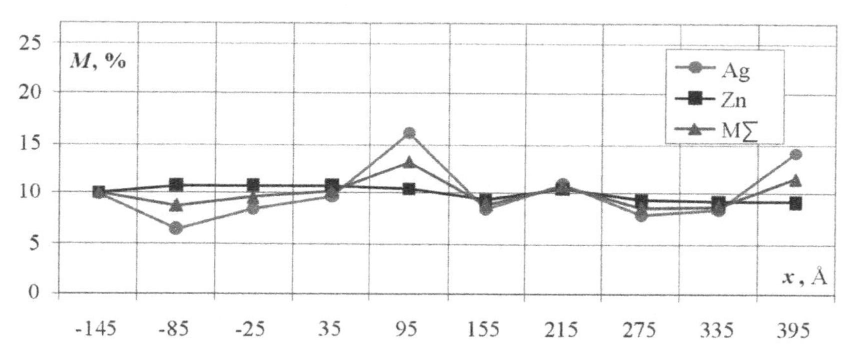

FIGURE 3.34 Distribution of nanocomposite mass through the volume rated for $t = 171$ ns. Lines with circle, square, and triangular markers show the distribution of Ag, Zn atoms, and total mass distribution, respectively.

Silver and zinc distribution through the nanoparticles is given in Figure 3.35. The value of total mass of nanoparticles containing a definite silver share is given along the ordinate axis. The change in the silver mass share in nanoparticles is given along the abscissa axis.

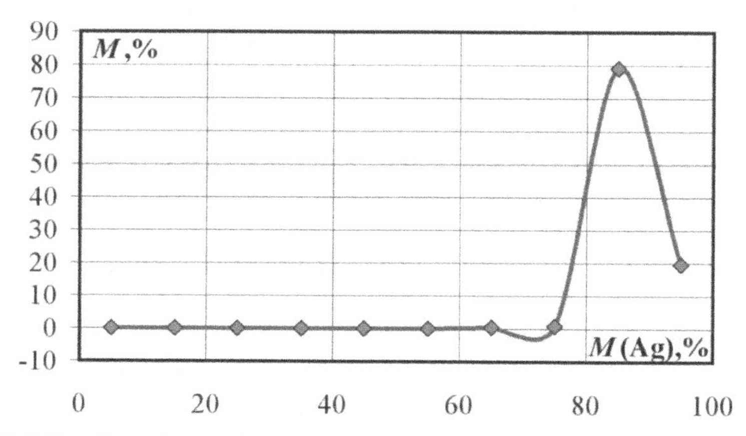

FIGURE 3.35 Dependence of mass share of composite material upon the relative mass of silver in nanoparticles.

In comparison with silver–copper nanoparticles, the silver–zinc ones have a more expressed maximum of mass distribution. The curve maximum (about 80% of nanocomposite mass) falls on 85% of silver share in nanoparticles and does not coincide with the initial mass share of silver on the contrary to the results obtained for Ag–Cu nanoparticles.

In conclusion to this section, the following main regularities of the composite nanoparticle formation should be enlisted:

1. The process of metal nanoparticle formation comprises two main stages: first—formation of nanoparticles from atoms; second—combining of nanoparticles. Stage duration is defined by the properties and composition of initial nanosystem.
2. Number of nanoparticles increases at the first stage of condensation and slowly decreases at the second, being the consequence of nanoparticle enlargement.
3. Total volume and surface of nanoparticles increase fast at the first stage of condensation, being practically constant at the second.
4. During condensation, the average diameter of nanoparticles increases.
5. Maximum of percent content of atoms in nanoparticles corresponds to the percent composition of the initial nanosystem.
6. Condensation rate does not sufficiently influence the percent content of atoms in composite nanoparticles.
7. Structure and homogeneity of nanoparticles depend upon the properties of initial materials.
8. Irregularity of composition arising in the modeling process can reach 5–10% from the mass of composite material, and is conditioned by the movement of large nanoparticles.

3.1.2 PROBABILISTIC LAWS OF DISTRIBUTION OF NANOPARTICLE STRUCTURAL CHARACTERISTICS

The calculation of macroscopic parameters of nanocomposites is a critical issue requiring the determination of power, structural-scale parameters, mechanical properties of nanoelements, as well as the investigation of formation and interaction processes of structural nanoelements constituting the composite. This problem as applicable to powder nanocomposites is discussed in a number of works of author.[155,157,160] According to the investigations, the main feature of nanoelements is as follows: when the characteristic size of nanoelements changes, their physical–mechanical characteristics (elasticity modulus, strength, deformation, and other parameters) can change by an order conditioned by the reconstruction (not only monotonous) of nanoelement atomic structure and shape.[108,125,127] Therefore, the nanoelement characteristic size is one of the main parameters determining the nanocomposite properties. It should be pointed out that nanoelement formation processes, determining the structure and properties, carried out on atomic and molecular levels, are probabilistic or stochastic. Thus, nanoelement parameters indicated always have a certain variation. So, the probabilistic analysis of the distribution of nanoelement parameters is quite a critical task.

Stochastic process of formation of nanoparticles is confirmed by the study of the stability of numerical solutions. Robustness of the results of modeling of composite nanoparticles is tested by coordinates, velocity, and parameters of nanoparticles. Among the composite mixture is chosen arbitrarily atom, and its trajectory is determined. After that, the initial position of the atom has changed by a small value ΔS_0, and carried out simulation of the behavior of the system with modified initial coordinates. On the basis of two predetermined trajectories of the test of the atom at each moment was calculated the amount of path displacement ΔS, which is the Euclidean norm between the original and the modified path. In Figure 3.36a, the results of studying the stability of solutions of the formation problem of composite nanoparticles for several initial change of coordinates are presented.

Analysis of the graphs shows that in the time interval of up to 10 pks, differences in the trajectories of the atom for $\Delta S_0 = 0.001$ A and $\Delta S_0 = 0.01$ A are small and at this temporary site solution can be considered as stable. For times greater than 10 pks and for $\Delta S_0 = 0.05$ A, the behavior of the curves indicates the instability of the system solutions.

The results of studying the stability of solutions for a variety changes in the initial speed of a single atom are shown in Figure 3.36b. In a sharp change of the curves, it can be seen that the solution is more sensitive to changes in the initial velocity than to coordinates, and is unstable in the initial velocity.

Nevertheless, a slight variation of the initial data does not affect the initial total solution of the problem, and system macro parameters remain unchanged. However, these parameters have a certain spread. In this regard, it is necessary to study the law of distribution of nanoparticle parameters.

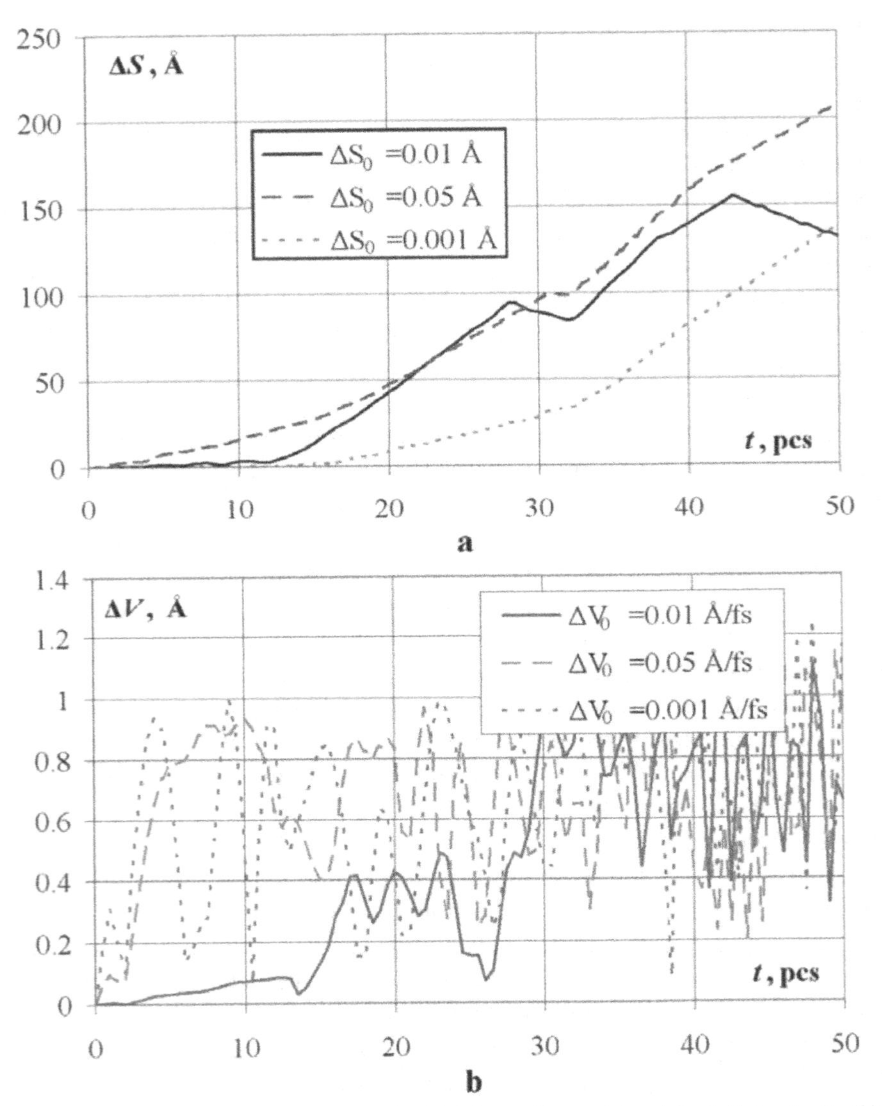

FIGURE 3.36 Stability of solutions to the initial coordinates (a) and the initial velocity (b).

3.1.2.1 PROBLEM DEFINITION AND THEORY

The aim of this chapter—probabilistic analysis using the technique of molecular dynamics of laws of distribution of nanoparticle parameters formed in the process of condensation from gaseous medium.

For the definition of the probabilistic laws of distribution of nanoparticle structural characteristics, the task of nanoparticle formation in gaseous medium was solved many times with one-type thermodynamic initial conditions. At the same time, initial conditions randomly differed in each numerical realization. At the initial time moment, the atoms were randomly distributed in the numerical cell with periodic boundary conditions. They were in a free state and were not grouped in nanoparticles. The distribution of initial velocities of the atoms was also set based on Maxwellian distribution, corresponding to the nanosystem initial temperature. The further cooling conditioned the condensation of free atoms in nanoparticles.

The selected data of some parameter of nanoparticles formed are obtained in the series of calculation experiments. Afterwards, the hypothesis on the type of the law of this parameter distribution is checked. Then the nanosystem distribution parameters at each moment in time are found based on the series of calculation experiments. Such properties as a number of nanoparticles in the volume unit, their average diameter, common volume, and surface area can serve as parameters. For the fixed moment in time, the distribution of a separate parameter X is expressed by selective random values $x_1, x_2, \ldots, x_N$. In general, the distribution law of the parameters is unknown, and its distribution becomes critical. The law of distribution of the selective parameters was estimated by Pirson criterion. [80]

Pirson criterion is based on the check of non-parametric hypothesis on the correspondence of the empiric distribution function to some theoretical law. For this, selective data of the parameter estimated were sequenced not by the decrease and grouped into the variation row $x_1, \leq x_2 \leq \ldots \leq x_{N,}$ where N—total number of selective data. The changing area of the random parameter X was split into m intervals $\Delta_1, \Delta_2, \ldots, \Delta_m$. The number of division intervals was determined by Sturgess heuristic formula m = 1 + 3.3221g N.[139]

The probabilities p_j (j = 1,2,...,m) of each selective value hitting each interval Δ_j are calculated in accordance with the assumed theoretical distribution law:

$$P_j = P\{X \in \Delta_j\} = \int_{\Delta_j} f(x)\, dx = F(x_j^+) - F(x_j^-), \quad (3.1)$$

where $f(x)$—distribution law of the assumed theoretical law; $f(x)$—integral probabilistic function; x_j^- and x_j^+—left and right boundaries of j -interval, respectively. The theoretical number of the values $\tilde{n}_j$ of random value X, in the interval Δ_j, is determined as $\tilde{n}_j = Np_j$. Empiric frequencies n_j are determined calculating selective values x_i, $i = 1, 2, \ldots, N$, which are in the interval Δ_j.

If the theoretical frequencies $\tilde{n}_j$ insufficiently differ from the empirical $\tilde{n}_j$, the hypothesis on the correspondence of the empirical distribution law to the theoretical one is accepted; otherwise the hypothesis is rejected. The known distributions can in turn be used as theoretical distribution laws: χ -square, Student, normal, etc.

The deviation degree between the empirical and theoretical frequencies is characterized by Pirson criterion.[80]

$$Kr = \sum_{j=1}^{m} \frac{\left(n_j - \tilde{n}_j\right)^2}{\tilde{n}_j} = \sum_{j=1}^{m} \frac{\left(n_j - Np_j\right)^2}{Np_j} = \sum_{j=1}^{m} \frac{n_j^2}{Np_j} - N. \quad (3.2)$$

In accordance with Pirson theorem, the criterion Kr has the distribution χ-square at $N \to \infty$ with the number of degrees of freedom $k = m - r - 1$. Here, m —the number of selection intervals and r —number of parameters of the distribution law assumed. For example, the normal law has two parameters—mathematical expectation and mean-square deviation. Consequently, the number of degrees of freedom when checking the hypothesis on the normality of the selection distribution law equals $k = m - 3$.

Thus with the significance level α, characterizing the reliability of the hypothesis accepted, and after defining the number of degrees of freedom k, the quantile of the distribution χ-square ($\chi^2_{\alpha,k}$) can be calculated, which is afterwards compared with the criterion value Kr. If $Kr \le \chi^2_{\alpha,k}$, the hypothesis on the coincidence of theoretical and empirical distribution laws is accepted. Otherwise, it is rejected. The obtained empirical function of the distribution of random value $F^*(x)$ is compared by Pirson criterion with various theoretical laws of distribution $F(x)$, given in Table 3.4.

The use of Pirson criterion requires the calculation of main parameters of the theoretical distribution law determined by their point estimates. For example, for the normal distribution law, it is necessary to find the estimates of two parameters: mathematical expectation and mean-square deviation. The point estimates of theoretical distribution laws are determined by the moment method.

The method essence consists in selecting as many empirical moments as many unknown distribution parameters have to be estimated. Usually, it is sufficient to estimate the moments of the first and second order, that is, the selective mathematical expectation and corrected dispersion:

$$M_x = \frac{1}{N} \sum_{i=1}^{N} x_i, \quad (3.3)$$

TABLE 3.4 Distribution Laws and Its Parameters.

Distribution law	Parameters of distribution	Distribution law	Mathematical expectation	Dispersion
Normal	μ—mathematical expectation, σ—mean-square deviation	$f(x)=\frac{1}{\sigma\sqrt{2\pi}}\exp\left(-\frac{(x-\mu)^2}{2\sigma^2}\right)$	μ	σ^2
γ-distribution	$k>0$, $\theta>0$—scale factor	$f(x)=x^{k-1}\frac{\exp(-x/\theta)}{\Gamma(k)\theta^k}$, $\Gamma(k)=\int_0^\infty x^{k-1}e^{-x}dx$	$k\theta$	$k\theta^2$
Student	$n>0$—number degrees of freedom	$f(x)=\frac{\Gamma((n+1)/2)}{\sqrt{n\pi}\Gamma(n/2)(1+x^2/n)^{(n+1)/2}}$	0	$\frac{n}{n-2}$
χ-square	$n>0$—number degrees of freedom	$f(x)=\frac{(1/2)^{n/2}}{\Gamma(n/2)}x^{n/2-1}\exp(-x/2)$	n	$2n$
Exponential	$\lambda>0$—intensity	$f(x)=\lambda\exp(-\lambda x)$	$\frac{1}{\lambda}$	$\frac{1}{\lambda^2}$

TABLE 3.4 *(Continued)*

Distribution law	Parameters of distribution	Distribution law	Mathematical expectation	Dispersion
Laplace distribution	$\alpha > 0$ —scale factor, β —shift factor	$f(x) = \frac{\alpha}{2}\exp(-\alpha\lvert x-\beta\rvert)$	β	$\frac{2}{\alpha^2}$
β -distribution	$\alpha > 0, \beta > 0$	$f(x) = \frac{x^{\alpha-1}(1-x)^{\beta-1}}{B(\alpha,\beta)}$, $B(\alpha,\beta) = \int_0^1 x^{\alpha-1}(1-x)^{\beta-1}\,dx$	$\frac{\alpha}{\alpha+\beta}$	$\frac{\alpha\beta(\alpha+\beta)^{-2}}{(\alpha+\beta+1)}$

$$D_x = \sigma_x^2 = \frac{1}{N-1}\sum_{i=1}^{N}(x_i - M_x)^2 . \qquad (3.4)$$

The selective moment estimates are equalized with theoretical moments constituting the equations expressing the dependence of distribution parameters upon the selective moments. As a result of solving the given equations, the estimates of the parent population parameters are found.

After finding out the law of distribution of nanoparticle parameters according to Pirson criterion, the bar chart of the probabilities was prepared and the empirical function of the random value $F^*(x)$ and theoretical law $F(x)$ were compared. The bar chart of the probabilities represents the joint graph of the distribution of theoretical probabilities p_j (3.4) and relative frequencies $p^*_j = n_j/N$. The values of integral theoretical and empirical functions at each interval were defined by the following formulas:

$$F_j = \sum_{l=1}^{j} P\{X \in \Delta_l\} = \sum_{l=1}^{j}\int_{\Delta_l} f(x)dx = F\left(\frac{x_j^+ + x_j^-}{2}\right), \qquad (3.5)$$

$$F^*_j = \sum_{l=1}^{j} p^*_l = \sum_{l=1}^{j}\frac{n_l}{N}. \qquad (3.6)$$

Additionally, the statistic function of the result distribution was calculated to estimate the normality of the distribution law of selective data.[163] The selective data of the parameter estimated are regulated by increase and grouped into the variation row $x_1 \le x_2 \le \ldots \le x_N$, where N—common number of selective data. The statistic distribution function is determined by the following formula:

$$\phi(x_1) = \frac{n(x_1)}{N+1}, \phi(x_i) = \phi(x_{i-1}) + \frac{n(x_i)}{N+1}, \ (i = 2, 3, \ldots, L), \qquad (3.7)$$

where $n(x_i)$—number of coinciding elements with the value x_i; L—number of elements of the variation row without recurrent values. This function represents a stepped line: the step width is determined by the value $x_{i+1} - x_i$, (i = 1, 2, …, $N-1$) the height of each next step is increased by $1/(N+1)$ in comparison with the previous one (Fig. 3.37). For repeating values of the variation row $x_i = x_{i+1} = \ldots = x_{i+k}$, the value of the statistic function of distribution increases by $1/(N+1)$, where k—number of coinciding elements. With a large quantity of observations N, the statistic function approaches stochastically the actual selective function of distribution.

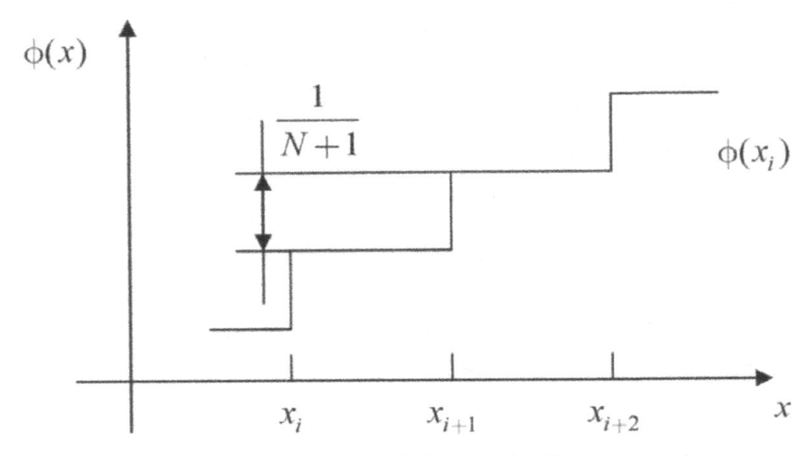

FIGURE 3.37 Statistic function of the variation row distribution.

With the help of statistic function of distribution and relation $\phi(x_i) = \Phi(z_i)$, the values of normally distributed value z_i are found. The function $\Phi(z_i)$ represents the distribution law of a normal random value with zero mathematical expectation and one mean-square deviation. The random value z is connected with the selective data by the relation $z_i = (x_i - M_x)/\sigma_x$. To check the distribution law normality, the points $(x_i z_i)$ corresponding to the selective data are marked in coordinate plane (x, z). If the points are marked along one straight line, the hypothesis on the normality of distribution law is accepted. Otherwise, the law of selective data distribution is not normal.

3.1.2.2 RESULTS AND DISCUSSION

The probabilistic analysis of the calculation of nanoparticle parameters formed from silver and copper atoms is presented below. The parameters of potentials for the metals investigated are given in Table 3.5.

The average density of nanoparticles Ro_{sr}, the average share of silver mass in nanoparticles $M(Ag)$, average diameter of nanoparticles D_{sr}, and the number of nanoparticles in the volume calculated N_{part} were considered as the properties of nanoparticles. The value of the volume calculated was selected in accordance with the requirement of representativeness of nanoparticle formation medium and equaled $W = 2.4 \times 10^{-23}$. The volume of selective data for the parameters estimated varied in the range from $N = 30$ up to 130. The nanoparticle formation was simulated in the time interval up to 10 ns.

TABLE 3.5 Parameters of Van der Waals Interactions.

Constants of Lennard–Jones potential	Ag	Cu
ε_i, 10^{-21} J	55,242	65,583
e_i^*, 10^{-10} m	2.968	2.624

The family of change curves of the average diameter of nanoparticles D_{sr} in time is shown in Figure 3.38c. In accordance with the calculation results, the beginning of metal atoms condensation is accompanied by the presence of small composite nanoparticles. After a while, the small nanoparticles are combined and, as a result, their average sizes grow. But despite random distributions of initial coordinates and velocities of metal atoms, the average diameter of nanoparticles fluctuates in a narrow range of data changes. It should be pointed out that at the final stages of atom and nanoparticle condensation corresponding to the simulation time interval $t = 6 \ldots 10$ ns, a more significant bundle of the average size of nanoparticles is observed.

Apart from the dimensional distribution of nanoparticles, such parameters as average density of nanoparticles (Fig. 3.38a), average share of silver mass in nanoparticles (Fig. 3.38b), and the number of nanoparticles in the volume calculated (Fig. 3.38d) are investigated. The change in the number of nanoparticles with time corresponds to the increase in their number at the initial moments and further decrease due to the combination of nanoparticles formed. The graphs of changes in average values of silver mass share and density of nanoparticles characterize the inner structure and composition of nanoparticles. The behavior of these values is mutually correlative, the parameters have great dispersion in comparison with the number of nanoparticles and their sizes.

To check the statistic hypotheses on the type of probabilistic law of nanoparticle parameters, the following distributions were considered: χ- square, Student, Laplace, normal, exponential, γ- and β- distributions. Pirson statistics was calculated for the final simulation stage $t = 10$ ns. With respect to all properties of nanoparticles (Ro_{sr}, M (Ag), D_{sr}, N_{part}), the value of Pirson criterion for the Student distribution, γ-, and β-distributions cannot be calculated since the probability theoretical functions are not defined with the distribution parameters obtained. Consequently, the hypotheses on the

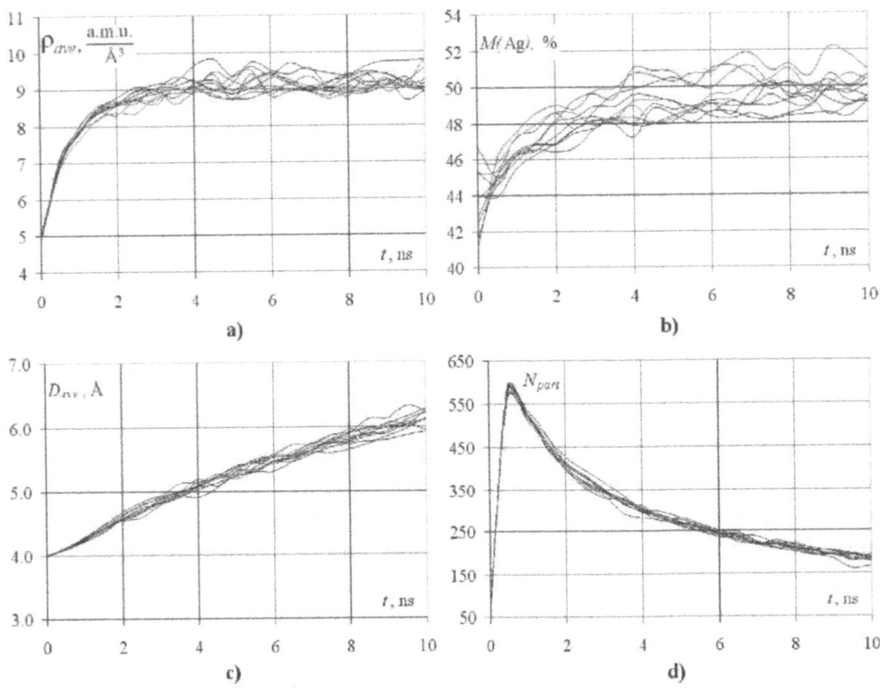

FIGURE 3.38 Family of change curves of: (a) average density of nanoparticles, (b) average share of silver mass in nanoparticles, (c) average diameter of nanoparticles, and (d) number of nanoparticles in the volume calculated.

correspondence of selective data to the probabilistic laws mentioned are rejected as unfounded. The analysis of Pirson statistics for the rest of the distribution laws shows that the criterion greatly depends on the volume of the data selected. With small volume of selection, none of the hypotheses on the law of distribution of nanoparticle properties is realized. Nevertheless, we can judge about the most probable distribution of the selection by the least criterion value. With the increase in the number of selective data, the decrease in Pirson statistics is observed and one of the hypotheses can be accepted.

During the statistic investigation with the help of Pirson criterion, it is found out that the average density of nanoparticles, average share of silver mass in nanoparticles, average diameter of nanoparticles and number of nanoparticles in the volume calculated have a normal law of distribution. Pirson statistics together with the quantile of the distribution of c-square for the parameters characteristics of nanoparticles with different selection volumes, as well as the values of distribution parameters estimated by the method of moments are given in the Table 3.6. The significance level during the estimation of hypotheses was $\alpha = 0.8$.

The data about Pirson statistics given in the Table 3.6 demonstrate that when the selection volume increases from $N = 30$ up to 70, the growth of the criterion value is observed. Such results are explained by the availability of low-probabilistic selective values that significantly influence the statistics value.

Besides when calculating Pirson criterion, it is recommended to split the selective data into odd number of intervals (for $N = 70$ the number of intervals corresponded to $m = 6$). For the selection volume $N = 130$, the statistics decreases considerably.

TABLE 3.6 Pirson Statistics of Normal Low for Different Selective Values Volume.

Nanoparticles parameter	$N = 30$		$N = 70$		$N = 130$		
	Statistics Kr	**Fractile** $\chi^2_{0.8,2}$	**Statis-tics Kr**	**Fractile** $\chi^2_{0.8,3}$	**Statistics Kr**	**Fractile** $\chi^2_{0.8,4}$	**Low parameters** μ/σ
Ro_{sr}	1.35	0.45	4.38	1.01	1.32	1.65	9.06 0.3
D_{sr}	1.8	0.45	8.79	1.01	1.62	1.65	6.14 0.12
M(Ag)	0.9	0.45	0.56	1.01	1.53	1.65	49.5 0.88
N_{part}	1.66	0.45	10.17	1.01	1.39	1.65	182.14 6.84

Taking into account the obtained parameters of normal distribution, the changes in the time of mathematical expectation of nanoparticle parameters are given in Figure 3.39: (a) average density of nanoparticles, (b) average share of silver mass in nanoparticles, (c) average diameter of nanoparticles, and (d) number of nanoparticles in the volume calculated. The gray area in the graph corresponds to the limiting error of parameter change. The limiting error was calculated based on the value of mean-square deviation: $\Delta_{пред} = 3\sigma_x$. Thus, the possible values of nanoparticle parameter X marked in gray in Figure 3.38 are determined by the relation $X_{возм}(t) = M_x(t) \pm \Delta_{пред}$ for the definite moment of time. (Ro_{sr}, M (Ag), D_{sr}, N_{part}) were in turn used as the variable X.

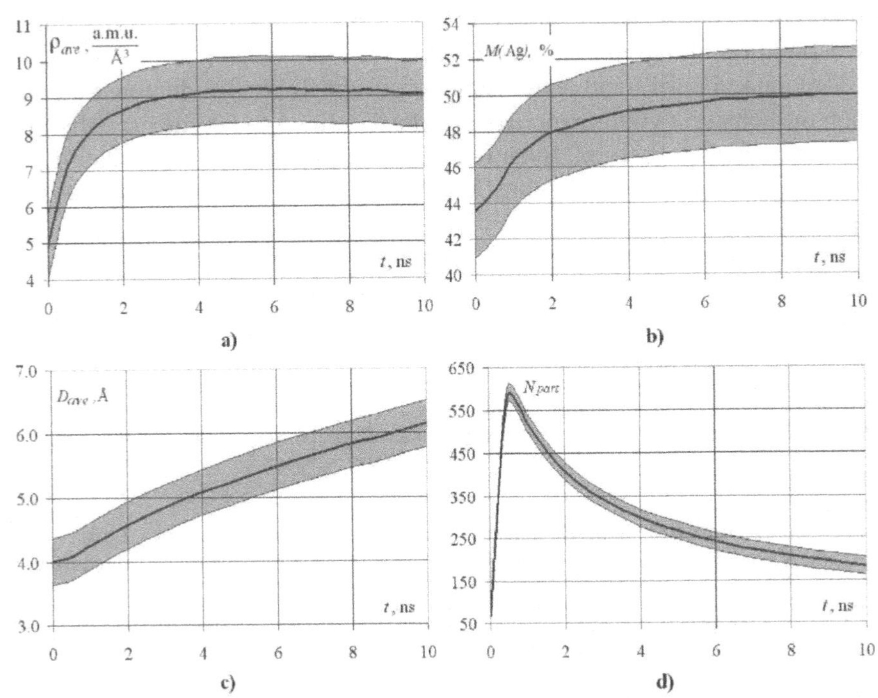

FIGURE 3.39 Change with the indication of the limiting error of nanoparticle parameters: (a) average density of nanoparticles, (b) average share of silver mass in nanoparticles, (c) average diameter of nanoparticles, and (d) number of nanoparticles in the volume calculated.

To compare the empirical function of the random value and standard normal distribution for all nanoparticle parameters investigated, we prepared the bar chart of probabilities given in Table 3.6. The sense of the bar chart consists in the values of probability of random value getting in each splitting interval for the moment of time t = 10 ns. The bar chart in Figure 3.40 depicts the empirical function of random value, the firm line—standard normal distribution law at the same selective mathematical expectation and mean-square deviation of nanoparticle parameter. Having analyzed the data in Figure 3.40, we see a slight deviation of probabilities from standard distribution law for the average density of nanoparticles, average share of silver mass in nanoparticles and number of nanoparticles in the volume calculated.

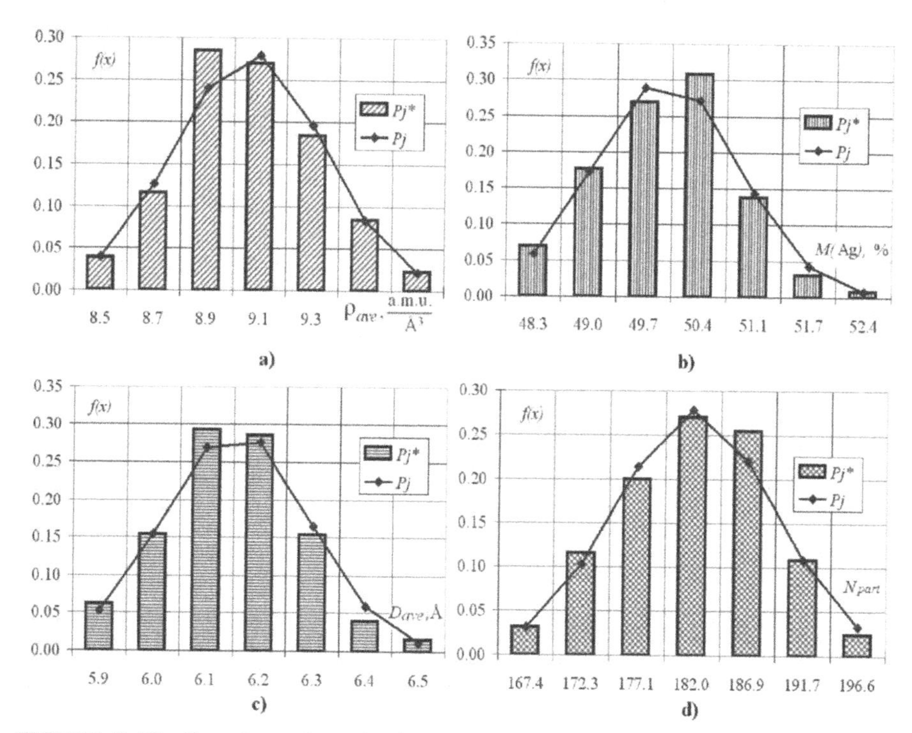

FIGURE 3.40 Bar charts for selective data corresponding to: (a) average density of nanoparticles, (b) average share of silver mass in nanoparticles, (c) average diameter of nanoparticles, and (d) number of nanoparticles in the volume calculated.

To continue with the analysis, we will consider theoretical functions of normal distribution $F(x)$ and function $F^*(x)$ characterizing the selective data given in Figure 3.41. The functions are arranged for four parameters of nanoparticles: Ro_{sr}, M (Ag), D_{sr}, N_{part}, besides the empirical distribution functions are depicted for the selection volume $N = 50$ (dot line with square markers) and $N = 130$ (dot line with triangular markers). The functions have similar mathematical expectations and mean-square deviations found by selective estimates. The curves of probability change are quite close to each other, have similar points of extremums and convexities of functions.

The comparison of the probabilistic function of normal distribution $F(x)$ and empirical functions $F^*(x)$ for the average density of nanoparticles, average share of silver mass in nanoparticles, average diameter of nanoparticles and number of nanoparticles in the volume calculated shows good correspondence of selective and theoretical data. At the same time, when the volume of selective data increases, the deviation between the theoretical and empirical dependencies becomes less noticeable. The best coincidence of functions is observed for the number of nanoparticles in the volume calculated.

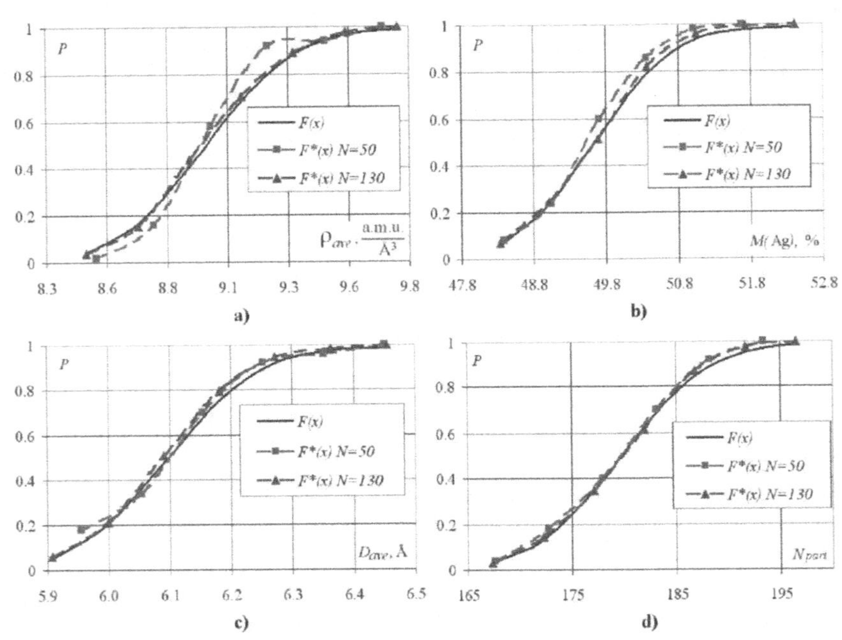

FIGURE 3.41 Comparison of theoretical and selective function of distribution probability for: (a) average density of nanoparticles, (b) average share of silver mass in nanoparticles, (c) average diameter of nanoparticles, and (d) number of nanoparticles in the volume calculated.

The hypothesis on the normality of the distribution of average density of nanoparticles Ro_{sr}, average share of silver mass in nanoparticles M (Ag), average diameter of nanoparticles D_{sr} and number of nanoparticles in the volume calculated N_{part} is confirmed by the arrangement of statistic function. In Figure 3.42, the selective data corresponding to the nanoparticle parameters and shown with round markers are compared with normally distributed random value z.

Approximation linear function is depicted with the firm line. The mathematical expectation and mean-square deviation of the random value z are determined for each parameter based on the formulas (3.3)–(3.4). In Figure 3.42, it is seen that the variation row for the investigated properties of nanoparticles in axes (Ds,z) is located along the straight line confirming the normality of the distribution law.

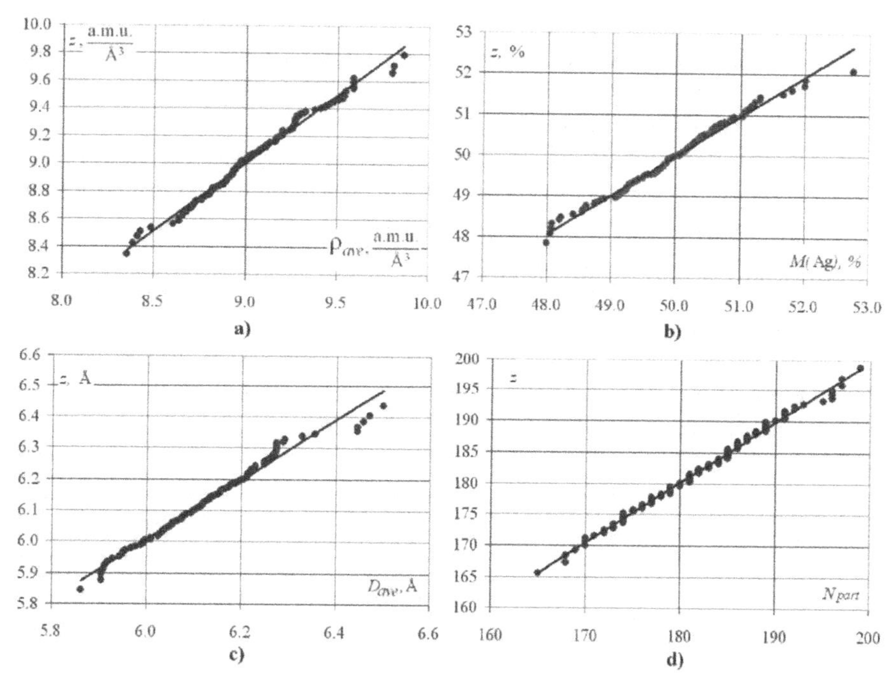

FIGURE 3.42 Statistic function of distribution for selective data corresponding to: (a) average density of nanoparticles, (b) average share of silver mass in nanoparticles, (c) average diameter of nanoparticles, and (d) number of nanoparticles in the volume calculated.

3.1.2.3 CONCLUDING REMARKS

1. Structural and numerical parameters of nanoparticles such as average density of nanoparticles, average share of silver mass in nanoparticles, average diameter and number of nanoparticles in the volume calculated in accordance with Pirson criterion at the significance level $\alpha = 0.8$ have the normal law of distribution. Mathematical expectations and mean-square deviations for the given parameters of nanoparticles were: $\mu = 9.06$ and $\sigma = 0.30$, $\mu = 49.96$ and $\sigma = 0.88$, $\mu = 6.14$ and $\sigma = 0.12$, $\mu = 182.14$ and $\sigma = 6.84$, respectively.
2. Calculations show that the check of the hypothesis on the correspondence of structural properties of nanoparticles to the theoretical law of distribution with the help of Pirson criterion requires a significant volume of selective data. For some properties of nanoparticles, the acceptance of the hypothesis on the distribution law became possible

with the selection volume $N = 50$. In general, for nanoparticle parameters investigated, the acceptance of the hypothesis on the normality of the law of distribution of selective data became possible with the selective data volume $N = 130$.

3. Pirson statistics for structural properties of nanoparticles nonlinearly depends on the selective data volume. With small selection volumes, the growth of statistics value is possible, with significant number of data Pirson statistics decreases.
4. The bar charts of the distribution of probabilities and comparison of probabilistic functions of normal distribution and selective empirical functions show good correspondence of selective and theoretical data. When the volume of selective data increases, the deviation between theoretical and empirical dependencies decreases.
5. Statistic functions of the variation row of nanoparticle parameters confirm Pirson hypothesis on the normal law of property distribution. For the average density of nanoparticles Ro_{sr}, average share of silver mass in nanoparticles M (Ag), average diameter of nanoparticles D_{sr} and number of nanoparticles in the volume calculated N_{part}, the statistic functions have a linear character and are well approximated by a straight line.

3.1.3 THE FORMATION OF MULTICOMPONENT NANOPARTICLES CONSISTING OF MOLECULES

3.1.3.1 THE TASK DEFINITION

Formation of multicomponent nanoparticles is necessary in many processes: forming of composites, composite coatings, manufacture of medicaments, plant nutrition from the gas phase, the formation of aerosol fire extinguishing, etc. Description of these processes is given in Chapter 1 and Fig. 1.1.

The problem of calculating the process of formation of such nanoparticles includes two stages. The first step is to calculate the molecular configurations; the second stage is the formation of nanoparticles composed of molecules.

The calculation of the configuration of the molecular formations requires "ab initio" calculations. The calculation of the process of the molecule joining-up in nanoparticles at cooling the gas mixture to normal temperature can be carried out with the help of the molecular dynamics methods.

Let us consider the results of calculations of the processes of nanoparticle condensation during operation of the aerosol fire extinguisher.

3.1.3.2 THE ANALYSIS OF THE CALCULATION RESULTS

The mix in the aerosol fire extinguisher includes seven basic components. The mass fractions of these components are presented in Figure 3.43. The gas phase mainly consists of the molecules of nitrogen and carbonic gas. The molecules of potassium carbonate (K_2CO_3) constitute the basic component of the solid phase.

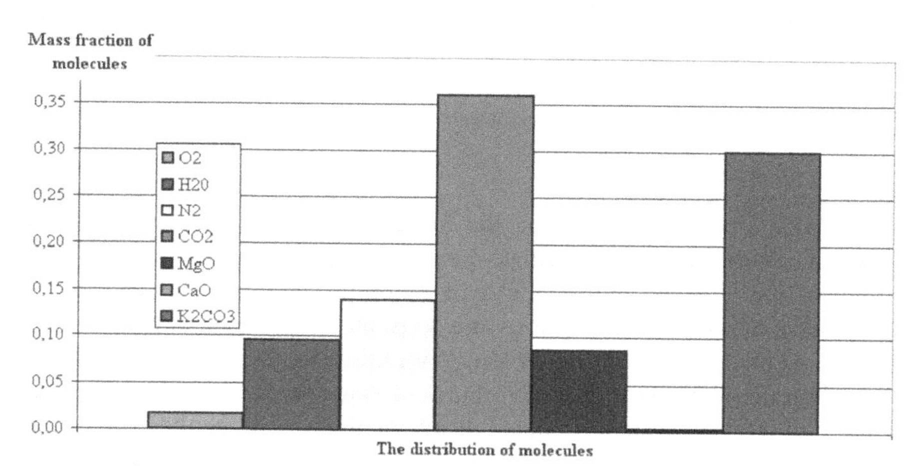

FIGURE 3.43 The main mass fractions of aerosol components.

For calculating the equilibrium configuration of the molecules of initial substances, the method of the first principles has been used. As a result of calculations, the equilibrium lengths of connections in molecules, the sizes of equilibrium corners and two-sided corners, and atomic charges of the molecules investigated were determined.

For each of seven types of the molecules forming the gas phase shown in Figure 3.43, the equilibrium configurations have been calculated. As an example, the appearance of a molecule of potassium carbonate with optimized geometry is shown in Figure 3.44. Figure 3.44 also contains the surface of electrostatic potential of a molecule of potassium carbonate, with the gray color corresponding to the positive potential +0.1 eV and the black one, to the negative potential −0.1 eV.

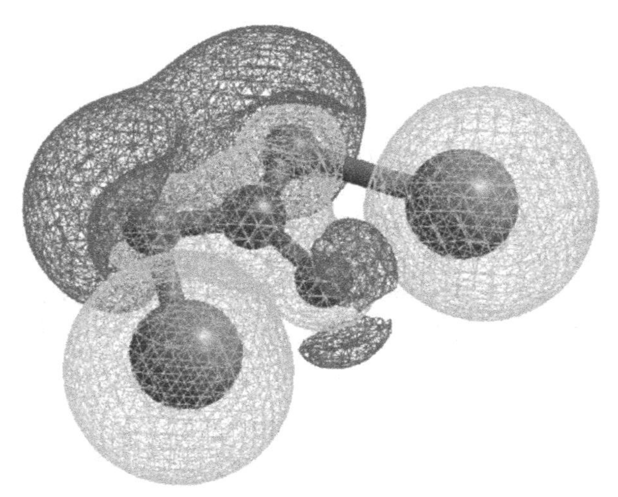

FIGURE 3.44 Structure of molecules, negative potential −0.1 eV and positive potential +0.1 eV of potassium carbonate.

Usually, in real experiments, the molecular system exchanges energy with the environment. For calculating such power interactions, special algorithms—"thermostats" are used. Use of a thermostat allows one to carry out calculation of molecular dynamics at a constant temperature of the environment or to change it by a certain law. With this aim, the stage of condensation of an aerosol mix after active burning was considered. Therefore, the temperature of the system modeled was kept slightly above the normal one (at a level of 310 K) due to the use of a Berendsen thermostat (Fig. 3.45a). As follows from this figure, with the Berendsen thermostat, the temperature was kept constant with an accuracy of 2–3%. Very important point of modeling is stabilization of the nanosystem pressure. Use of the "barostat" algorithms makes it possible to model the behavior of the system at constant pressure. The simplest of them is the Berendsen barostat in which the pressure is kept stationary by means of scaling the settlement cell. The positions of particles in systems are modified on each time step according to the factor of barostat scaling. The dependence of the change in the modeled volume of the system is presented in Figure 3.45b.

The process of condensation of molecules into nanoparticles is reflected by the dynamics of change in the relative weight of grouped molecules (Fig. 3.46). The line with square symbols corresponds to the relative weight of condensed K_2CO_3 molecules, with rhombic symbols, to H_2O, with triangular symbols, to MgO, and with round symbols, to N_2.

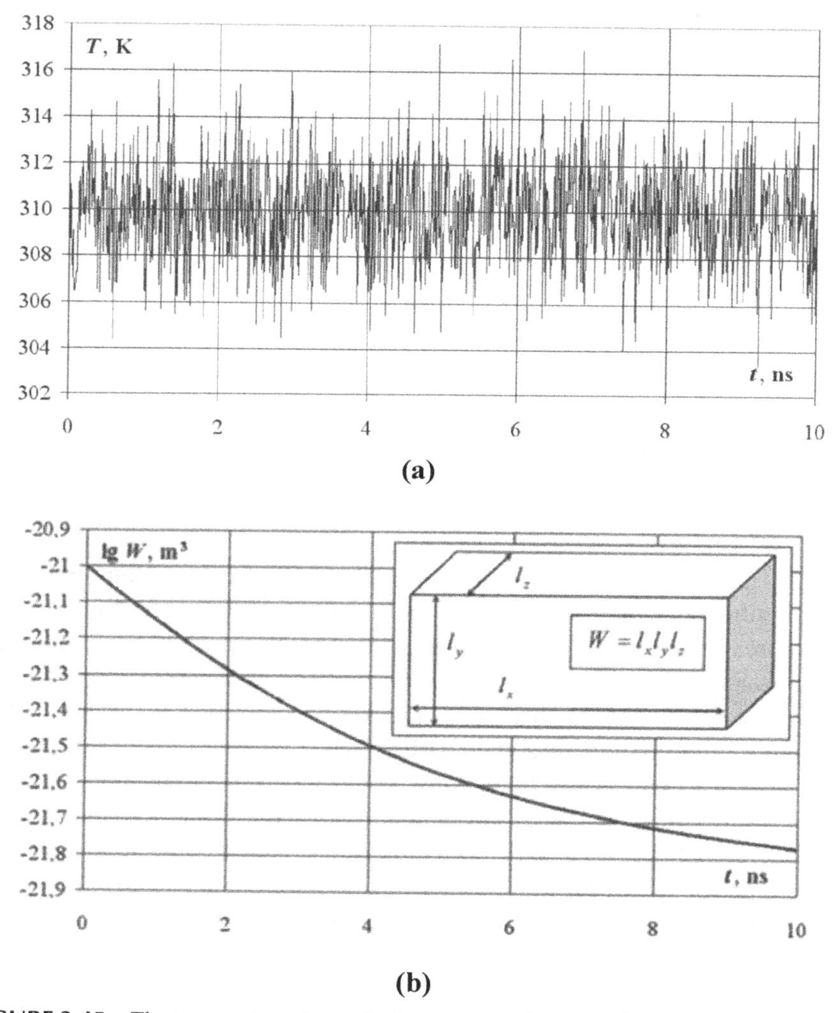

FIGURE 3.45 The temperature change in the system during modeling (a) and change in the volume of settlement during modeling (b).

Curves describing the changes in the amount of condensed molecules of oxygen and calcium oxide were not obtained, since the molecules of these substances were in a gaseous state. Carbonic gas and nitrogen were condensed into nanoparticles least actively. Such an effect point to the fact that the CO2, O2, and N2 molecules were present in the volume investigated as compound components of the mix of air with combustion products which under normal conditions do not unite into nanoparticles.

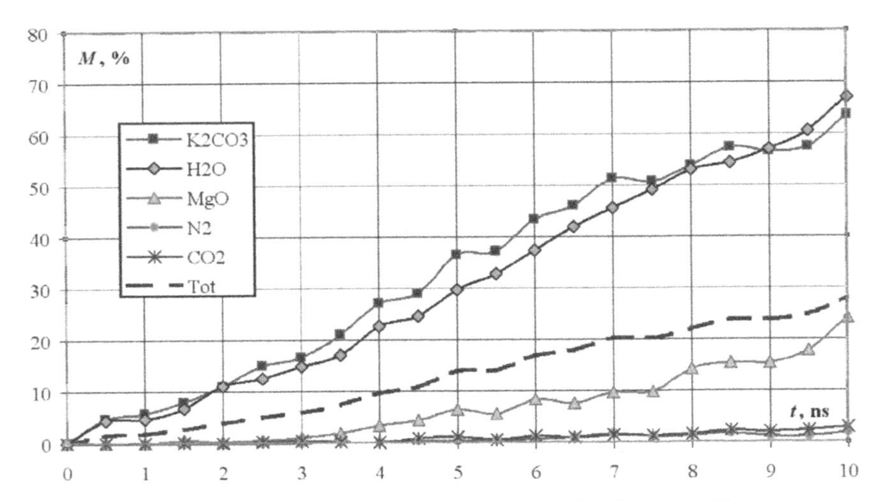

FIGURE 3.46 Change in the mass fraction of nanosystem molecules grouped into nanoparticles.

The quantity of the molecules of calcium oxide in the volume modeled is insignificant, which corresponds to their mass fraction in Figure 3.43. Therefore, they constitute a part of the nanoparticles formed in small quantity. The analysis of Figure 3.47 shows that the association of molecules into nanoparticles during the entire stage of condensation is observed. The molecules of potassium carbonate participate in condensation most actively. They have the greatest weight and are the centers of condensation. Because of the small values of the forces of interaction, the unstable nanoparticles can lose molecules from their structure, and break up into nanostructures smaller in size.

In Figure 3.47, the change in the number of nanoparticles in a unit volume at the logarithmic scale is given. The analysis of this figure shows that the amount of nanoparticles is stably reduced due to the merging and association of nanostructures. The asymptotic behavior of a curve is observed: the quantity of nanoparticles in a unit volume varies more slowly eventually.

The weight of the nanoparticles formed grows, the mobility decreases, which causes a decrease in the number of nanoparticles in a unit volume. In using the mesodynamics method, the increase in the spatial scale of the nanosystem because of association of several symmetrically displayed settlement areas was obtained. It has allowed carrying out an analysis of the processes of nanoparticle formation at big time scales. The mesodynamics considers only generated nanoparticles; the influence of the gas phase is determined by the magnitude of casual forces. Therefore, a substantial increase in the computing productivity has been obtained in comparison with the molecular dynamics modeling due to reduction in the number of objects in the system.

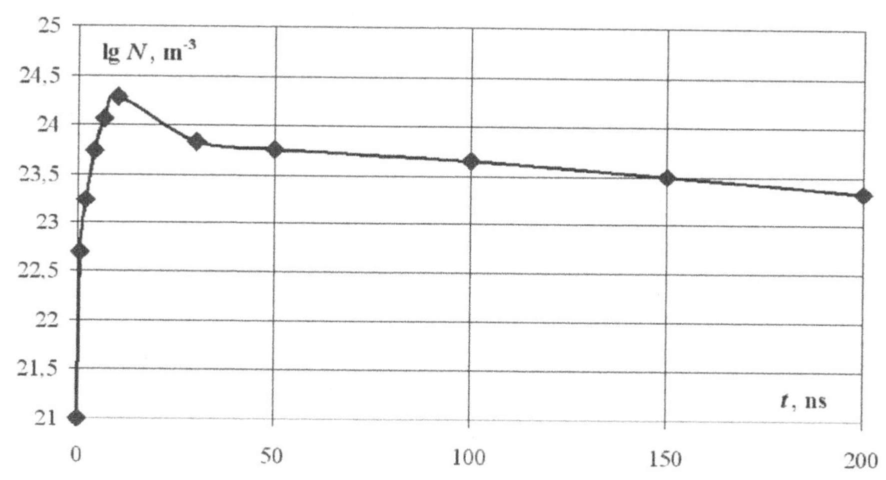

FIGURE 3.47 Change in the number of nanoparticle in a unit volume obtained in modelling by the molecular dynamics and mesodynamics method.

The size distribution of nanoparticles in various moments of time is presented in Figure 3.48. For the size of a nanoparticle, the maximal distance between two atoms of a nanoparticle was taken. In the figure, the dependences the amount of nanoparticles on the diameter in 1, 5, and 10 ns are given. At the beginning of the process of condensation, small nanoparticles are formed. By the moment of 1 ns, the number of nanoparticles with the diameter of up to 10 Å prevails, and the number of large particles is insignificant. Eventually nanoparticles are integrated; therefore, the number of particles of small diameter decreases. However, there are larger nanoparticles having the size of more than 20 Å whose formation was not observed at the earlier stages of condensation.

The example of the nanoparticles formed by 10 ns is given in Figure 3.49. It is composed of 62 atoms, including 12 water molecules, two molecules of carbon dioxide, one molecule of magnesium oxide and three molecules of potassium carbonate. The nanoparticle has a no spherical shape, with its diameter being 20 Å.

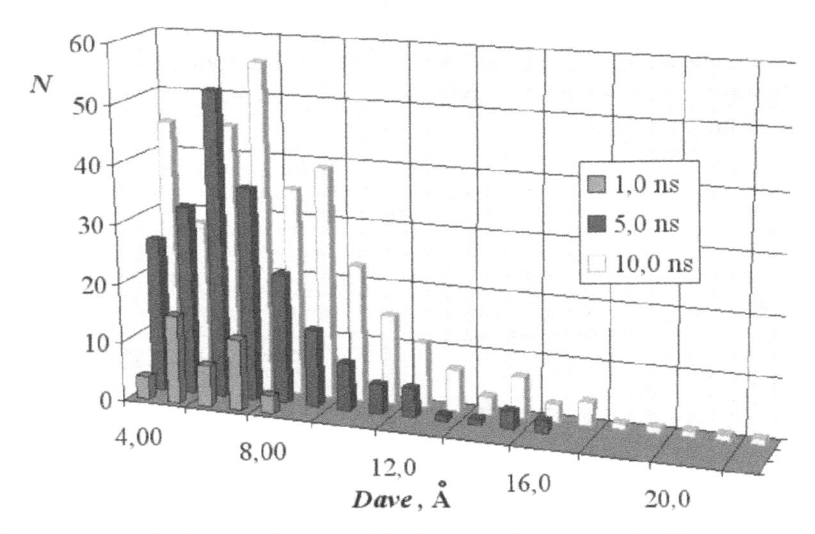

FIGURE 3.48 Distribution of nanoparticle sizes at various moments of time.

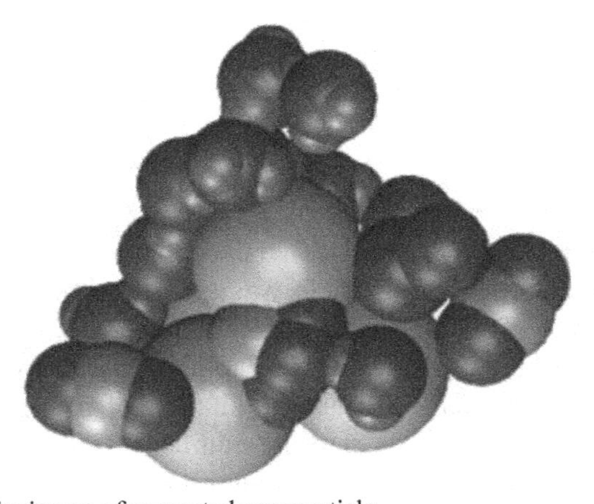

FIGURE 3.49 The image of generated nanoparticle.

In Figure 3.50a, the change in the total volume of nanoparticles at the logarithmic scale for the initial stage of condensation is shown. The sharp increase in the total volume up to 1 ns corresponds to the active condensation of free atoms of an aerosol mix of nanoparticles. Transition of all free atoms in a condensed state characterizes the smoother growth of the number of nanoparticles corresponding to the gradual integration of the already generated nanoparticles. The dynamics of the total volume of nanoparticles is approximated by a logarithmic function (the continuous line without symbols in Figure 3.50a). In the box, the kind of approximate function is given, and also its square-law error.

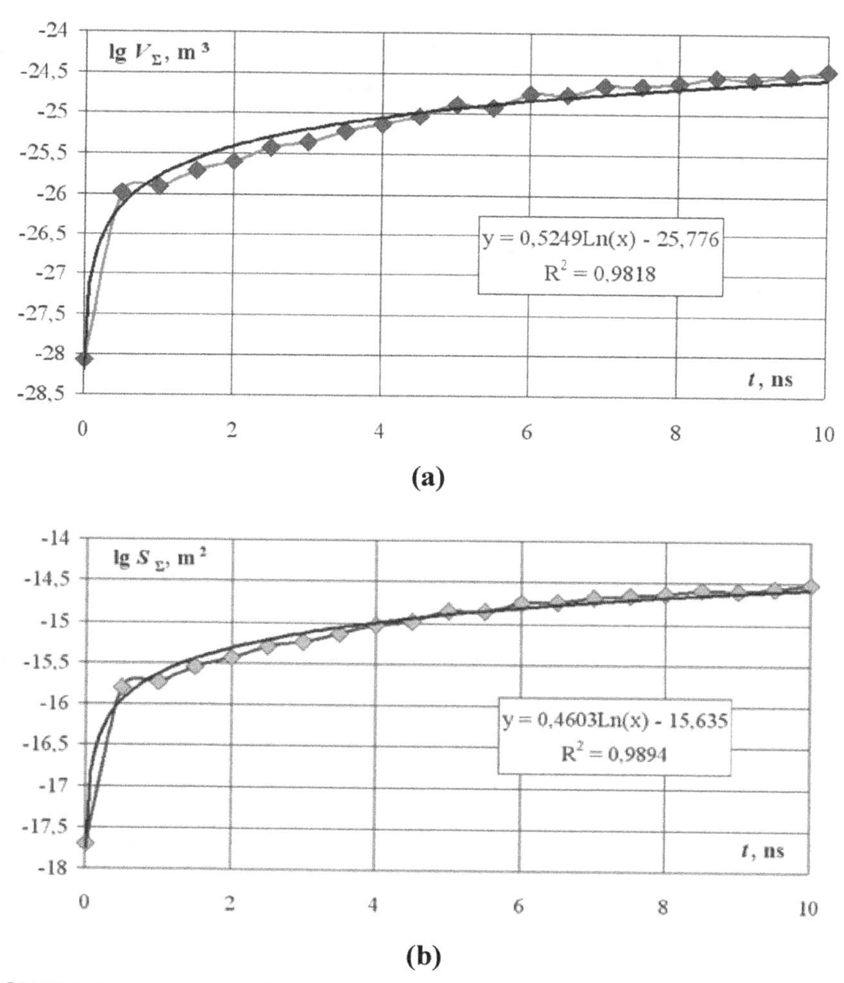

FIGURE 3.50 Change in the total volume of condensed nanoparticles in time (a) and change of a total surface of nanoparticles.

The behavior of the dependence of the total surface of nanoparticles on time (Fig. 3.50b) is similar to the dynamics of the total volume of nanoparticles. The significant gain in the given parameter at the initial stage of condensation testifies to origin of a large number of nanoparticles. It is coordinated in Figure 3.47. The activity of interacting nanoparticles depends on their total surface; therefore, the analysis of this characteristic is rather important.

In using aerosols, of great significance is the problem of uniformity of the distribution of initial materials over the volume in vestigated. The criterion of uniformity is the distribution of the weight of substance over the volume of the settlement cell. This distribution is shown in Figure 3.51 at the moment of 0 ns and 10 ns from the beginning of condensation.

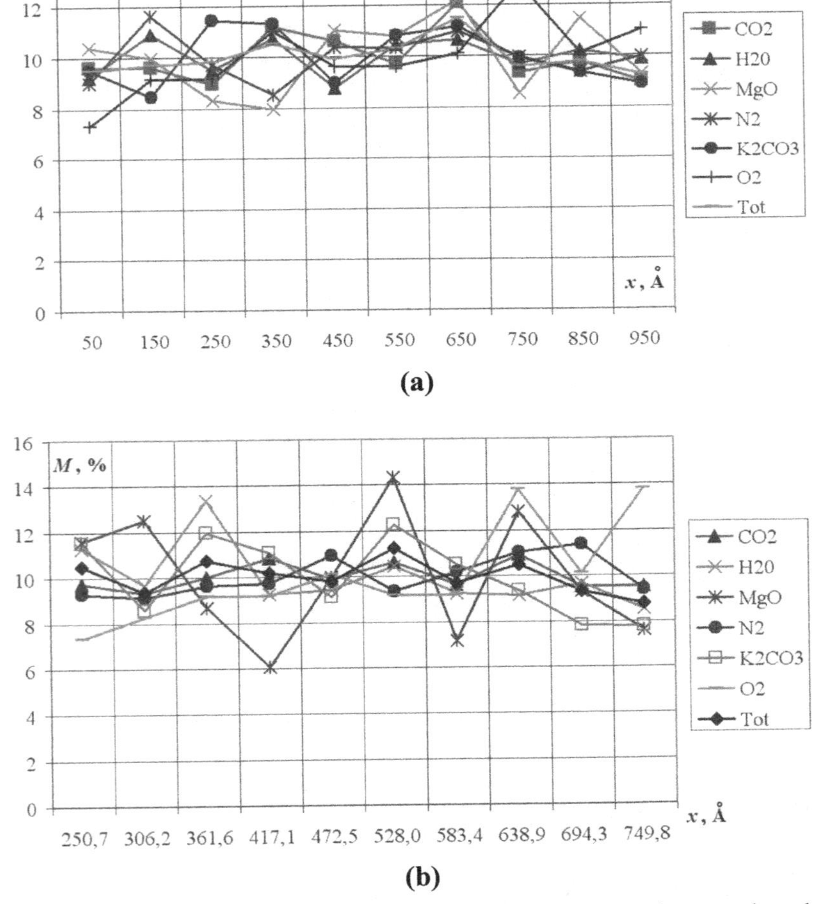

FIGURE 3.51 Distribution of the weight of nanoaerosol over the settlement volume before the process of condensation (a) and after condensation lasting for 10 ns (b).

At the initial moment, the aerosol mix as atoms and molecules is distributed in regular intervals over the settlement volume. Insignificant fluctuations of each component of the nanoaerosol do not deviate from the average value by more than 2%. During interaction, the system components are redistributed over the settlement volume and the uniformity of this distribution changes insignificantly. For some components of the mix in 10 ns of condensation, the deviation from the uniform distribution achieves 4–6%. Nevertheless, the given fluctuations are not great, and it is possible to define the absence of the connection of the centers of condensation as nanoparticles and their congestions from the arrangement in the nanosystem modeled.

3.2 MODELING NANOPARTICLE DISPERSION

3.2.1 MODELING NANOPARTICLE DISPERSION BY THE FINITE ELEMENT METHOD

In this section, the results are presented for the numerical calculations of the dispersion of powder particles of different forms and at different types of dynamic loading and unloading Figure 3.52.

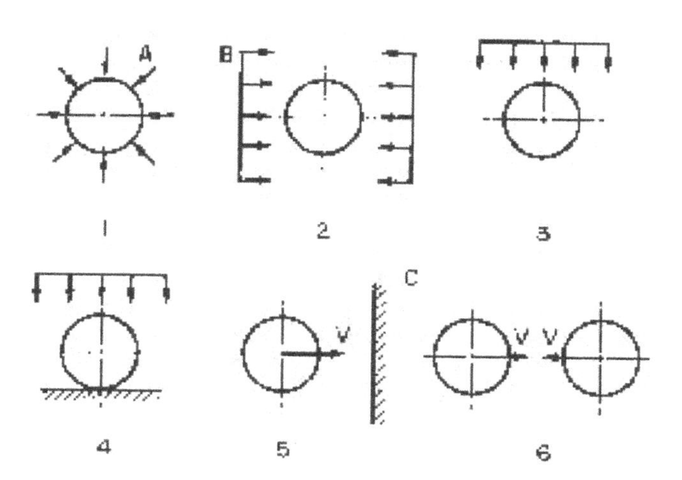

FIGURE 3.52 The schemes of loading of powder particles: 1—instantaneous release of high pressure, 2—loading by two shock waves, 3—loading by one shock wave, 4—loading of a particle by a shock wave near a rigid surface, 5—a particle impact on the surface of a solid, and 6—colliding particles.

3.2.1.1 DISPERSION OF PARTICLES AT PRESSURE RELEASE

This section concerns with the investigation of the processes of the destruction of particles of different forms at pressure release. In particular, Figure 3.53 shows the patterns of the sequential destruction of a particle of a conical form. The analysis of the destruction patterns shows that at the first stage of destruction, the surface fragments in the regions of the largest concentration of stresses separate from the particle. Then the destruction front propagates from the initial destruction region deep into the particle. At a certain moment, a main crack appears in the particle (Fig. 3.53-3). It propagates with the propagation of the general front of the destruction leaving it behind. At the end of the destruction process, the crack splits the particle into two large fragments (Fig. 3.53-6).

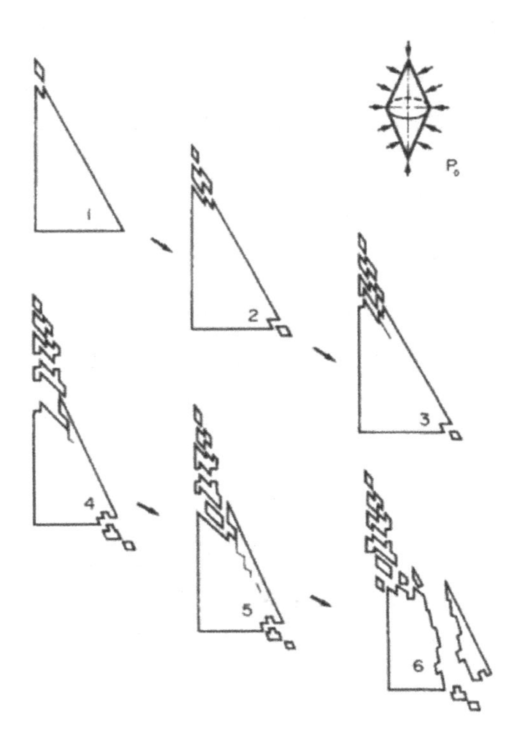

FIGURE 3.53 A conical particle destruction.

The kinetics of the destruction is of importance. Figure 3.54 displays the plot of the destruction front propagation in the conical particle with time.

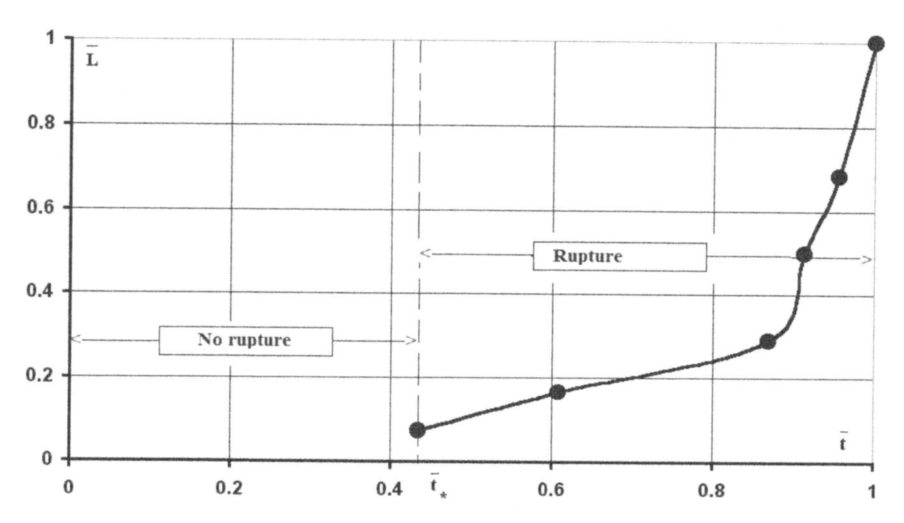

FIGURE 3.54 The development of the destruction front in a conical particle with time.

Here, on the horizontal axis, the time $\tilde{t}$ is laid off in the relationship with the time t_* of the particle destruction; and on the vertical axis, the length $\tilde{L}$ of the destruction front penetration within the particle depth is laid off in the relationship with its specific size H:

$$\tilde{t} = \frac{t}{t_*};\ \tilde{L} = \frac{L}{H}$$

It can be seen that the first destruction does not appear immediately after particle unloading but in a definite time t_* that is required for the wave of unloading to pass through the particle and the maximal tensile stress, which is equal to ultimate stress at tension, to appear. After the appearance of the first "destructed" finite element, a gradual enlargement of the destruction area takes place. At the moment, $\tilde{t} = 0.9$, the destruction velocity sharply increases, which leads to the split of the particle.

The pattern of the destruction of the particle, in which the regions of stress concentration are absent, is of interest. Figure 3.55 shows the patterns of destruction of a spherical particle.

Its destruction does not start from the surface after loading with high pressure followed by immediate pressure release. The first area of the destruction appears inside the particle at some distance from the surface and propagates farther under the particle surface without appearing on its surface. As a result, a surface layer separates from the sphere and later breaks into small fragments. The calculations show that when the initial level of pressure is sufficiently high, this pattern of destruction is repeated. Deep in the particle, another area of destruction is formed, which propagates equidistantly to the destruction surface, which was formed at the first stage. Then, the next surface layer of the particle is separated. The process is repeated as long as there is enough accumulated energy in the particle to form another new area of destruction. When the energy drops to a certain level, the particle is not destructed any more, though it still oscillates, being compressed and stretched in sequence with time.

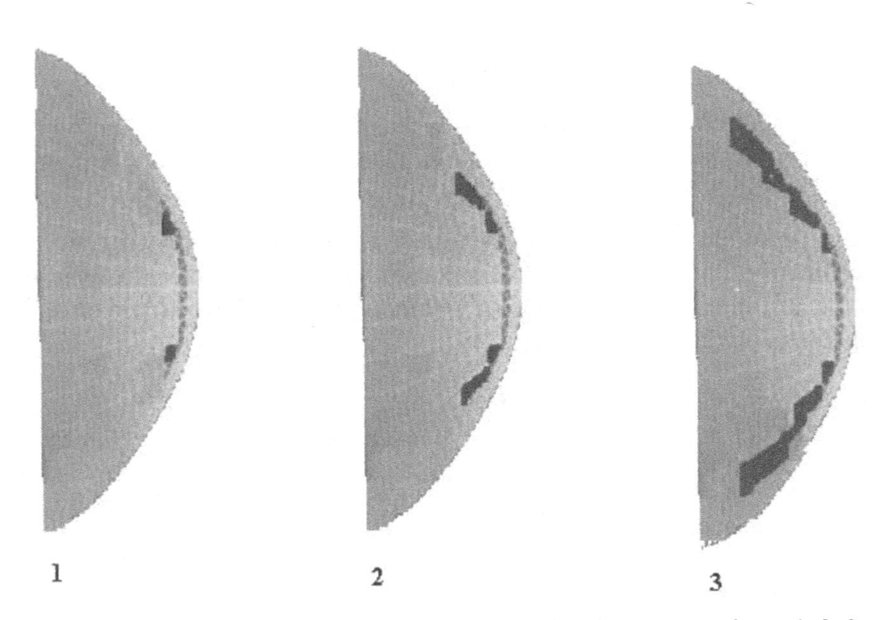

FIGURE 3.55 The destruction of a spherical particle after the pressure release: 1, 2, 3—sequential development of the destruction front.

Figure 3.56 shows the plots of the maximal normal stress distribution along the radius of the spherical particle, which clearly show the above destruction mechanism. At the initial moment (Fig. 3.56-1), the stress along the radius is constant and equal to the pressure that compresses the particle. After unloading (Fig. 3.56-2), the stress along the particle radius is alternate. In the central part of the particle, it is negative and equal to the initial pressure of compression. On the particle surface, it is zero. At a certain distance from the surface, the stress is positive. If the positive maximal stress reaches the ultimate tensile stress, destruction will start. During dynamic unloading, the positive maximum of the stress shifts deep into the particle, which causes the development of the particle surface-layer detachments that are successive in time.

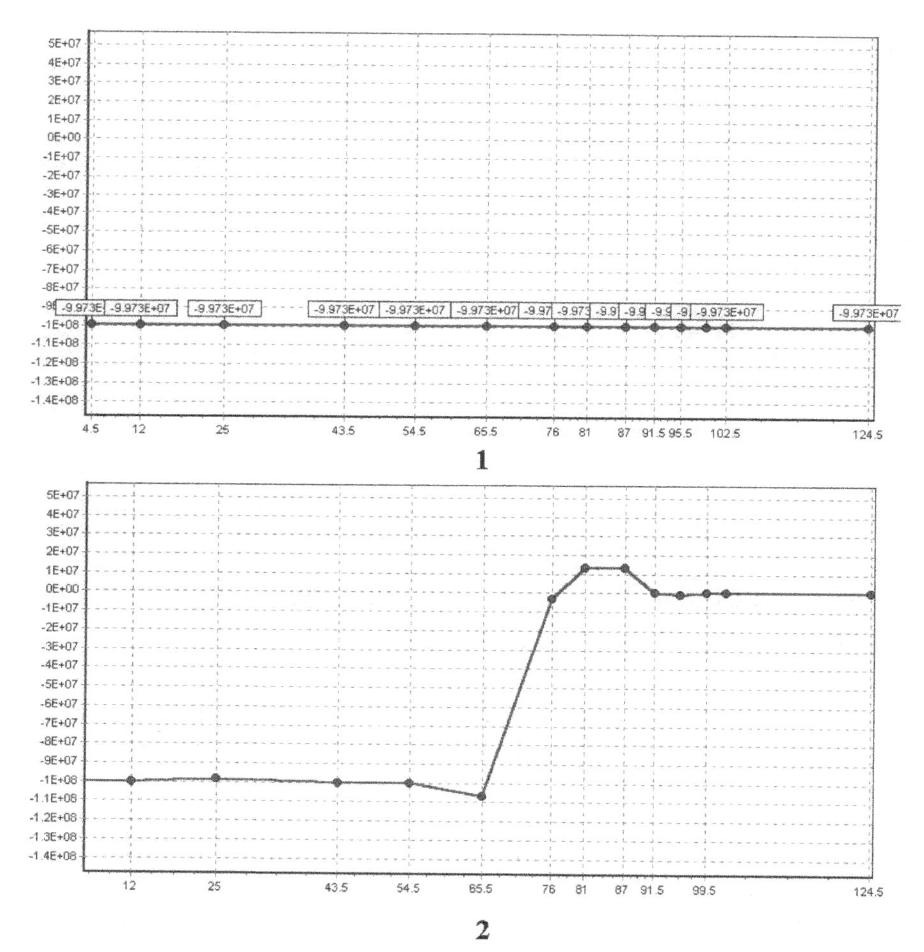

FIGURE 3.56 Distribution of maximal normal stresses along the spherical particle radius: 1—stresses at the initial moment; 2—stresses in a particle after unloading.

It should be noted that the calculation results for the destruction of the particles of a rod like or spherical form obtained in the previous chapter based on the expansion of the solution in terms of their forms and oscillation frequencies are confirmed by the finite-element calculation. The destruction patterns obtained by these methods are similar: according to the calculations, each time the destruction appears deep in the particle at some distance from its free surface propagating equidistantly to it.

It should be noted that the rate of the ambient pressure decrease essentially influences the destruction of the particles. The calculations show that if the pressure decrease is slow enough for the stress field to "follow" the change of the external pressure, the elastic energy is released quasi-statically and does not cause the particle fragmentation.

The influence of the imperfections in the particle form on the destruction processes is of importance. Normally, a powder particle form differs from a canonical form since there are various defects on its surface (Fig. 3.57).

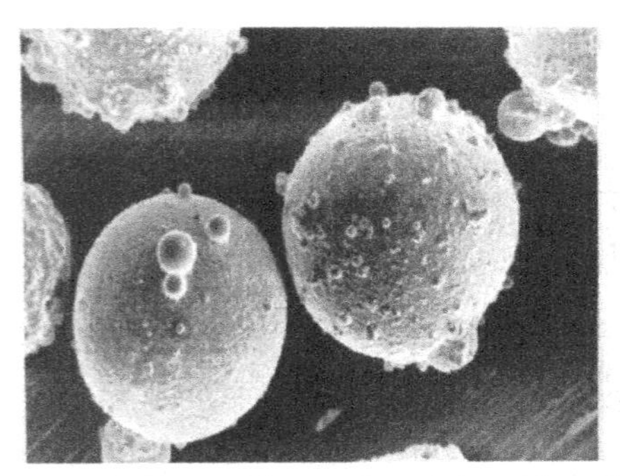

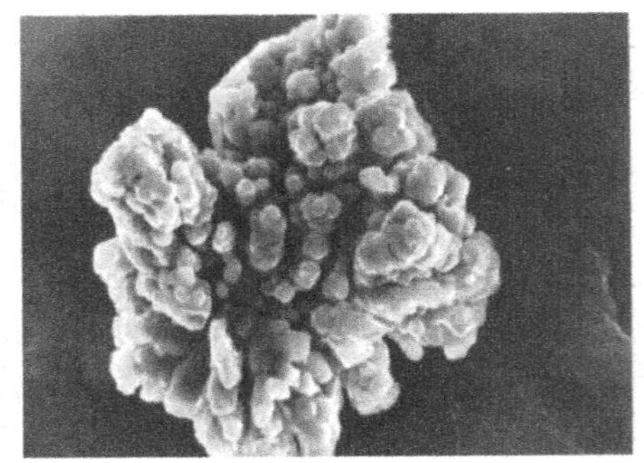

FIGURE 3.57 Photos of particles with surface defects.

To analyze the destruction of such particles, the calculation was carried out for the particle with a protuberance. Figure 3.58 shows a finite element model for this particle.

The analysis of the calculation results shows that the destruction area is localized in the region of the connection of the particle and the surface formation. At first, the destruction occurs on the free surface in the region of the connection of the particle and the surface formation. Then, the finite elements are "destructed" in the direction of the line of their connection. As a result, the surface defect completely departs from the particle. After their separation, the particle is destructed by the above-mentioned mechanisms.

Thus, the calculations show that it is surface defects that are the initiators of the destruction.

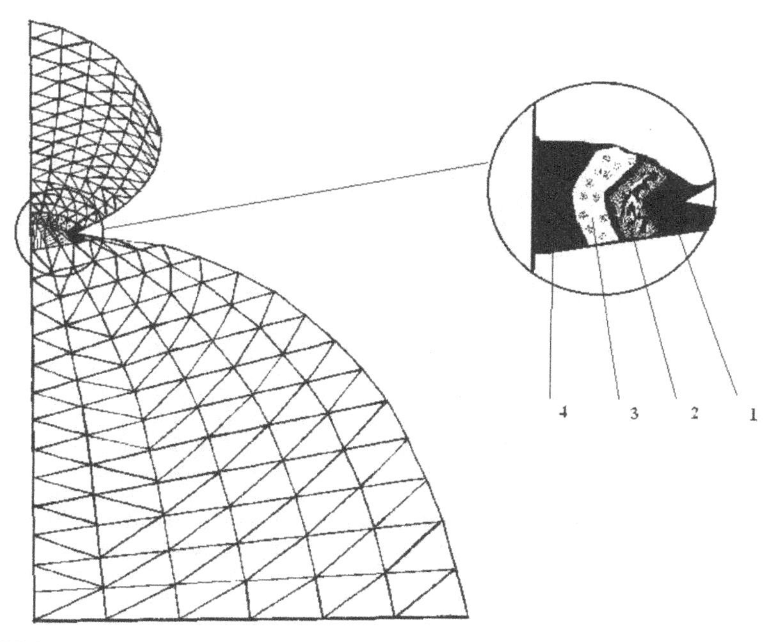

FIGURE 3.58 The calculated finite-element scheme and the destruction of a particle with a surface defect. The regions of destruction are shown sequentially in the circle (1–4).

3.2.1.2 DISPERSION OF PARTICLES IMPACTING ON AN OBSTACLE

In Figure 3.59, one can see a destruction pattern of the particle impacting on an obstacle.

The impact happens on a small area at the top of the particle. At the left of Figure 3.59, there is a scheme with the destruction area appearing at the first moments of the impact. One can see that the destruction takes place deep in the particle, in the region, which is situated under the center of the contact area. From this point, the destruction propagates first in the direction perpendicular to the direction of the impact and later "destructed" elements appear which are directed at an angle close to 45° relative to the impact. The number of the destructed elements in the above direction gradually increases. At the end of the dispersion, this area traverses the whole particle splitting it into two parts. Thus, unlike the destruction of particles caused by the pressure release when several particles of different fraction are formed, the destruction caused by impacting on an obstacle leads to the split of the particle into two large parts.

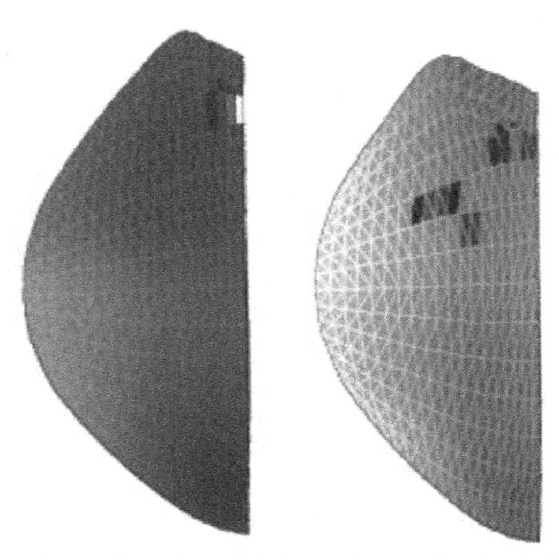

FIGURE 3.59 The destruction of a particle impacting on a barrier. At the left, the beginning of destruction is shown and at the right, the final picture of destruction.

3.2.1.3 DISPERSION OF PARTICLES AT THE INTERACTION WITH SHOCK FRONT

Figure 3.60 displays a finite-element scheme of the calculation of the dispersion of a particle at the interaction with a shock wave. In Figure 3.61a, a pattern of the particle destruction is shown at the interaction with a shock wave. At the initial moment, an internal destruction of the particle is observed in some depth from the direction of the shock. Then, the destruction area increases without going out on the surface and is compressed in the direction of the shock. In this case, the particle is compressed from the direction of the shock, which does not allow the internal destruction to go out onto the surface.

Similar to the case of impacting on an obstacle, "destructed" elements appear which are situated at a certain angle to the normal of the shock wave. This area of destruction develops and transverses the particle causing its splitting into two parts. It should be noted that this mode of the particle destruction is confirmed by the experimental investigations of the sample destruction at a high-speed impact (Fig. 3.61). It can be seen that similar to the above calculations, in the central part of the sample, the area appears which is perpendicular to the impact axis. On the sample edges, one can observe cone-shape cleaved surface directed at a certain angle to the initial destruction area.

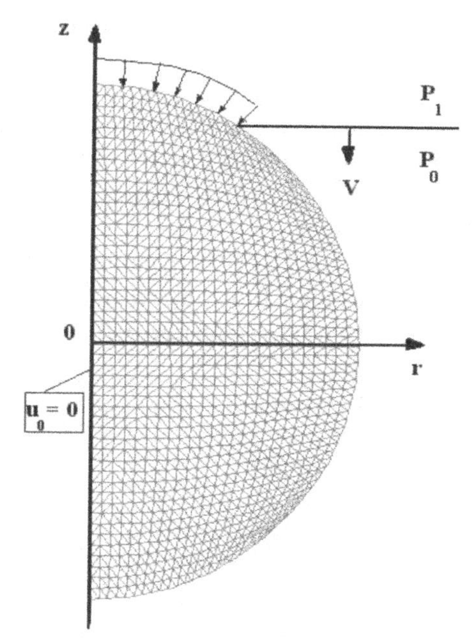

FIGURE 3.60 The calculation scheme of the interaction of a particle and a shock wave.

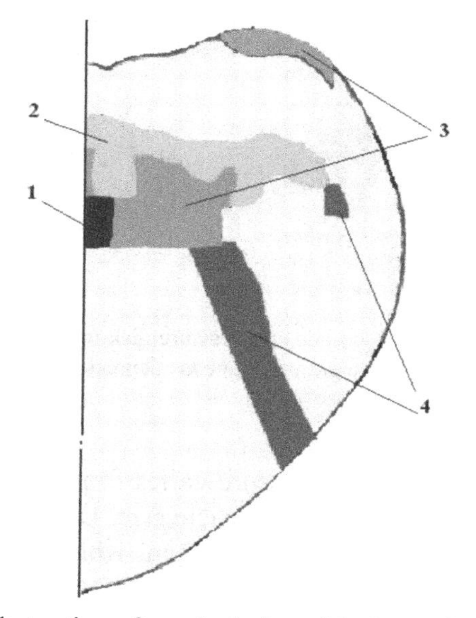

FIGURE 3.61 The destruction of a spherical particle interacting with a shock wave (a calculation pattern of destruction). Regions (1–4) designate the regions of sequential destruction.

3.2.2 MODELING NANOPARTICLE DISPERSION BY THE MOLECULAR DYNAMICS METHOD

Note that the use of molecular dynamics in the study of dynamic behavior of nanoparticles allows to obtain results that are not observed in the simulation of the processes by the continuum mechanics. As an example, consider dispersing of a disk nanoparticles.

Figure 3.62 shows the sequence (1–5) dynamic fracture disk nanoparticles consisting of 2000 atoms after full compression and high instantaneous depressurization (side and front views, respectively).

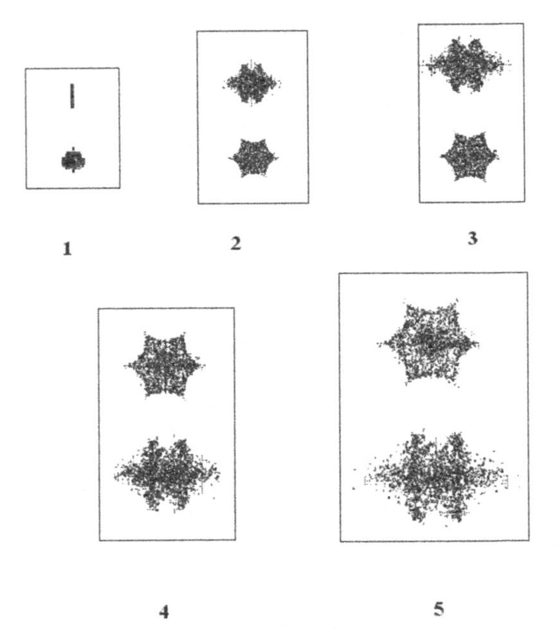

FIGURE 3.62 The pictures (1–5) dynamic fracture disk nanoparticles consisting of 2000 atoms after full compression and high instantaneous depressurization (side and front views, respectively).

At time $t = 0$, the external pressure varies from maximum to zero. Due unloading starts increasing in nanoparticle size. During extension, it passes through various forms of stable equilibrium, which, as shown by calculations are not necessarily spherical. Therefore, on the surface of the particles appear edges and corners (Fig. 3.62-2). It is they determine the evolution of the particle dispersion process, giving rise to symmetrical structure, without, however, the full spatial symmetry. Atoms move predominantly in six directions in the plane of the disk particles and in the two directions along the perpendicular to this plane. Thus, initially axisymmetric particle breaks in violation of the axial symmetry.

Symmetry breaking with the "explosion" of a particle cannot be seen clearly. Figure 3.63 shows the destruction of the spherical particles after full compression and fast unloading.

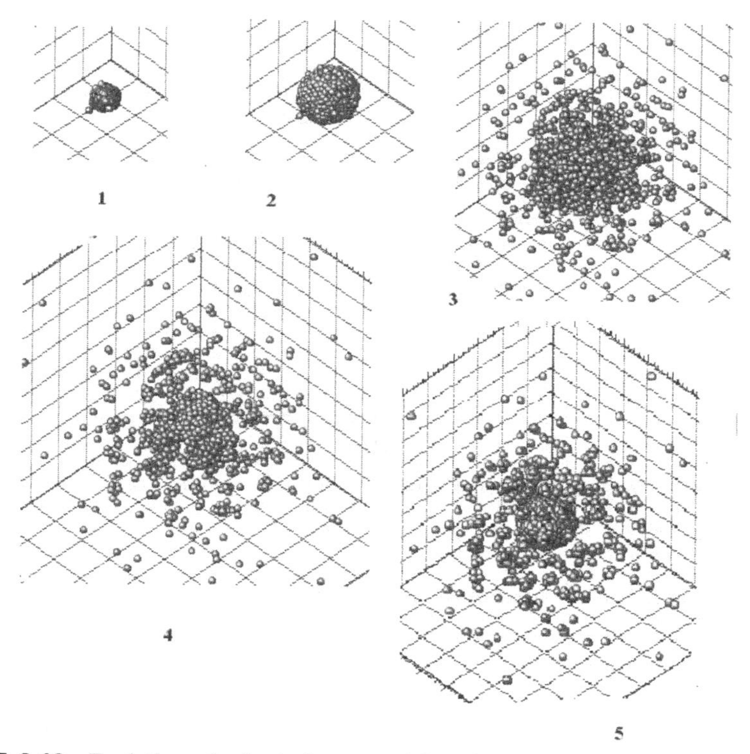

FIGURE 3.63 Evolution of spherical nanoparticles after 20% compression and sudden removal of the compressive load. The numbers correspond to the time points (1–0; 2–0.1; 3–200; 4–500; 5–10,000). 10^{-13} s, respectively.

It is seen that in this case, from the surface, atoms are separated symmetrically in all directions, while maintaining the central "core," which is not destroyed. The atoms, separated from the particle, forming smaller particles that surround a central atom group.

Figure 3.64 shows the nanoparticle consisting of 1000 atoms, before and after the impulse loading, directed radially to nanoparticle.

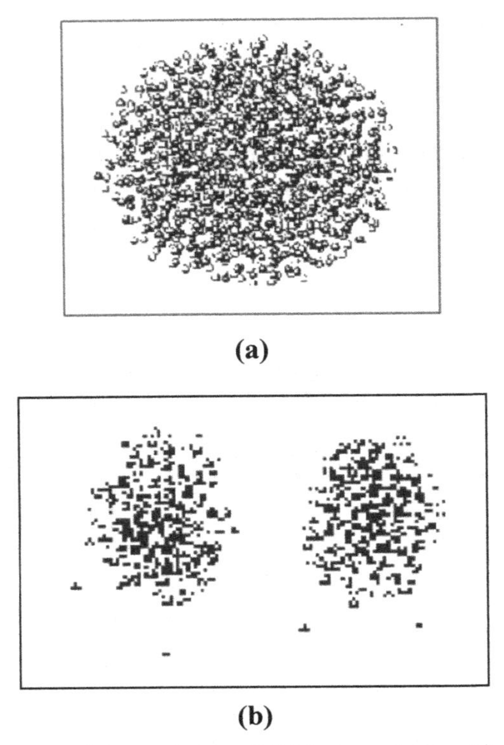

FIGURE 3.64 A nanoparticle consisting of 1000 atoms, before (a) and then (b) pulsed loading particles.

It is seen that after loading, nanoparticle is divided into two smaller-size stable particles. Part of the atoms, thus, there is allocated and independently. After the separation, small nanoparticles tend to get close, but do not merge.

KEYWORDS

- **numerical simulation**
- **nanoparticles**
- **"up-down" processes**
- **"bottom-up" processes**
- **structural characteristics**

CHAPTER 4

NUMERICAL SIMULATION OF NANOSYSTEM FORMATION

CONTENTS

ABSTRACT

The problems of numerical simulation of nanosystem formation are considered. First, the interactions of nanoelements are described. Different types of nanoelement interaction are described: adhesion, fusion, and absorption. Stable and unstable forms of nanoparticle system equilibrium are investigated. The changes in the strength of powder nanomaterials depending upon the nanoparticle diameter are evaluated. The task for forming nanocomposites with different pressing schemes is presented. The complete pressing cycle is considered: loading and unloading of powder material. The evolution of technological stresses in the material being pressed and changes in the material parameters are discussed in detail. Simulation of electrodeposition of metal matrix nanocomposite coatings is presented. Simulation of formation of superficial nano-hetero structures: nanowhiskers and quantum dots are described in detail.

4.1 SIMULATION OF NANOELEMENTS INTERACTIONS AND SELF-ASSEMBLING

4.1.1 DEFINITION OF PROBLEM

The properties of a powder nanocomposite are determined by the structure and properties of the nanoelements, which form it. When the characteristics of powder nanocomposites are calculated it is also very important to take into account the interaction of the nanoelements since the changes in their original shapes and sizes in the interaction process and during the formation (or usage) of the nanocomposite can lead to a significant change in its properties and a cardinal structural rearrangement. In addition, the experimental investigations show the appearance of the processes of ordering and self-assembling leading to a more organized form of a nanosystem.[3,44,56,83,84,123,191] The nanoelements assembly from different type of materials into a superlattice can provide a new way to construct precisely controlled nanocomposites (metamaterials).[102]

In general, three main types of the nanoelements assembly processes can be distinguished. The first process is due to the regular structure formation at the interaction of the nanostructural elements with the surface where they are situated; the second one arises from the interaction of the nanostructural elements with one another; the third process takes place because of the influence of the ambient medium surrounding the nanostructural elements. The ambient medium influence can have "isotropic distribution" in the space or it can be presented by the action of separate active molecules connecting nanoelements to one another in a certain order. The external action significantly changes the original shape of the structures formed by the nanoelements. For example, the application of the external tensile stress leads to the "stretch" of the nanoelement system in the direction of the maximal tensile stress action; the rise in temperature, vice versa, promotes a decrease in the spatial anisotropy of the nanostructures.[44] Note that in the self-organizing process, parallel with the linear displacements, the nanoelements are in rotary movement. The latter can be explained by the action of forces moment caused by the asymmetry of the interaction force fields of the nanoelements, by the presence of the "attraction" and "repulsion" local regions on the nanoelement surface, and by the "non-isotropic" action of the ambient as well.

The above phenomena play an important role in nanotechnological processes. They allow developing nanotechnologies for the formation of nanostructures by the self-assembling method (which is based on self-organizing processes) and building up complex spatial nanostructures consisting of different nanoelements (nanoparticles, nanotubes, fullerenes, super-molecules, etc.).[93,98,117] However, in a number of cases, the tendency toward self-organization interferes with the formation of a desired nanostructure. Thus, the nanostructure arising from the self-organizing process is, as a rule, "rigid" and stable against external actions. For example, the "adhesion" of nanoparticles interferes with the use of separate nanoparticles in various nanotechnological processes, the uniform mixing of the nanoparticles from different materials and the formation of nanocomposite with desired properties. In connection with this, it is important to model the processes of static and dynamic interaction of the nanostructure elements. In this case, it is essential to take into consideration the interaction force moments of the nanostructure elements, which causes the mutual rotation of the nanoelements.

The investigation of the above dependences based on the mathematical modeling methods requires the solution of the aforementioned problem on the atomic level. This requires large computational aids and computational time, which makes the development of economical calculation methods urgent. The objective of this work was the development of such a technique.

In this part, in development of author researches of formation and movement of nanoparticles, the methods of numeric simulation within the framework of molecular dynamics were used for calculating the interactions of nanostructural elements. Method offered is based on the pair wise static interaction of nanoparticles potential built up with the help of the approximation of the numerical calculation results using the method of molecular dynamics. Based on the potential of the pair wise interaction of the nanostructure elements, which takes into account forces and moments of forces, the method for calculating the ordering, and self-organizing processes has been developed. The dependence of the ultimate stress limit of the monodisperse powder nanocomposite on the diameter of the constituent nanoparticles has been calculated. The investigation results on the self-organization of the system consisting of two or more particles are presented and the analysis of the equilibrium stability of various types of nanostructures has been carried out.

Let us consider the realization of the above procedure taking the calculation of the metal particles interaction as an example.

The potentials of the atomic interaction of Morse (4.1) and Lennard-Johns (4.2) were used, in the calculations

$$\Phi(\vec{\rho}_{ij})_m = D_m\left(\exp(-2\lambda_m(\left|\vec{\rho}_{ij}\right| - \rho_0)) - 2\exp(-\lambda_m(\left|\vec{\rho}_{ij}\right| - \rho_0))\right) \quad (4.1)$$

$$\Phi(\vec{\rho}_{ij})_{LD} = 4\varepsilon\left[\left(\frac{\sigma}{\left|\vec{\rho}_{ij}\right|}\right)^{12} - \left(\frac{\sigma}{\left|\vec{\rho}_{ij}\right|}\right)^{6}\right], \quad (4.2)$$

where D_m, λ_m, ρ_0, ε, σ are the constants of the materials studied.

At the first stage of the problem, the coordinates of the atoms positioned at the macromaterial lattice points (Fig. 4.1-1) were taken as the original coordinates. During the relaxation process, the initial atomic system is rearranged into a new "equilibrium" configuration (Fig. 4.1-2) in accordance with the calculations based on equations of MD, which satisfies the condition when the system potential energy is approaching the minimum (Fig. 4.1, the plot).

When the forces of the interaction of two nanoparticles of the same size were calculated, their parameters after free relaxation were used.

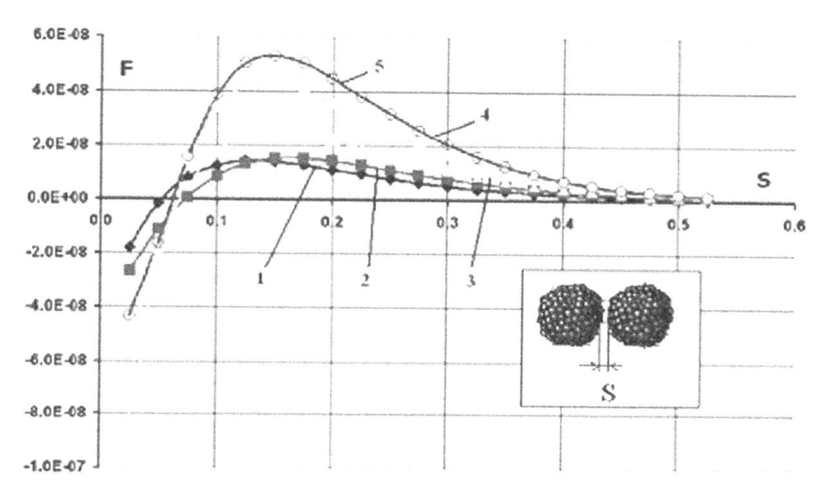

FIGURE 4.1 The initial crystalline (1) and cluster (2) structure of the nanoparticle consisting of 1331 atoms after relaxation; the plot of the potential energy U [Nm]change for this atomic system in the relaxation process (n —number of iteration in time).

4.1.2 STATICS INTERACTION OF NANOPARTICLES

Figure 4.2 shows the calculation results demonstrating the influence of the sizes of the nanoparticles on their interaction force. One can see from the plot that the larger nanoparticles are attracted stronger, that is, the maximal

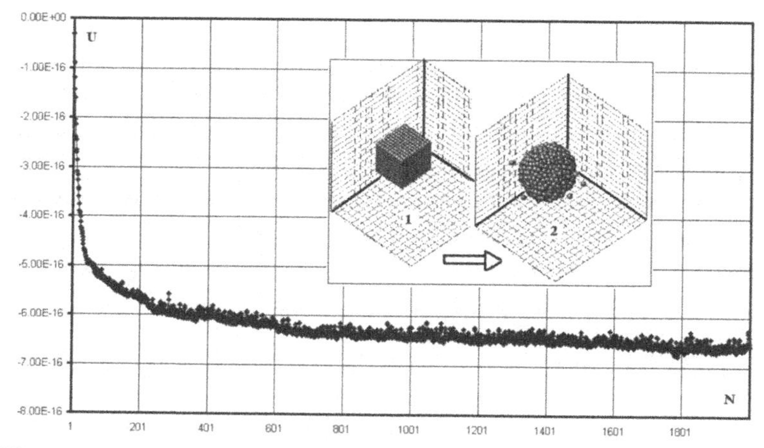

FIGURE 4.2 The dependence of the interaction force F [N] of the nanoparticles on the distance S [nm] between them and on the nanoparticle size: 1— d = 2.04; 2—d = 2.40; 3—d = 3.05; 4—d = 3.69; 5—d = 4.09 nm.

interaction force increases with the size growth of the particle. Let us divide the interaction force of the nanoparticles by its maximal value for each nanoparticle size, respectively. The obtained plot of the "relative" (dimensionless) force (Fig. 4.3) shows that the value does not practically depend on the nanoparticle size since all the curves come close and can be approximated to one line.

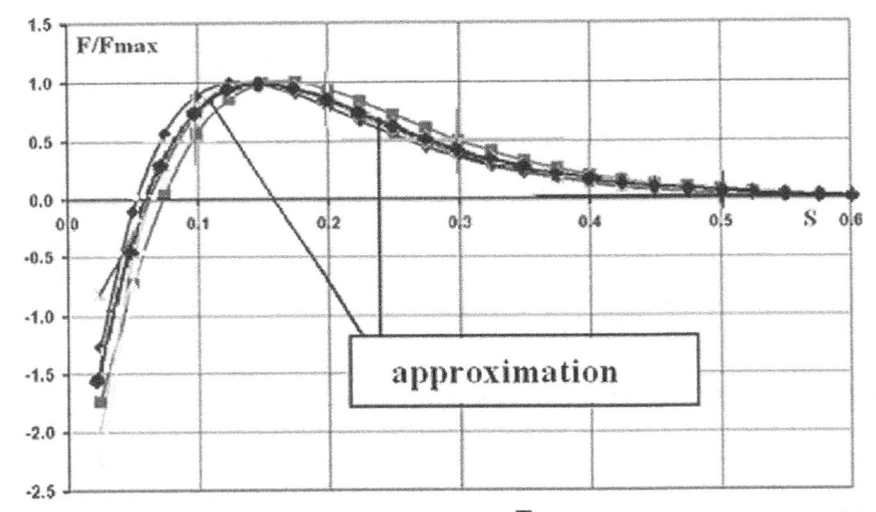

FIGURE 4.3 The dependence of the "relative" force $\overline{F}$ of the interaction of the nanoparticles on the distance S [nm] between them.

Figure 4.4 displays the dependence of the maximal attraction force between the nanoparticles on their diameter that is characterized by nonlinearity and a general tendency toward the growth of the maximal force with the nanoparticle size growth.

The total force of the interaction between the nanoparticles is determined by multiplying of the two plots (Figs. 4.3 and 4.4).

Using the polynomial approximation of the curve in Figure 4.5 and the power mode approximation of the curve in Figure 4.6, we obtain

$$\overline{F} = (-1.13S^6 + 3.08S^5 - 3.41S^4 - 0.58S^3 + 0.82S - 0.00335)10^3, \tag{4.3}$$

$$F_{max} = 0.5 \cdot 10^{-9} \cdot d^{1.499}, \tag{4.4}$$

$$F = F_{max} \cdot \overline{F}, \tag{4.5}$$

where d and S are the diameter of the nanoparticles and the distance between them [nm], respectively; F_{max} is the maximal force of the interaction of the nanoparticles [N].

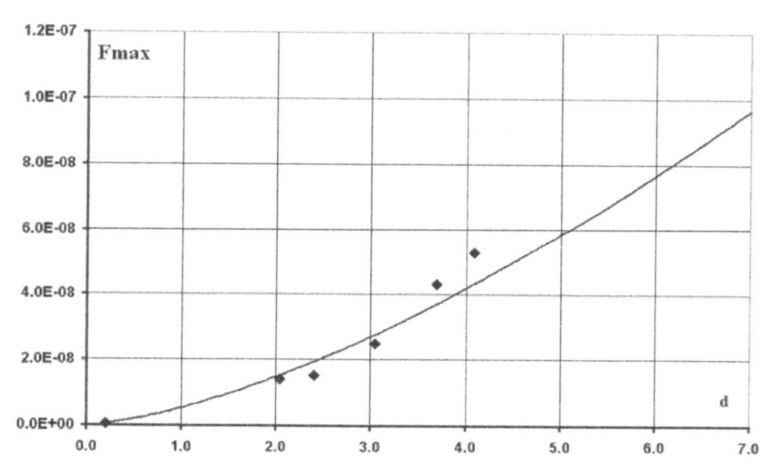

FIGURE 4.4 The dependence of the maximal attraction force F_{max} [N] the nanoparticles on the nanoparticle diameter d [nm].

Dependences (4.3)–(4.5) were used for the calculation of the nanocomposite ultimate strength for different patterns of nanoparticles' "packing" in the composite (Fig. 4.5).

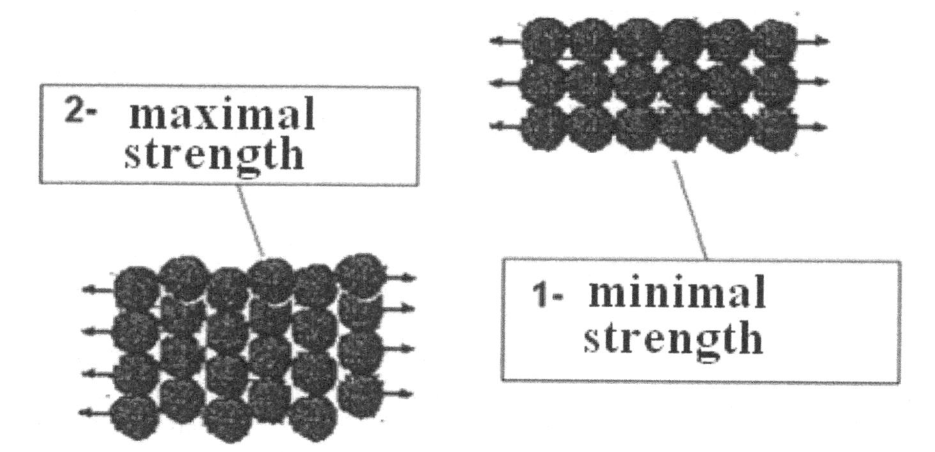

FIGURE 4.5 Different types of the nanoparticles' "packing" in the composite.

Figure 4.6 shows the dependence of the ultimate strength of the nanocomposite formed by monodisperse nanoparticles on the nanoparticle sizes. One can see that with the decrease of the nanoparticle sizes, the ultimate strength of the nanomaterial increases, and vice versa. The calculations have shown that the nanocomposite strength properties are significantly influenced by the nanoparticles' "packing" type in the material. The material strength grows when the packing density of nanoparticles increases. It should be specially noted that the material strength changes in inverse proportion to the nanoparticle diameter in the degree of 0.5, which agrees with the experimentally established law of strength change of nanomaterials (the law by Hall–Petch):[52]

$$\sigma = C \cdot d^{-0.5}, \tag{4.6}$$

where $C = C_{max} = 2.17 \cdot 10^4$ is for the maximal packing density; $C = C_{min} = 6.4 \cdot 10^3$ is for the minimal packing density.

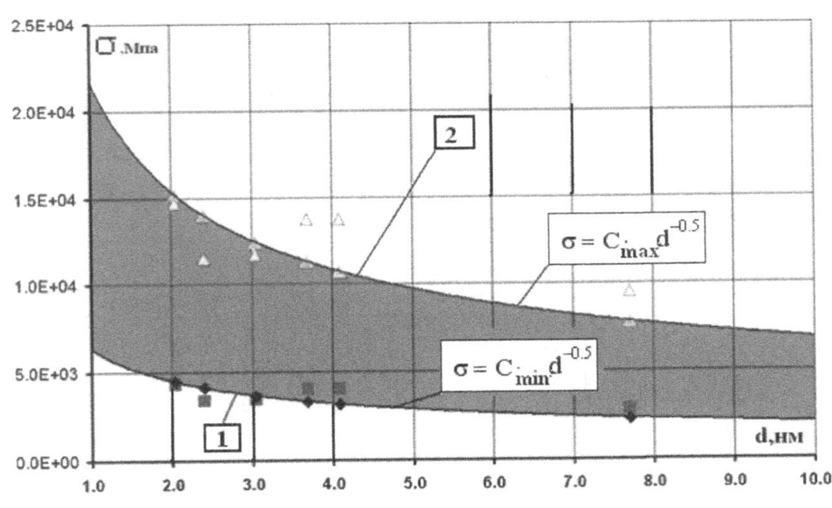

FIGURE 4.6 The dependence of the ultimate strength σ (MPa) of the nanocomposite formed by monodisperse nanoparticles on the nanoparticle sizes d (nm).

The electrostatic forces can strongly change force of interaction of nanoparticles. For example, numerical simulation of charged sodium (NaCl) nanoparticles system (Fig. 4.7) has been carried out. Considered ensemble consists of eight separate nanoparticles. The nanoparticles interact due to van der Waals and electrostatic forces.

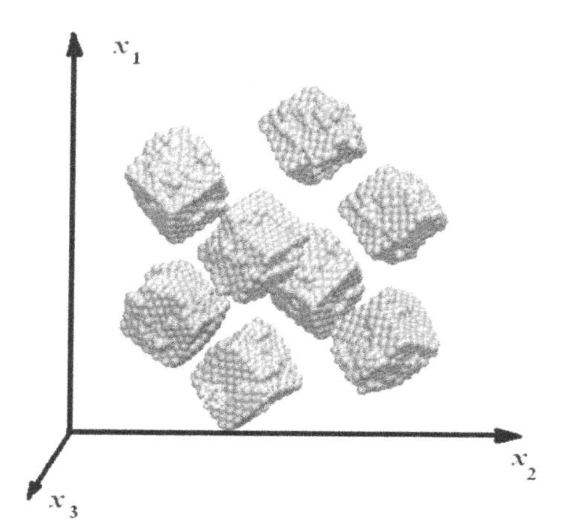

FIGURE 4.7 Nanoparticles system consists of eight nanoparticles NaCl.

Results of particles center of masses motion are introduced in Figure 4.8 representing trajectories of all nanoparticles included into system. It shows the dependence of the modulus of displacement vector $|R|$ on time. One can see that nanoparticle moves intensively at first stage of calculation process. At the end of numerical calculation, all particles have got new stable locations, and the graphs of the radius vector $|R|$ become stationary. However, the nanoparticles continue to "vibrate" even at the final stage of numerical calculations. Nevertheless, despite of "vibration," the system of nanoparticles occupies steady position.

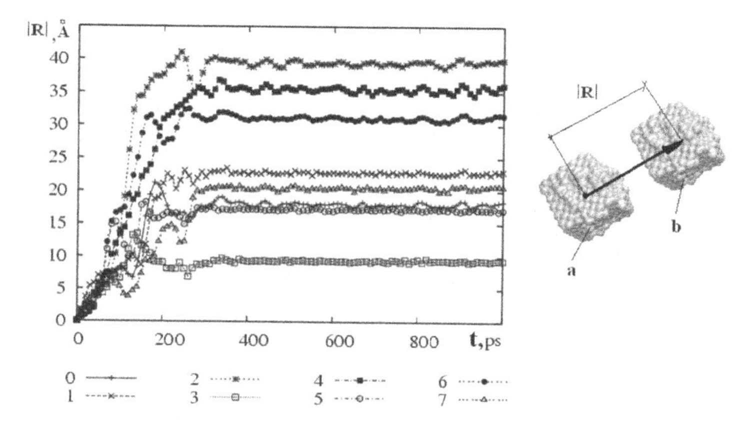

FIGURE 4.8 The dependence of nanoparticle centers of masses motion $|R|$ on the time t; a, b—the nanoparticle positions at time 0 and t, accordingly; 1–8—are the numbers of the nanoparticles.

However, one can observe a number of other situations. Let us consider, for example, the self-organization calculation for the system consisting of 125 cubic nanoparticles, the atomic interaction of which is determined by Morse potential (Fig. 4.9).

FIGURE 4.9 The positions of the 125 cubic nanoparticles: (a) initial configuration; (b) final configuration of nanoparticles.

As you see, the nanoparticles are moving and rotating in the self-organization process forming the structure with minimal potential energy. Let us consider, for example, the calculation of the self-organization of the system consisting of two cubic nanoparticles, the atomic interaction of which is determined by Morse potential. Figure 4.10 displays possible mutual positions of these nanoparticles. The positions, where the principal moment of forces is zero, corresponds to pairs of the nanoparticles 2–3; 3–4; 2–5 (Fig. 4.10), and defines the possible positions of their equilibrium.

Figure 4.11 presents the dependence of the moment of the interaction force between the cubic nanoparticles 1–3 (Fig. 4.10) on the angle of their relative rotation. From the plot follows that when the rotation angle of particle 1 relative to particle 3 is $\pi/4$, the force moment of their interaction is zero. At an increase or a decrease in the angle the force moment appears. In the range of $\pi / 8 < \theta < 3\pi / 4$ the moment is small. The force moment rapidly grows outside of this range. The distance S between the nanoparticles plays a significant role in establishing their equilibrium. If $S > S_0$ (where S_0 is the distance, where the interaction forces of the nanoparticles are zero), then the

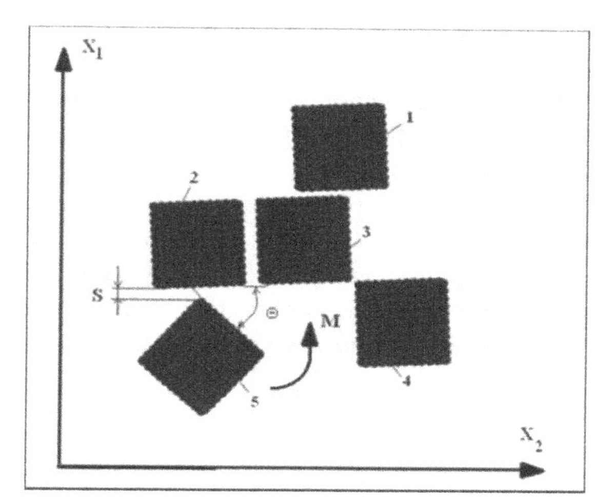

FIGURE 4.10 Characteristic positions of the cubic nanoparticles.

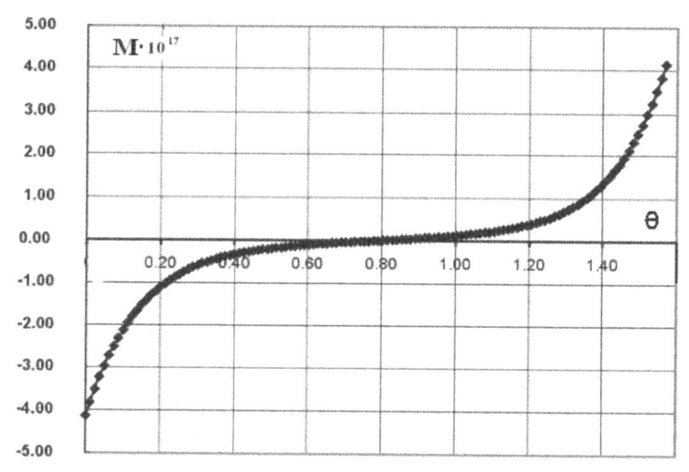

FIGURE 4.11 The dependence of the moment M [Nm] of the interaction force between cubic nanoparticles 1–3 (see in Fig. 4.9) on the angle of their relative rotation θ [rad].

particles are attracted to one another. In this case, the sign of the moment corresponds to the sign of the angle θ deviation from $\pi / 4$. At $S < S_0$ (the repulsion of the nanoparticles), the sign of the moment is opposite to the sign of the angle deviation. In other words, in the first case, the increase of the angle deviation causes the increase of the moment promoting the movement of the nanoelement in the given direction, and in the second case, the angle deviation causes the increase of the moment hindering the movement of the nanoelement in the given direction. Thus, the first case corresponds to the unstable equilibrium of nanoparticles, and the second case—to their stable equilibrium. The potential energy change plots for the system of the interaction of two cubic nanoparticles (Fig. 4.12) illustrate the influence of the parameter S. Here, curve 1 corresponds to the condition $S < S_0$ and it has a well-expressed minimum in the $0.3 < \theta < 1.3$ region. At $\theta < 0.3$ and $\theta > 1.3$, the interaction potential energy sharply increases, which leads to the return of the system into the initial equilibrium position. At $S < S_0$ (curves 2–5), the potential energy plot has a maximum at the $\theta = 0$ point, which corresponds to the unstable position.

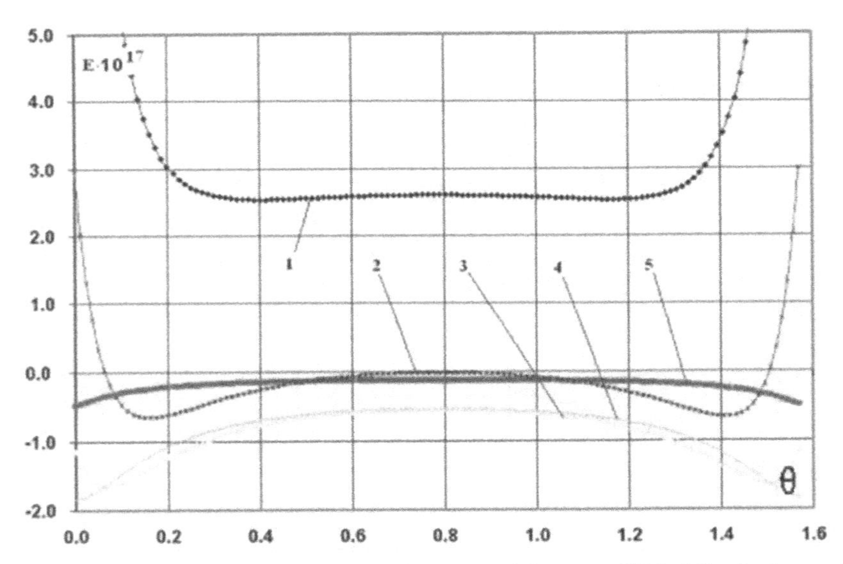

FIGURE 4.12 The plots of the change of the potential energy E [Nm] for the interaction of two cubic nanoparticles depending on the angle of their relative rotation θ [rad] and the distance between them (positions of the nanoparticles 1–3, Fig. 4.10).

The carried out theoretical analysis is confirmed by the works of the scientists from New Jersey University and California University in Berkeley who experimentally found the self-organization of the cubic micro-particles of plumbum zirconate-titanate (PZT):[143] the ordered groups of cubic micro-crystals from PZT obtained by hydrothermal synthesis formed a flat layer of particles on the air–water interface, where the particle occupied the more stable position corresponding to position 2–3 in Figure 4.10.

Thus, the analysis of the interaction of two cubic nanoparticles has shown that different variants of their final stationary state of equilibrium are possible, in which the principal vectors of forces and moments are zero. However, there are both stable and unstable stationary states of this system: nanoparticle positions 2–3 are stable, and positions 3–4 and 2–5 have limited stability or they are unstable depending on the distance between the nanoparticles.

Note that for the structures consisting of a large number of nanoparticles, there can be a quantity of stable stationary and unstable forms of equilibrium. Accordingly, the stable and unstable nanostructures of composite materials can appear. The search and analysis of the parameters determining the formation of stable nanosystems is an urgent task.

It is necessary to note, that the method offered has restrictions. This is explained by change of the nanoparticles form and accordingly variation of interaction pair potential during nanoparticles coming together at certain conditions.

The merge (accretion[4]) of two or several nanoparticles into a single whole is possible (Fig. 4.13). Change of a kind of connection cooperating nanoparticles (merging or coupling in larger particles) depending on its sizes, it is possible to explain on the basis of the analysis of the energy change graph of connection nanoparticles (Fig. 4.14). From Figure 4.14 follows, that, though with the size increasing of a particle energy of nanoparticles connection E_{np} grows also, its size in comparison with superficial energy E_S of a particle sharply increases at reduction of the sizes nanoparticles. Hence, for finer particles energy of connection can appear sufficient for destruction of their configuration under action of a mutual attraction and merging in larger particle.

Spatial distribution of particles influences on rate of the forces holding nanostructures, formed from several nanoparticles, also. In Figure 4.14 the chain nanoparticles, formed is resulted at coupling of three nanoparticles (from 512 atoms everyone), located in the initial moment on one line. Calculations have shown, that in this case nanoparticles form a stable chain. Thus, particles practically do not change the form and cooperate on "small platforms."

In the same figure the result of connection of three nanoparticles, located in the initial moment on a circle and consisting of 256 atoms everyone is submitted. In this case particles incorporate among themselves "densely," contacting on a significant part of the external surface.

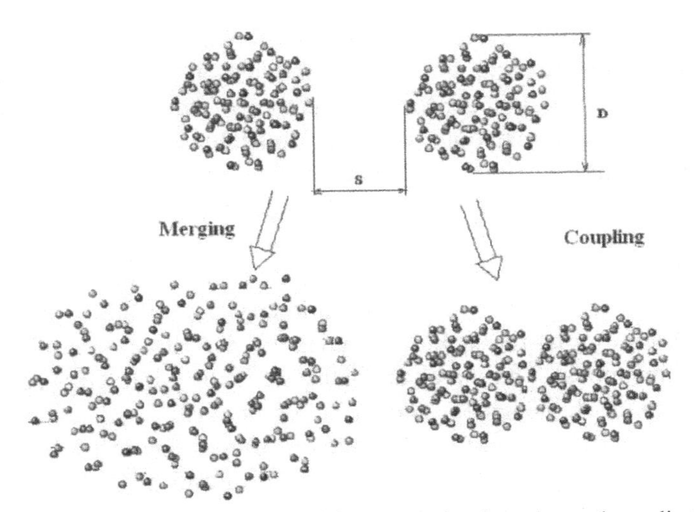

FIGURE 4.13 Different type of nanoparticles connection (merging and coupling).

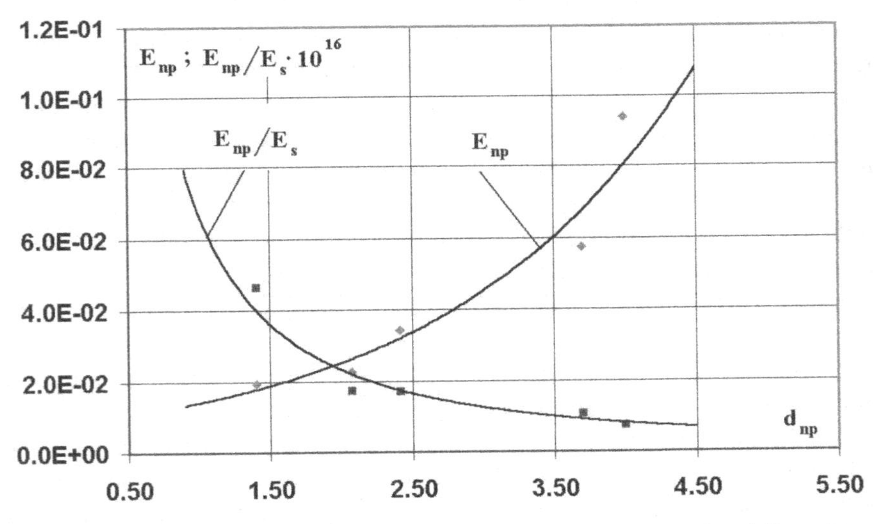

FIGURE 4.14 Change of energy of nanoparticles connection E_{np} [Nm] and E_{np} ration to superficial energy E_S depending on nanoparticles diameter d [nm]. Points designate the calculated values. Continuous lines are approximations.

Distance between particles at which they are in balance it is much less for the particles collected in group ($L^0_{3np} < L^0_{2np}$). It confirms also the graph of forces from which it is visible, that the maximal force of an attraction between particles in this case (is designated by a continuous line) in some times more, than at an arrangement of particles in a chain (dashed line) $F_{3np} > F_{2np}$ (Fig. 4.15).

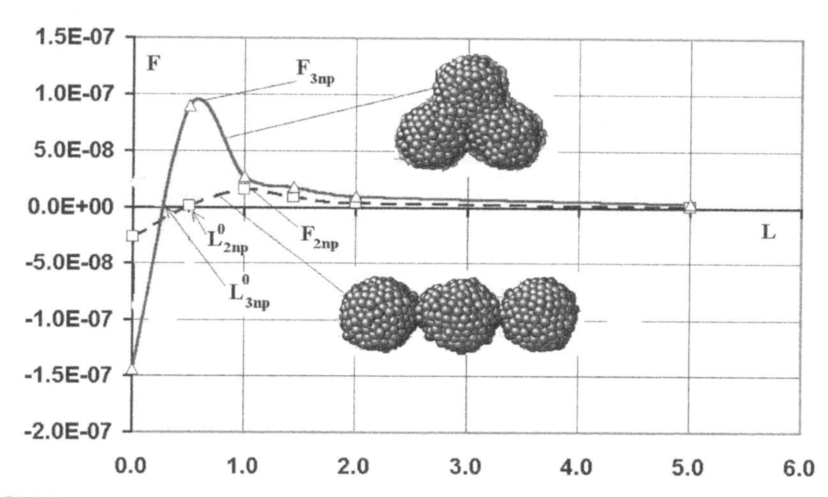

FIGURE 4.15 Change of force F [N] of three nanoparticles interaction, consisting of 512 atoms everyone, and connected among themselves on a line and on the beams missing under a corner of 120 degrees, accordingly, depending on distance between them L [nm].

Experimental observation of spatial structures formed by nanoparticles (Fig. 4.16), confirms that the nanoparticles are collected in compact objects. The internal atomic structure of the area of nanoparticles connection is noticeably different from the structure of free nanoparticle.

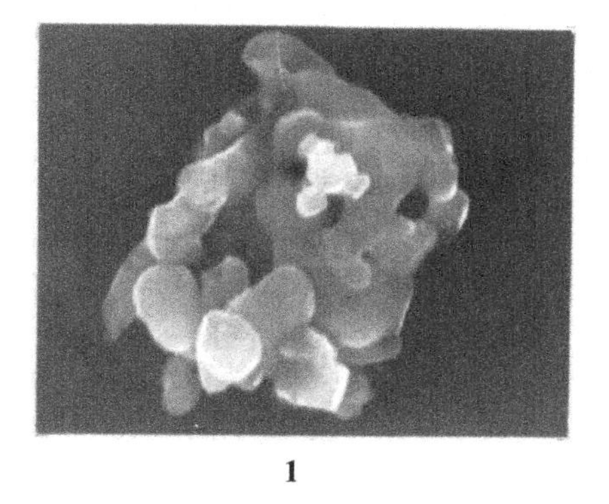

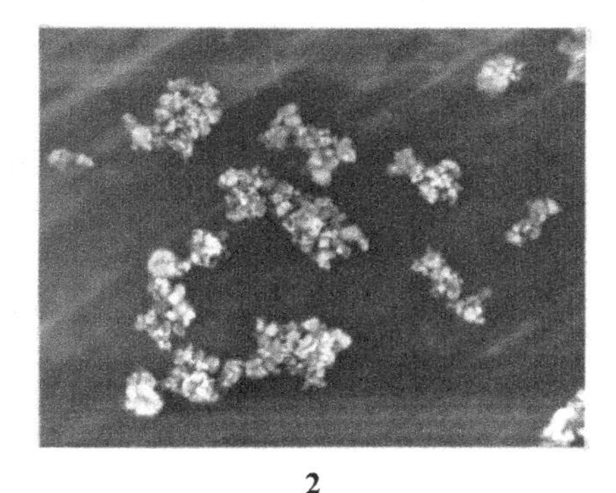

1 2

FIGURE 4.16 Photos of nanoparticles formed from the merger or the adhesion of the particles of smaller size (scale: (1) 1:5000 and (2) 1:1000).

Type of static interaction of nanoparticles depends strongly on the temperature. Figure 4.17 presents the picture of the interaction of nanoparticles with the environment temperature changes according to Figure 4.18. You can see that the temperature increase is observed consistently change the appearance of nanoparticles connection: coupling–merging—dissociation of nanoparticles.

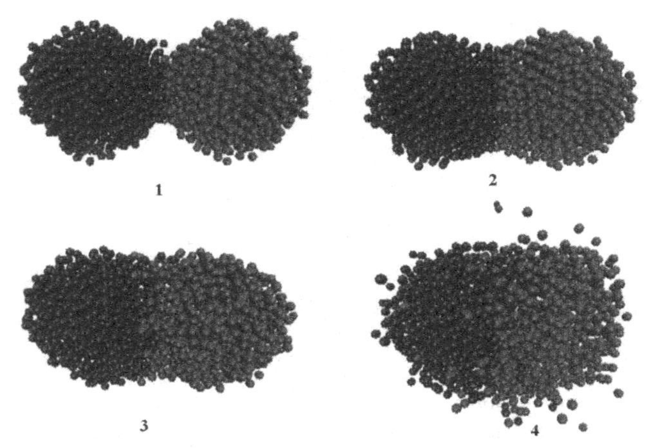

FIGURE 4.17 Changing the type of connection nanoparticles with increasing temperature.

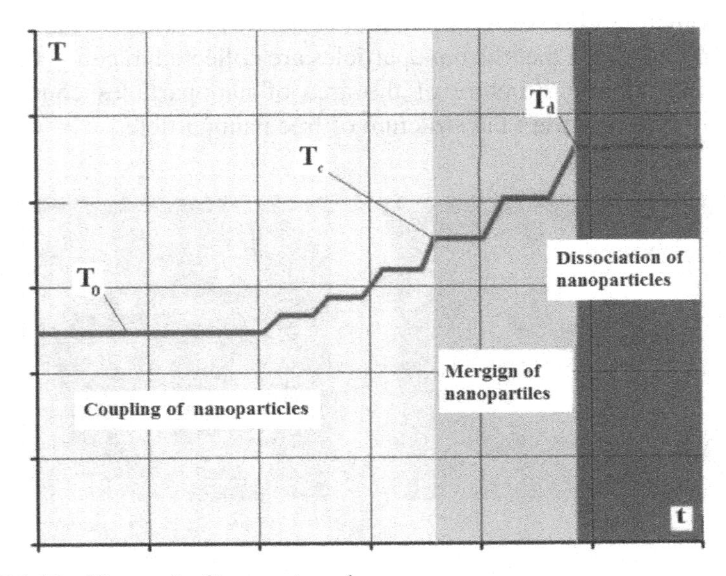

FIGURE 4.18 The graph of temperature change.

4.1.3 DYNAMIC INTERACTION OF NANOPARTICLES

This section discusses the problems of the dynamics of nanoparticles and nanoparticle interaction with individual atoms.

Figure 4.19 presents sequentially the picture of the interaction nanoparticles consisting of 12 atoms, with three atoms moving to the particle.

Initially, the particle is stable; velocity of atoms in it at the initial time is zero. Calculations showed that the result of the interaction depends strongly on the kinetic energy of the atoms approaching. At low initial speed of the atoms, the nanoparticle captures these atoms and gradually reconstructed forming a new larger nanoparticle.

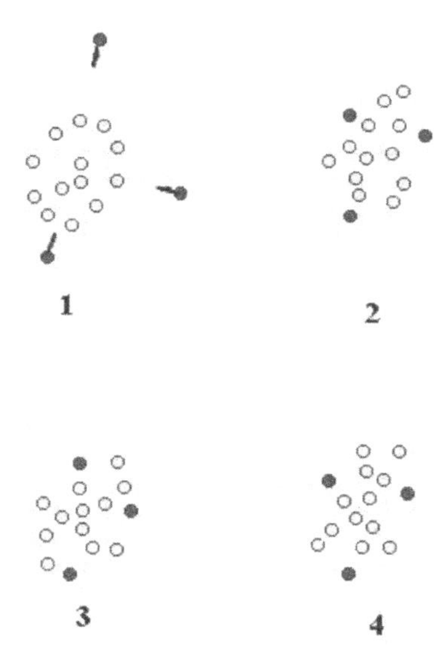

FIGURE 4.19 The pictures of forming a new nanoparticle in time (1–4) with attached external environment atoms; ○—the atoms of nanoparticle and •—the atoms of the environment.

The stability of the new particle graph shows changes over time of its energy (Fig. 4.20). We can see that the energy of the new nanoparticles is constant in time, since the curves and full of potential energy have a horizontal asymptote.

By increasing the speed of movement of the atoms from the external environment the result of this interaction changes. Figure 4.21 shows an example of the destruction of the nanoparticle under the influence of three atoms, approaching it with prescribed initial velocities.

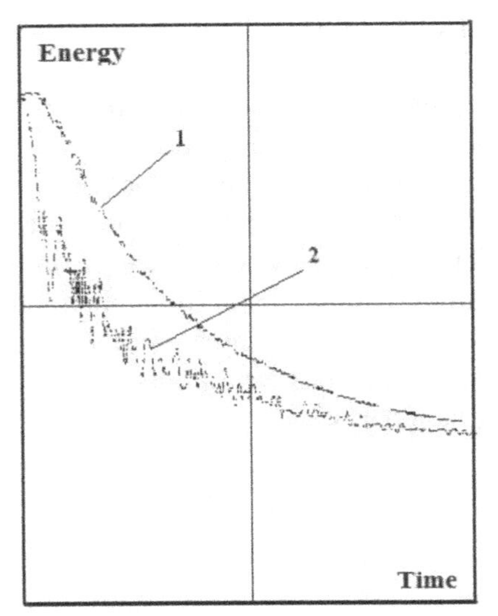

FIGURE 4.20 The change in time the energy of "stable" groups of atoms: 1—the total energy of the system and 2—potential energy of the system.

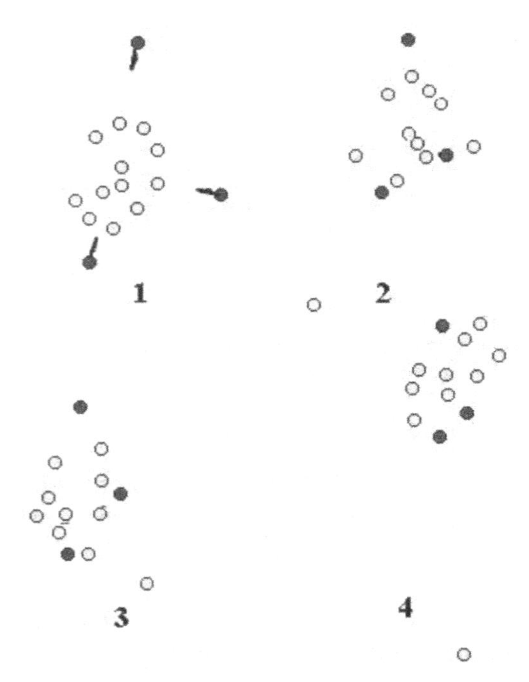

FIGURE 4.21 Pictures of destruction of nanoparticle in time (1–4) under the action of the environment: ○—atoms of nanoparticle and ●—ambient atoms.

As follows from the calculations the environment atoms knock the atoms constituting nanoparticle, flying through it. The change in time of energy such system of atoms (Fig. 4.22) shows that its structure is unstable, as evidenced by the absence of a horizontal asymptote at the graph.

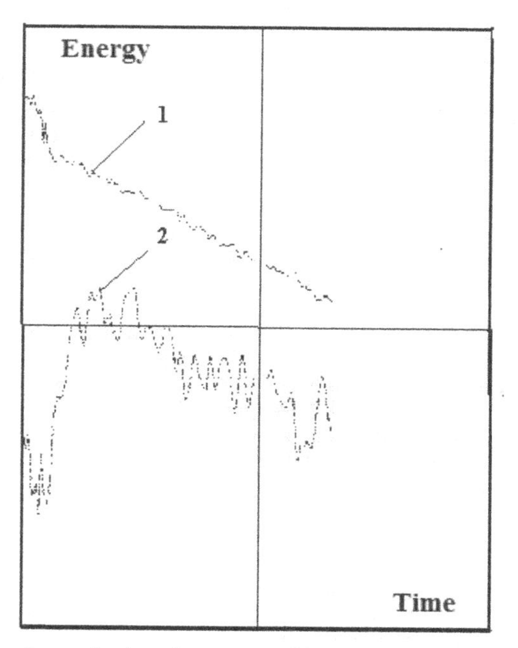

FIGURE 4.22 The change in time the energy of "unstable" groups of atoms: 1—the total energy of the system and 2—the potential energy of the system.

Analysis of the interaction between the nanoparticles also allows us to conclude significant role in this process the initial energy of the particle motion. Various processes at interaction of the nanoparticles, moving with different speed, are observed: the processes of agglomerate formation, formation of larger particles at merge of the smaller size particles, absorption by large particles of the smaller ones, dispersion of particles on separate smaller ones or atoms.

For example, in Figure 4.23 the interactions of two particles are moving toward each other with different speed are shown. At small speed of moving is formed steady agglomerate (Fig. 4.23 (right)). In a Figure 4.23 (left) is submitted interaction of two particles moving toward each other with the large speed. It is visible, that steady formation in this case is not appearing and the particles collapse.

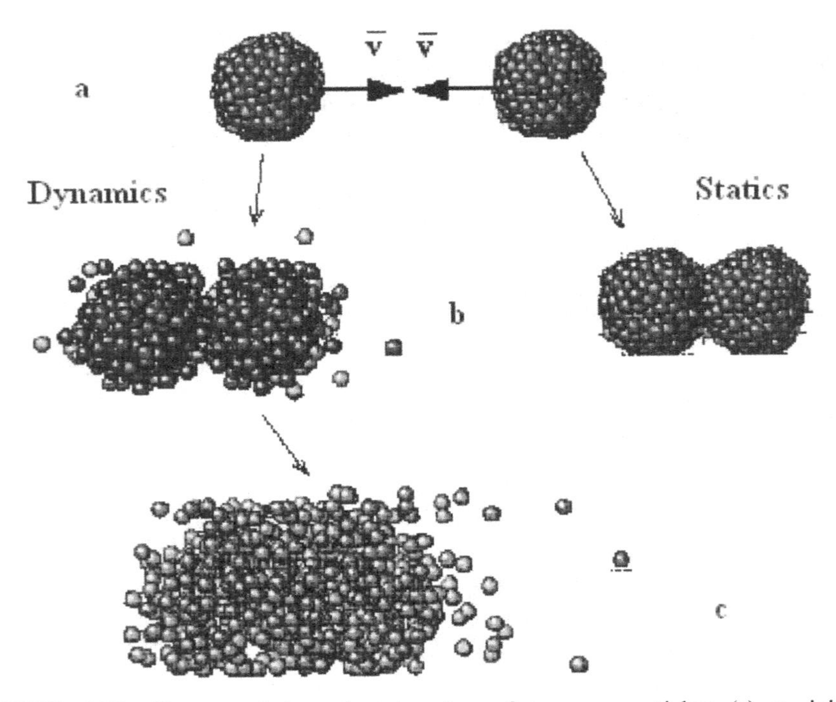

FIGURE 4.23 Pictures of dynamic interaction of two nanoparticles: (a) an initial configuration nanoparticles; (b) nanoparticles at dynamic interaction; and (c) the "cloud" of atoms formed because of dynamic destruction two nanoparticles.

The nature of the interaction of nanoparticles also depends strongly on the ratio of the size of nanoparticles. In Figure 4.24 the pictures of the interaction of two zinc nanoparticles with different are sizes presented. As in the previous case, the nanoparticles move from the initial position (1) toward each other. At low initial velocity at contact nanoparticles are joined and form a stable conglomerate (2).

By increasing the speed, the larger nanoparticle absorbs smaller one, and formed a single nanoparticle (3,4). With further increase in the speed of nanoparticle, the smaller particles penetrate rapidly into a large and destroy it (5).

These examples show that the use of dynamic compaction process to form nanocomposites requires proper selection of load mode, providing the integrity of the nanoparticles. At high energy of dynamic loading instead of the nanocomposite dispersion corresponding to the initial size of the nanoparticles, the nanocomposite can be obtained with a much larger grain, which significantly changes the properties of the composite.

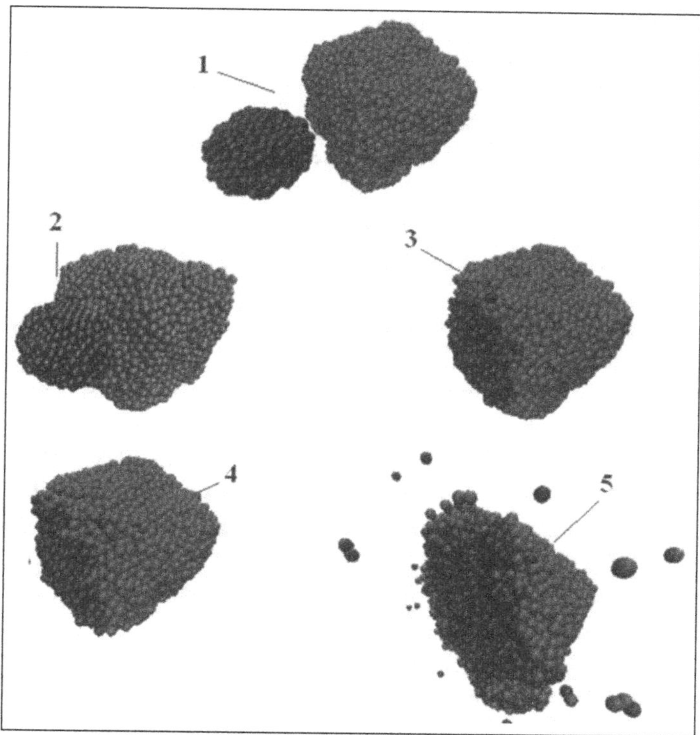

FIGURE 4.24 Pictures of the interaction of two zinc nanoparticles: (1) initial configuration of the nanoparticles; (2) nanoparticle compound; (3,4) large nanoparticle absorption of smaller size particle; and (5) destruction of nanoparticles upon impact.

4.2 SIMULATION OF ELECTRODEPOSITION OF METAL MATRIX NANOCOMPOSITE COATINGS

4.2.1 INTRODUCTION

Composite materials represent a multiphase system which is consisted of two or more components, that can be classified as reinforcing elements and as a binder matrix. The properties of the composite material are determined by the ratio of the parameters of reinforcing elements and binder matrix, as well as of technology their production. As a result of combining of utility of matrix and reinforcing elements it is formed the new complex properties of the composite, which is not only reflects the original properties of its components, but also includes properties that do not had isolated components.

The composite coating with improved and unique operational characteristics, such as wear resistance, cracking resistance, anti-friction properties, corrosion resistance, radiation resistance, and high adhesion to the substrate can be produced by this technology.

There are many kinds of traditional techniques for formation of surface layers with improved physicochemical properties. The most widely used the surface hardening, surface strengthening, and various methods of chemical and thermal machining, for example, the carburizing, nitriding, boriding, and etc.

In recent time the methods influencing on the work piece surface by beams of particles (ions, atoms, and clusters) or high-energy quantum (ion-plasma surface treatment and laser machining) are used extensively. The methods of gas-phase deposition of composite coatings at atmosphere pressure or at vacuum environment have a significant development. Also the thermal sprayed coating methods receive a powerful development in connection with practical application of technique of plasma and detonation spraying of powdered materials.

A new branch of knowledge is the process of composite electrochemical coatings formation.[141]

Metal matrix composite electrochemical coatings (MMEC) are prepared from the suspensions, representing electrolyte solutions with additives of certain quantity of a superfine powder. The particles are adsorbed onto cathode surface in combination with metal ions during electrocodeposition (ECD) process and the metal matrix composite coating is formed. MMEC consists of galvanic metal (dispersion phase) and particles (dispersed phase). The ECD process is schematically displayed in Figures 4.25 and 4.26. Initially, all electrolyte inclusions are considered as impurities that degraded the quality of the electroplated metal. Efforts were made to prevent incorporation of undesirable particles by enclosing soluble anodes in a bag to prevent dissolved anode material from being codeposited with the plated metal at the cathode or by periodic filtration of electrolyte. Filters were used to remove from the plating solution unwanted suspended particles which could cause dull, rough, and poorly adherent deposits.[24]

The first application of MMEC dates back to the beginning of twenty centuries. The nickel matrix with sand particles composite was utilized as anti-slip coatings on ship stairs. Also in 1928 Fink and Prince investigated the possibility of using ECD process to produce self-lubricating copper-graphite coatings on part of car engines. Despite this the systematical scientific investigation of the ECD process occurred only in the early 1960s, immediately after the electrophoretic deposition method has industrial application for coating metallic substrates by charged particles. In the electrophoretic deposition process, the suspended charged particles moves toward and deposit onto the substrate surfaces as the result of an applied electric field. The ECD process was developed with the intent of increasing the versatility of the electrophoretic deposition process by combining it with electroplating.

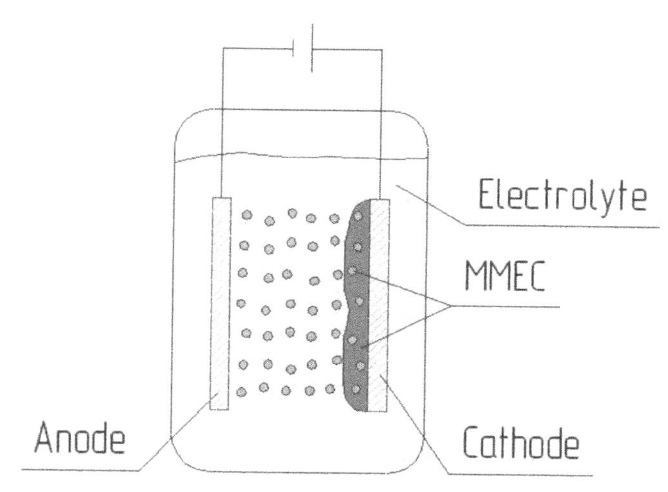

FIGURE 4.25 The ECD process.

There are the following steps of the ECD process: (a) the particles in suspension obtain a surface charge, (b) the charged particles and metal ions are transported through the liquid by the application of an electric field (electrophoresis), convection and diffusion, (c) the particles and metal ions are adsorbed onto the electrode surface, and (d) the particles adhere to the electrode surface through van der Waals forces, chemical bonding, or other forces and, simultaneously, adsorbed metal ions are reduced to metal atoms. Metal matrix are encompassed the adsorbed particles and thus the MMEC is formed.

The particles of metal including those that cannot be electrochemically deposited from water-based electrolytes, synthetic diamond, ceramic or organic materials with dimensions smaller than 100 um are usually used as dispersed phase.

Also the nanomaterials can be employed as the dispersed phase. In accordance with received opinion the nanomaterials are particles with dimensions smaller than 100 nm, also this material is known as ultra-dispersed or nanophase material.

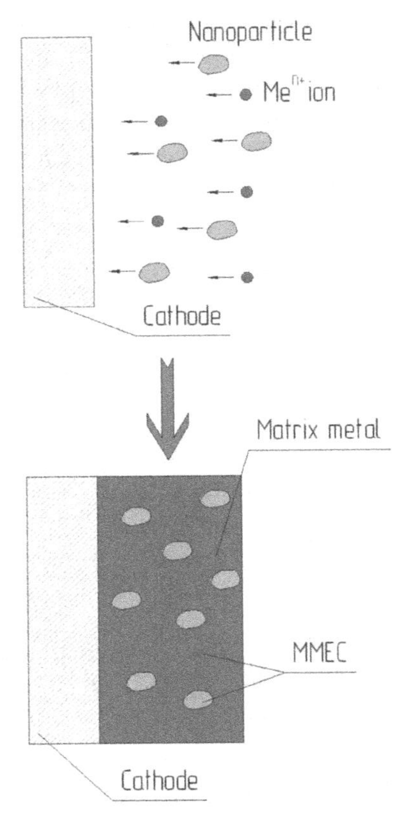

FIGURE 4.26 The process of MMEC formation.

Recent investigations have highlighted that ECD process is an attractive approach for the preparation of nanocomposite coatings.

The advantages of ECD process over other coating methods are the coating uniformity both thickness and chemical composition even for covered detail with complex shapes, reduction of waste often in comparison with dipping or spraying techniques, low levels of environment contamination, the ability to continuously process parts.[2] In addition, ECD process avoids the problems related to high temperature or high pressure processing, also in this method is not necessary to realize the deposition process in vacuum or in an atmosphere of shielding gas.

The low processing temperature (around room temperature) minimizes the inter diffusion or undesirable chemical reactions.[3] The film thickness can be accurately controlled by regulation of the consumed charge.

However, there are a number of problems, such as the no uniformity of the distribution of nanoparticles in electrolyte volume by reason of agglomeration and sedimentation of particles.

The particles of oxides Al_2O_3, ZrO_2, TiO_2, SiO_2, Cr_2O_3, various allotropic form C, carbides SiC, WC, TiC, nitrides Si_3N_4, polymers (polystyrene, PTFE), and number of other materials are employed as dispersed phase. Diameter of particles as usual varies from 4 to 800 nm. The metals as Cu, Ni, Co, Cr, Zn, Ag, Fe, Au, As, or their alloys can be used as dispersion phase.

The concentration of particles suspended in solution are usually varied from 2 to 200 g/L producing composites with typically 1 ± 10 vol.% of embedded particles. It has been reported that the incorporation reached up to 50 vol.% when gravity sedimentation was additionally used.

The ECD process is used to produce both soft magnetic materials for sensor implementation, and hard-magnetic materials. Moreover, the incorporation of magnetically hard particles of barium ferrites in an electrodeposited Ni or Ni-alloy matrix was found to significantly increase the coercivity of the resulting coatings.

The applications of MMEC include wear and abrasion resistant surfaces, lubrication, high hardness tools, dispersion-strengthened alloys, and for protection against oxidation and hot corrosion. ECD process has been used to produce high surface area cathodes which have been used as electro catalysts for hydrogen electrodes in industrial water electrolysis.

The process of the ECD finds application in such areas of the industry, as automobile production, building, electric power production, and also in airspace industry along with oil and gas production.

MMEC based on nickel and aluminum oxides are being applied in different technological fields with high demands on friction and corrosion resistance. The ECD of SiC nanoparticles into Cu matrix leads to a more appreciable grain refinement and, as a consequence, the nanocomposite coating present a very high micro hardness, 61% higher than pure copper coating, and an increase of 58% of the abrasion resistance.

On the ECD process and hence the structure, the morphology and the properties of the composite coatings is affected next electrodeposition parameters like the electrolysis conditions (composition and agitation of the electrolytic bath, presence of additives, temperature, and pH), the electrical profile and the particle properties (type, size, shape, surface charge, concentration and dispersion in the bath), interaction between particles and the electrolyte ions and the character and velocity of fluid motion.

It has been shown that the surface charge of the nanoparticles is an important parameter for the ECD process. An interesting result was that negatively charged particles are more easily codeposited compared to positively charged ones.

The intensity of electrolyte agitation during ECD process is the important factor that affects on the uniformity of nanoparticles distribution in the bulk of electrolyte and on the nanoparticles mass transfer to the electrode surface.

The properties of the MMEC are also strongly influenced by the parameters of the applied voltage. For example, the application of pulsed current deposition provide the deposition of coatings with improved properties such as wear resistance, corrosion resistance surface roughness, and grain size in comparison with the direct current deposition. Also, the application of pulsed current in the ECD process allows significant increase the particles content in the MMEC. The particles content in the MMEC is a key parameter that determines the properties of the coating such as wear resistance, high temperature corrosion, oxidation resistance, and the coefficient of friction. Another important factor is the uniformity of the nanoparticles distribution in the MMEC.

There are a number of types of electrochemical cells which are used in experimental study of the ECD process, such as the parallel plate electrodes, rotating disk or cylindrical electrodes, impinging jet electrode (IJE).The IJE provides the selective and high-speed deposition.

The rotating cylinder electrode (RCE) is one of the most common geometries for different types of studies,[116] such as metal ion recovery, alloy formation, corrosion, effluent treatment, and Hull cell studies. RCEs are also particularly well suited for high mass transport studies in the turbulent flow regime.

In this work the results of mathematical modeling of the ECD process of system Cu-Al_2O_3 on a RCE are presented. The mathematical model describes the mass transfer of electrolyte ions and held in suspension of nanoparticles throughout the volume of the electrolyte, the electrode processes, adsorption and desorption of electroactive ions and nanoparticles on electrode surface and turbulent flow of rotating liquid in electrochemical cell.

In recent years a number of numerical modeling were conducted to study the turbulent flow on RCE. For instance Perez and Nav[116] investigated the copper electrodeposition process on RCE with various applied current density and electrode rotation speed using the standard turbulent model k-epsilon (k-ε) and tertiary current distribution boundary conditions for mass transfer of ions. But in their work the mass transfer of copper ions was investigated only within a tiny stagnant steady diffusion layer. The influence of turbulent flow on the steel corrosion was investigated in Ref. 10 where authors were also used the standard turbulent model k-ε to modeling the flow. In our work we have found that the Low Reynolds k-ε model is more suitable for modeling of the coupled electrochemical processes. The first time ever was found that in the system with the RCE the unsteady diffusion layer is formed and consequently it is necessary to modeling the mass transfer of electroactive ions and nanoparticles thorough the volume of the electrochemical cell.

4.2.2 PROBLEM FORMULATION

The basis for this investigation is provided by the fundamental works of Stojak and Talbot,[141,142] where the process of deposition of the Cu–Al_2O_3 MMCC on a RCE was experimentally investigated. In their work the experiments were conducted using a three-electrode system, consisting of a RCE, a concentric stationary electrode, and a saturated calomel reference electrode placed in a Luggin capillary. The water solution of 0.1M CuSO4 and 1.2M H_2SO_4 was used as electrolyte. The experiments were conducted at various electrode rotational rates: 500, 1000, and 1500 rpm and concentration of nanoparticles: 39, 120, and 158 g/L. The geometric parameters of electrochemical cell are presented in Figure 4.27. The computational domain (Fig. 4.28) is the 2D-axisymmetric space consisting of electrolytic solution, RCE as cathode and anode which is fixed on external cylinder.

The hydrodynamics of fluid depends on rotational speed of RCE, electrochemical cell geometric parameters and kinematic viscosity of electrolyte and is possible to characterize by the Reynolds number according to eq 4.7.

$$\mathrm{Re} = \frac{\Omega \cdot r_i \cdot (r_o - r_i)}{\tilde{o}} \tag{4.7}$$

where Ω is RCE rotating velocity [rad/s], r_i(=6 mm] is the radius of the inner cylinder, r_o (=24 mm) is the radius of the outer cylinder, υ(=1.155 mm^2/s) is the kinematic viscosity. The value of the Reynolds numbers for the deposition conditions of 500, 1000, and 1500 rpm are, respectively, equal to 4896, 9792, and 14,688. For the system with RCE the critical Reynolds number is equal to 200.[49] Consequently, the flow of electrolyte has a complex and turbulent nature at all investigated modes and the hydrodynamic of electrolyte turbulent flow should be accounted in the simulation process.

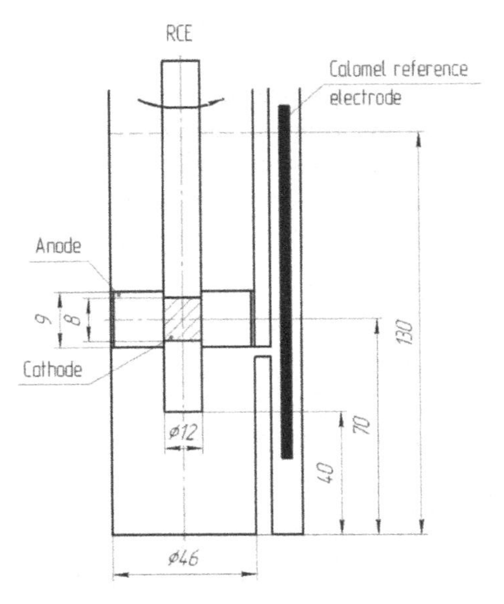

FIGURE 4.27 The geometric of cell.

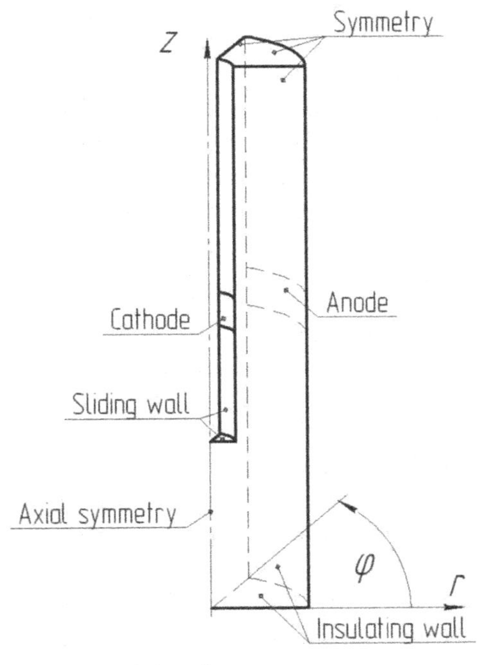

FIGURE 4.28 The computational domain.

4.2.3 HYDRODYNAMIC MODEL

Mathematical modeling of the turbulent flow of rotating fluid is still the intractable problem. Application of methods of direct numerical simulation (DNS) and large-eddy simulation (LES) is limited due to the necessity to use a lot of computational power. For this reason the k-ε Reynolds Averaged Navier-Stokes (RANS) turbulence model is most commonly used to describe the turbulent motion of the fluid in the numerical modeling of technological processes in engineering calculations.[1] However, this model can be used only to simulate the turbulent flow at some distance from the fixed or moving boundaries and is not suitable for the simulation of rotating fluid. In this model the special wall function is used to describe the flow near the walls since the motion of the fluid near the surface of the border is very different from the motion in the volume. This wall function is represented the analytical expression which depend on flow and grid mesh parameters. For this reason the computational domain is assumed to be displaced some distance from the wall. The bulk concentrations of electroactive ions at immediate vicinity of electrode surface are necessary to determine in modeling of electrochemical process with account of concentration polarization near the electrode surfaces. Consequently, the k-ε RANS model is not suited for modeling of ECD process. In this work we used to describe the turbulent fluid flow the Low Reynolds Number Turbulence Model (low Re k-ε RANS) wherein the wall function is not used. There are many different kinds of low Re k-ε RANS model. The most common used models are the Abe-Kondoh-Nagano (k-ε AKN), the Chang-Hsieh-Chen (k-ε CHC), the Launder-Sharma (k-ε LS), and the Yang-Shih (k-ε YS). The Abe-Kondoh-Nagano (k-ε AKN) model is used in this work as basic model according to the review paper,[74] where the above mentioned models were compared.

The system of equations which describe the fluid motion is represented below:

RANS:

$$\rho\frac{\partial \mathbf{U}}{\partial t}+\rho(\mathbf{U}\cdot\nabla)\mathbf{U}= \nabla\left[-P+(\mu+\mu_t)\left(\nabla\mathbf{U}+(\nabla\mathbf{U})^{\mathrm{T}}\right)-\frac{2}{3}\rho\kappa\mathbf{I}\right]+\boldsymbol{F}, \quad (4.8)$$

where **U** is the averaged velocity field, P is the averaged pressure, ρ is the electrolyte density, μ is the dynamic viscosity depends only on the physical properties of the electrolyte, μ_T is the turbulent eddy viscosity (6) which is supposed to emulate the effect of unresolved velocity fluctuations the eddy viscosity, k is the turbulent kinetic energy, ε is the turbulent dissipation rate, and F is the volume forces vector.

The continuity equation is:

$$\nabla\ U = 0 \tag{4.9}$$

The transport equations for turbulent kinetic energy, k, and turbulent dissipation rate,ε, as well as eddy viscosity equation are represented below:

$$\rho\frac{\partial k}{\partial t} + \rho(\mathbf{U}\cdot\nabla) = \nabla\cdot\left[\left(\mu + \frac{\mu_T}{\sigma_k}\right)\nabla k\right] + P_k - \rho\varepsilon \tag{4.10}$$

$$\begin{gathered}\rho\frac{\partial \varepsilon}{\partial t} + \rho\mathbf{U}\cdot\varepsilon = \\ \nabla\left[\left(\mu + \frac{\mu_t}{\sigma_\varepsilon}\right)\nabla\varepsilon\right] + C_{\varepsilon 1}\frac{\varepsilon}{\kappa}P_k - f_\varepsilon C_{\varepsilon 2}\rho\frac{\varepsilon^2}{\kappa}\end{gathered} \tag{4.11}$$

$$\mu_T = \rho f_\mu C_\mu \frac{k^2}{\varepsilon} \tag{4.12}$$

where $C_{\varepsilon 1}$, $C_{\varepsilon 2}$, C_μ, σ_k, and σ_ε are the model constants, P_k is the production term of turbulent kinetic energy (4.13) and f_ε, and f_μ are the damping functions (4.14 and 4.15).

$$P_k = \frac{\mu_T}{2}\left|\nabla U + (\nabla U)^T\right|^2 \tag{4.13}$$

$$f_\varepsilon = \left(1 - e^{-l^*/14}\right)^2 \cdot \left(1 - 0.3e^{(R_t/6.5)^2}\right) \tag{4.14}$$

$$f_\varepsilon = \left(1 - e^{-l^*/14}\right)^2 \cdot \left(1 + \frac{5}{R_t^{3/4}} e^{(-R_t/200)^2}\right) \tag{4.15}$$

where l^* is the dimensionless number characterizing the distance to the closest wall (4.16), R_t is the turbulent Reynolds number (4.17),

$$l^* = \frac{\rho u_s l_W}{\mu}, \tag{4.16}$$

$$R_t = \frac{\rho k^2}{\mu\varepsilon}, \tag{4.17}$$

where u_ε is the Kolmogorov velocity scale (4.18), l_w is the distance to the closest wall (4.19).

$$u_\varepsilon = \left(\frac{\mu\varepsilon}{\rho}\right)^{1/4}, \tag{4.18}$$

$$l_w = \frac{1}{G} - \frac{l_{ref}}{2}, \tag{4.19}$$

where l_{ref} is the characteristic length, which is depended by the model geometry and usually be equal a half of minimal side of rectangle encompassing the model geometry. In this work l_{ref} is taken equal half of radius of external cylinder. The G is the solution of modified Eikonal equation.[74]

$$\nabla G \cdot \nabla G + \sigma_w G(\nabla \cdot \nabla G) = (1 + 2\sigma_w)G^4 \tag{4.20}$$

where σ_w is dimensionless constant range from 0 to 0.5. In this work σ_w is equal 0.1.

The initial condition for equitation is

$$G = \frac{2}{l_{ref}} \tag{4.21}$$

The boundary conditions for modified Eikonal equation

On insulating and sliding wall boundaries:

$$G = \frac{2}{l_{ref}} \tag{4.22}$$

The homogeneous Neumann condition (4.23) is used on the other boundaries

$$\nabla G \cdot \mathbf{n} = 0 \tag{4.23}$$

Equation 4.20 with initial (4.21) and boundary conditions (4.22 and 4.23) must be solved before starting to solve the main differential equation system.

Following constants are used in this model:

$$C_{\varepsilon 1} = 1.5;\ C_{\varepsilon 2} = 1.9;\ C_\mu = 0.09;\ \sigma_k = 1.4;\ \sigma_\varepsilon = 1.5. \tag{4.24}$$

4.2.4 ELECTROLYTE SUBSTANCES MASS TRANSFER

Kinetics of electrode processes depends on the concentrations of ions and particles near the surface of the working electrode, which are known only in the initial moment of the process.[111] They are equal to the concentrations in the bulk solution in well-stirred solution. However, at initial time of deposition process the concentrations of all substances in the electrode-electrolyte surface are changed. Mass transfer layer is formed near the surface of the electrode, the thickness of which depends on the concentration distribution in the bulk solution and the parameters of fluid flow. In a well-stirred solution of electrolyte the concentrations of active substances is considered to be not change outside of this layer and are equal to their bulk values.

Three mechanisms of mass transfer in the bulk of electrolyte: diffusion, convection, and migration, are well known. The mass transfer by diffusion occurs either to the electrode surface or from it, depending on the concentration gradient of the substance. Mass transfer by convection is defined by the terms of hydrodynamic fluid. This process can be created artificially by stirring, but it can also occur *in vivo* by reason of changes in fluid density. Migration is the third mechanism of mass transfer of substances. It is a result of electrostatic forces, which act only on charged particles and ions in contradistinction from diffusion and convection, which act on the mass transfer of all ions and particles. According to the three mass transfer mechanisms the total substance flow $\mathbf{N}_\Sigma$ is assumed to can be divided into flows of diffusion $\mathbf{N}_d$, migration $\mathbf{N}_m$ and convection $\mathbf{N}_c$

$$\mathbf{N}_\Sigma = \mathbf{N}_d + \mathbf{N}_m + \mathbf{N}_c \tag{4.25}$$

The mathematical modeling of mass transfer is performed by means of diffusion-convection equations and is investigated throughout the volume of the electrochemical cell. The electrode processes are described based on the tertiary current distribution boundary condition,[11] which is considered the osmic resistance drop in electrolyte solution, the over potential on the electrode and ion activity near electrode surface.

The mass transfer equations of electrolyte ions and suspended nanoparticles are defined by the law of conservation of mass:

$$\frac{\partial c_i}{\partial t} + \nabla \cdot \mathbf{N}_i = 0, \tag{4.26}$$

where subscript i denote Cu^{2+} ions and p denote some nanoparticles; $\mathbf{N}_i$ is the mass flux density of ion or nanoparticle and defined by the Nernst–Planck equation (4.21).

$$\mathbf{N}_{\mathrm{i}} = -D_{\mathrm{i}}\nabla c_{\mathrm{i}} - z_{\mathrm{i}}m_{l}Fc_{\mathrm{i}}\nabla\phi_{l} + c_{\mathrm{i}}\mathbf{u}, \tag{4.27}$$

where D_i is the diffusion coefficient, c_i is the volume concentration, z_i is the charge number, um_i is the electrophoretic mobility, F is the Faraday constant, φ_l is the electrolyte potential, and $\boldsymbol{u}$ is the vector of electrolyte velocity.

The diffusion coefficient of nanoparticles was determined by using the Einstein's equation (4.28). The diffusion coefficients of Cu^{2+} ions which were found in the experimental work[142] are used in this work.

$$D_p = \frac{k \cdot T}{6\pi \cdot \mu \cdot r_p} \tag{4.28}$$

where k is Boltzmann constant, T is the electrolyte temperature and r_p is the radius of nanoparticles.

Electrophoretic mobility of Cu^{2+} ions and nanoparticles specials are defined by the equitation:

$$m_i = \frac{D_i}{R \cdot T}, \tag{4.29}$$

where R is the universal gas constant.

The Poisson's equation is used to define the electrolyte potential, φ_l, and to close the system of equations.

$$-\nabla \cdot \mathbf{i}_l = 0, \tag{4.30}$$

$$\mathbf{i}_l = \sigma_l \nabla \varphi_l, \tag{4.31}$$

where $\mathbf{i}_l$ is the current density, σ_l is the electrolyte conductivity The current density depends on the mass flux densities of ions via (4.32) since ions are both mass and charge carriers.

$$\mathbf{i}_l = F \cdot z_i \cdot \mathbf{N}_i, \tag{4.32}$$

The electrochemical reaction occurring on electrode surfaces is described by eq 4.33.

$$Cu^{2+} + 2e^- \Leftrightarrow Cu_m^0, \tag{4.33}$$

where Cu^{2+} denote the Cu^{2+} electrolyte ions, while Cu_m^0 denote the metal form of copper atoms adsorbed on the cathode and $\Delta\varphi_{eq}$ is the standard electrode potential of reaction (27).

The rates of cathode and anode reactions are defined in accordance with the Butler–Volmer theory by the following equations:

$$i_n = i_0 \left(\frac{c_i}{c_i^b} \right)^v \cdot \left[\exp\left(\frac{\alpha_a F}{RT} \eta_s^a \right) - \exp\left(\frac{-\alpha_c F}{RT} \eta_s^c \right) \right], \tag{4.34}$$

where c_i^b is the bulk concentration of Cu^{2+} ions; α_a, α_c are, respectively, the transport coefficients of anodic and cathode reactions; η_s^a, η_s^c are, respectively, the over potentials on anode and cathode electrodes.

The over potential is the driving forces for electrochemical reactions, its value can be defined by eq 4.35.

$$\eta_s^k = V_s^k - \varphi_l^k, \tag{4.35}$$

where superscript k denotes the electrode (a—anode; c—cathode), V_s^k is the potential of respective electrode, φ_l^k is the potential of electrolyte near the surface of respective electrode.

The cell voltage, U, is the difference between the electric potentials of anode and cathode electrodes.

$$U = V_s^a - V_s^c, \tag{4.36}$$

The adsorption/desorption processes of nanoparticles on cathode surface are described by (4.37)

$$P \underset{r_p^d}{\overset{r_p^a}{\Leftrightarrow}} P_a, \tag{4.37}$$

where P denote the Al_2O_3 nanoparticles suspended in electrolyte volume, while P_a denote the adsorbed nanoparticles Al_2O_3 and r_p^a, r_p^d are, respectively, the rates of adsorption and desorption processes. The assumption that the desorption process is not going on are used in the work. The isotherm adsorption (4.38) is used to describe the process of nanoparticles adsorption

$$r_p^a = k_p^a c_p RT, \tag{4.38}$$

where k_p^a is adsorption coefficient of nanoparticles.

Initial conditions for system of equations are following:

$$c_p = c_{p0},\ c_p = c_{p0}, \tag{4.39}$$

$$U = f(t),\ \mathbf{u} = 0,\ p = 0, \tag{4.40}$$

for turbulent kinetic energy, k_0

$$k_0 = \left(\frac{\mu}{\rho(0.1 \cdot l_{ref}} \right)^2 \tag{4.41}$$

for turbulent dissipation rate, ε_0,

$$\varepsilon_0 = \frac{C_\mu \cdot k_0^{1.5}}{0.1 \cdot l_{ref}}, \tag{4.42}$$

boundary conditions for system of equations are following on electrode surfaces:

$$\mathbf{N}_i = \frac{1}{F \cdot z_i} \mathbf{i}_l,\ \mathbf{N}_\mathrm{p} = r_p^a \cdot \mathbf{n}, \tag{4.43}$$

on insulating wall:

$$\mathbf{N}_\mathrm{i} = \mathbf{N}_\mathrm{p} = 0,\ \varphi = 0 \tag{4.44}$$

$$\mathbf{u} \cdot \mathbf{n} = 0, k = 0, \varepsilon = \frac{2\mu k}{\rho l_w^2} \tag{4.45}$$

on sliding wall:

$$u_r = 0, u_\varphi = \omega \cdot r, k = 0, u_r = 0, \varepsilon = \frac{2\mu k}{\rho l_w^2}, \tag{4.46}$$

where ω is the RCE rotational speed;
symmetry boundaries:

$$\begin{gathered} \mathbf{u} \cdot \mathbf{n} = 0, \\ (-p\mathbf{I} + \mu\,(\nabla\mathbf{u} + (\nabla\mathbf{u})^T)) \cdot \mathbf{n} = 0, \\ \nabla k \cdot \mathbf{n} = 0, \\ \nabla\varepsilon \cdot \mathbf{n} = 0. \end{gathered} \tag{4.47}$$

The closed system of equations is received. The model consists of nine partial differential equations (4.8–4.11, 4.26, and 4.30). To solve this system of equations a sufficient number of initial and boundary conditions has been given. The finite element method[196] is used in this work to solve this system of differential equations.

4.2.5 RESULTS AND DISCUSSIONS

First of all, let us consider the results of mathematical simulation without the hydrodynamic modeling of rotating turbulent flow. These results are presented in Figures 4.29–4.32.

From dependence (Figs. 4.29 and 4.30) of current density against electrode rotation frequency and nanoparticles concentration it is possible to conclude that current density is continual increased with increase of electrode rotation frequency. This dependence is corresponded to convective diffusion theory. Contrariwise, the current density is decreased with the increase of nanoparticles concentration in bulk of electrolyte. This can be explained by decrease of electrolytic solution conductivity, which is decreased with increase of nanoparticles concentration and decrease of cathode surface-active area owing nanoparticles adsorption. For example, 5 nm nanoparticles concentration augmentation with 39–158 g/L reduces current density from 2.8 to 3% and 50 nm nanoparticles concentration augmentation with 39–158 g/L reduces current density from 5.5 to 10.7%.

The analysis of nanoparticles weight content dependences from bulk nanoparticles concentrations and electrode rotating speed (Figs. 4.31 and 4.32) has shown that the maintenance of particles in a composite coating rises with augmentation of their concentration in electrolyte for nanoparticles of all radiuses. As an example the augmentation of 5 nm nanoparticles concentration with 39–158 g/L leads to weight content augmentation from 273 to 289%, depending on rotating speed of an electrode.

For verification of the received results as well as for definition suitability of a mathematical model the modeling results have been compared with the data of experimental work (Stojak and Talbot, 1999), which underlies this paper. Thus, for 25 nm with concentration of 158 g/L, and the electrode rotational speed 500 rev/min, the nanoparticles weight content in the experimental work is 4.2% and in this work is 3.49%, with accuracy of modeling is 16.9%.

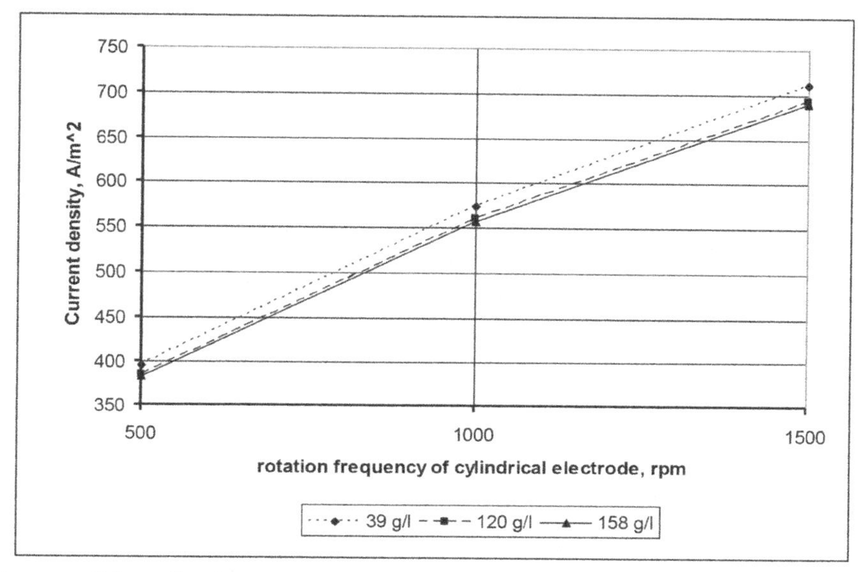

a) nanoparticles radius 5 nm.

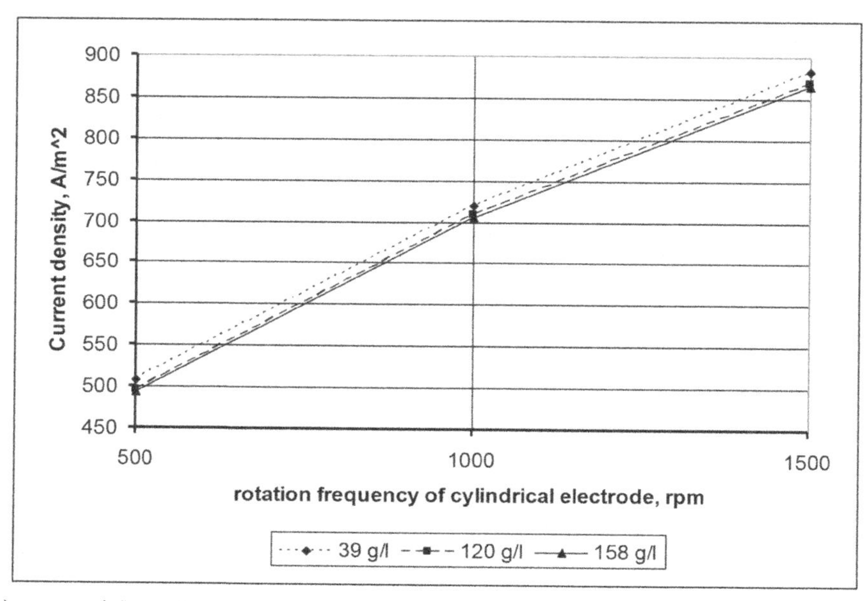

b) nanoparticles radius 25 nm.

FIGURE 4.29 The graph of dependences of current density from electrode rotational frequency and various nanoparticles concentrations.

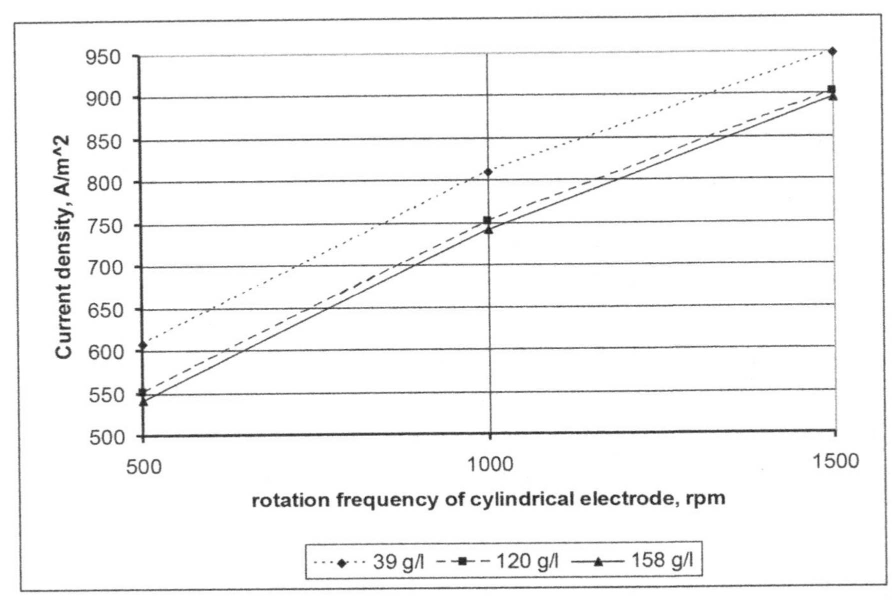

a) nanoparticles radius 50 nm.

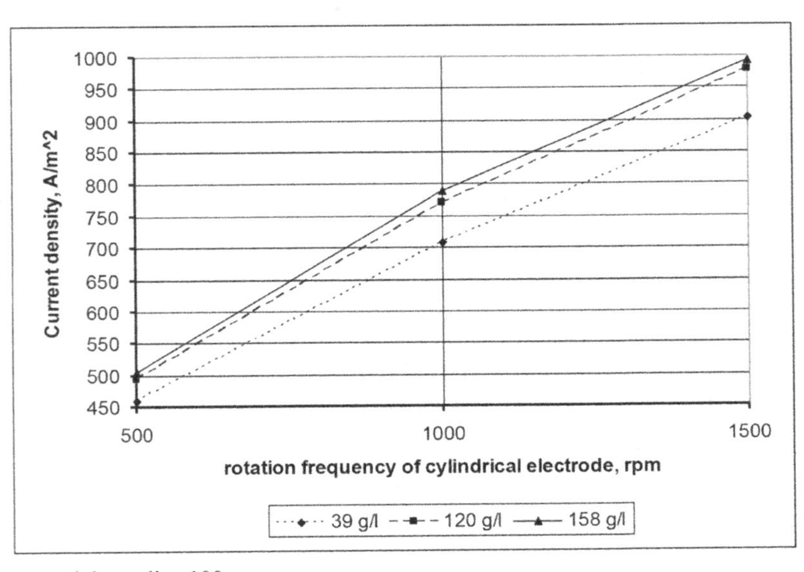

b) nanoparticles radius 100 nm.

FIGURE 4.30 The graph of dependences of current density from electrode rotational frequency and various nanoparticles concentrations.

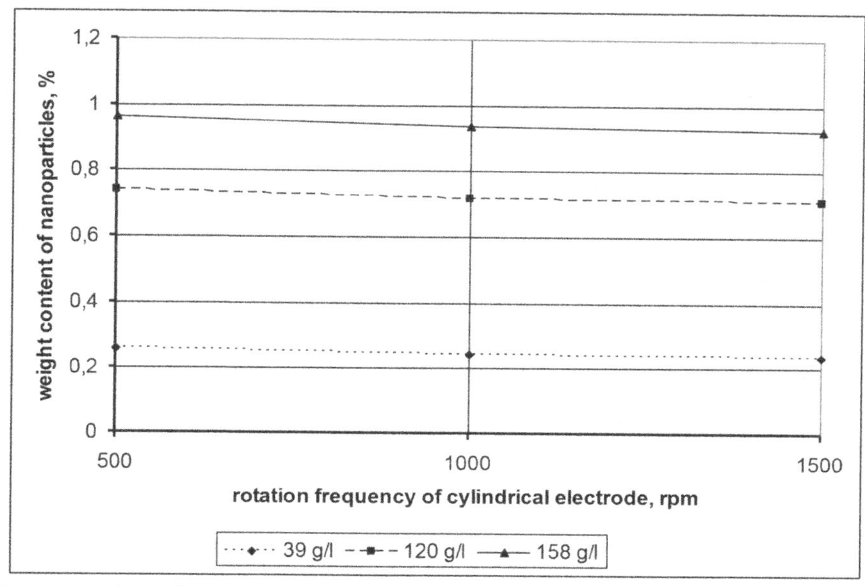

a) nanoparticles radius 5 nm.

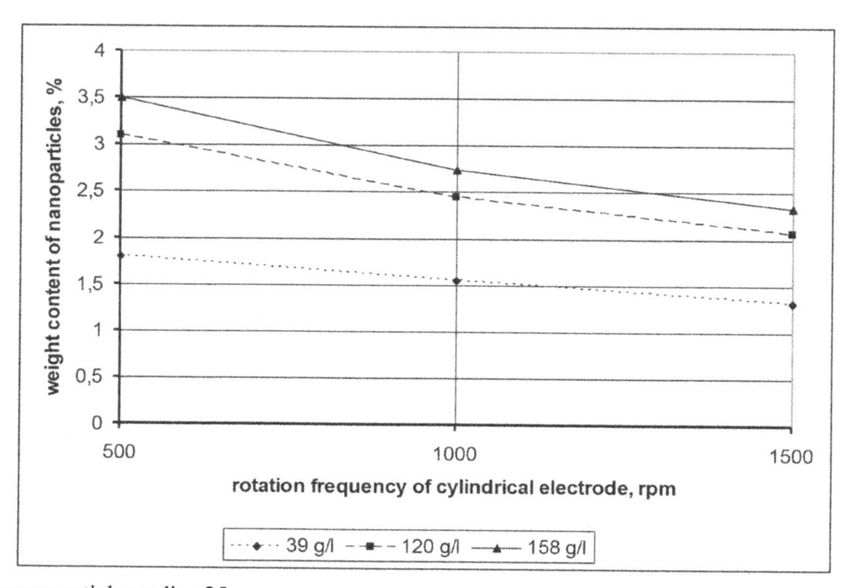

b) nanoparticles radius 25 nm.

FIGURE 4.31 The graph of dependences of nanoparticles weight content from electrode rotational frequency and various nanoparticles concentrations.

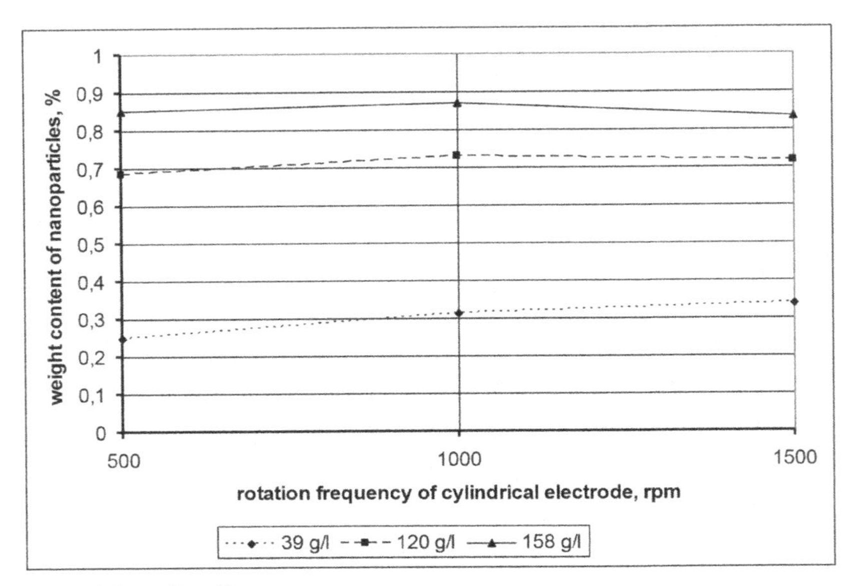

a) nanoparticles radius 50 nm.

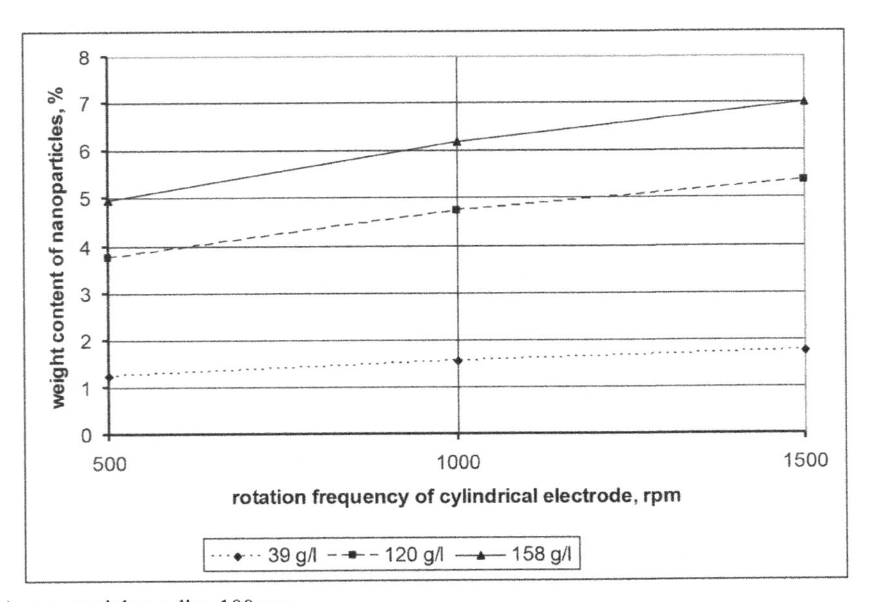

b) nanoparticles radius 100 nm.

FIGURE 4.32 The graph of dependences of nanoparticles weight content from electrode rotational frequency and various nanoparticles concentrations.

The hydrodynamic modeling of rotating turbulent flow arising in RCE system was done to compare the results of modeling with the well-known experimental data.[9] The standard k-ε RANS turbulence model and low Re k-ε RANS described in this book were used to determine the more appropriate model. The configuration of system with rotating inner and stationary outlier concentricity cylinders with the geometry parameters r_i=52.5 mm, r_o=59.46 mm, a $L = 208.8$ mm is depicted in Figure 4.33. The inner cylinder is rotating with the rotating velocity Ω. The modeling was done with a 2D-axisymmetric computational domain during 100 s at slowly increasing Reynolds number from 0 to 2000.

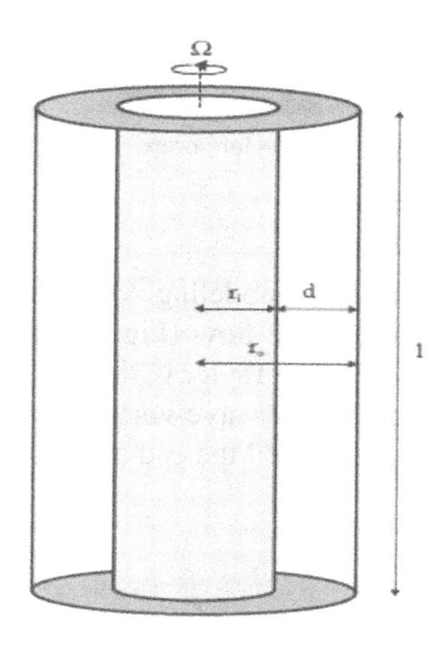

FIGURE 4.33 The geometry of model.

Andereck et al.[9] found that laminar Couette flow (Fig. 4.34a) is stable while the Reynolds number of system will not reach the first critical Reynolds number (Re = 120), after that the flow is changing to the new form and the toroidal Taylor vortex flow (Fig. 4.34b) is established. The Taylor vortices are appeared in all volume between cylinders and the order of their disposition relative to the rotation axis is not been changing through time. The Taylor vortices flow is stable in a narrow range of Reynolds number from 120 to 175. The flow is changing to the new form again when the Reynolds number of system reaches its second critical value. At this new form the vortices are started to migrate relative to the rotation axis and the wave-vortex flow (Fig. 4.34c) is established.

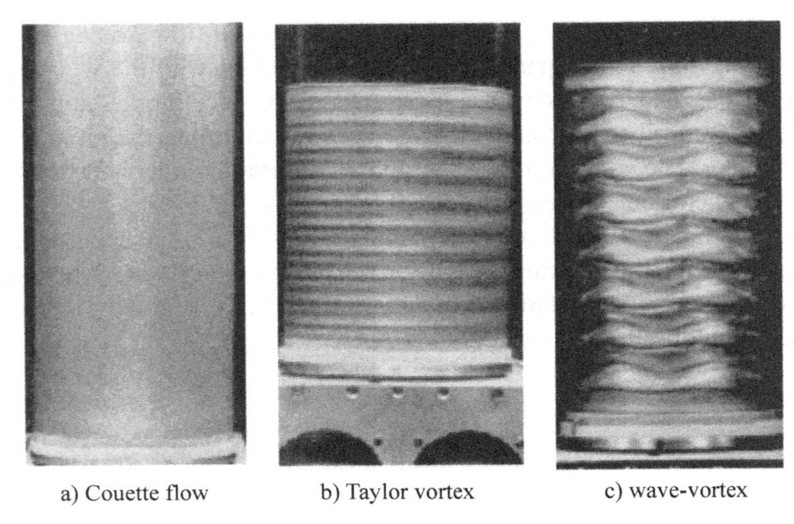

FIGURE 4.34 The experimental date.[9]

The results of mathematical modeling with low Re k-ε RANS model revealed that the laminar Couette flow (Fig. 4.35a) is stable at Reynolds number Re from 0 to 200–250, the Taylor vortex flow (Fig. 4.35b) is stable at Re from 250 to 1000, after that the wave-vortex flow (Fig. 4.35c) is formed. The wave-vortex flow is stable until the end of simulation.

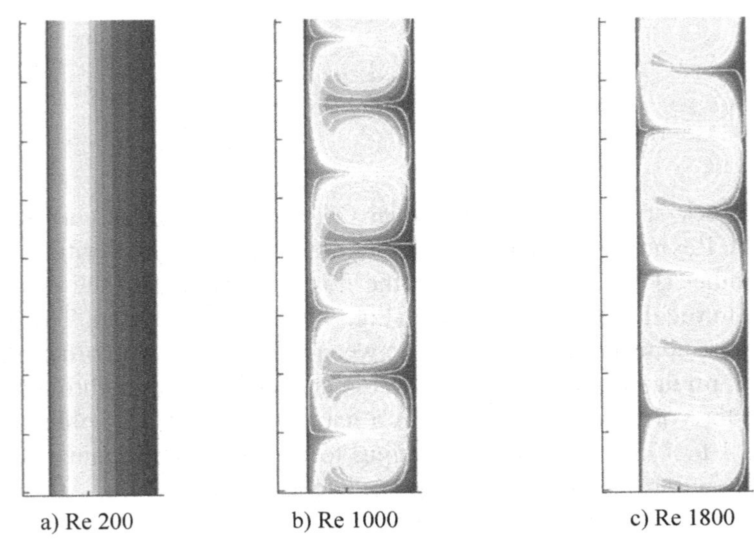

FIGURE 4.35 Fields of liquid velocity vector at various Re, m/s.

The results of mathematical modeling showed that the standard k-ε RANS turbulence model is not appropriated to describe and predict the turbulent rotating flow because the previously mentioned flow changing were not predicted by this model. At this model the quasi-Taylor vortices flow with only one vortex is appeared after laminar Couette flow. The results of modeling with the standard k-ε RANS turbulence model are obviously diverging with the experimental date.

The mathematical simulation of turbulent flow of electrolyte in RCE cell without considering the mass transfer of electrolyte ion and suspended nanoparticles was also performed in order to examine the flow dynamics at various Reynolds numbers. The rotation speed was gradually increased from 0 to 2000 RPM during 200 s at this simulation. It was found that the flow of the electrolyte at all investigated modes has a complex and turbulent nature. At low Reynolds numbers ($Re < 400$) the flow can be characterized as laminar Couette flow. The flow changing is occurred at Re from 400 to 900, respectively, at 41 and 92 rpm. However, at this Re the Taylor vortices are not formed in contradistinction from the case of fluid motion between two concentric cylinders but a lot of small vortices are formed and have been moved chaotic relative to the axis of rotation of RCE.

Hydrodynamic mathematical modeling of the electrodeposition of copper on RCE was conducted at three various rotational speed of the inner electrode 500, 1000, and 1500 rpm in accordance with the experimental conditions. The process of ECD of Cu–Al_2O_3 nanoparticles was simulated at three various concentrations of nanoparticles 39, 120, and 158 g/L only at 1500 rpm. The computational time was limited by 105 s. At this time the rotational velocity of RCE and cell voltage varies from the open circuit value (+0.037 mV vs SCE) to −0.5 V at a scan rate of 5 mV/s according to Figure 4.36.

The result of hydrodynamic mathematical modeling of Cu electrodeposition found that the flow has a developed turbulence character at all investigated rotation speed (Fig. 4.37) in accordance with the results which were mentioned above. It also was defined that the zones with high and low fluid velocity are continuously formed near the electrode surface.

The following results apply to all investigated modes: when the voltage between the electrodes reaches the magnitude of −0.4 V relative the calomel reference electrode the concentration of Cu^{2+} ions near the cathode surface is reduced to values close to zero. Consequently, strong concentration polarization is occurring at this cell voltage and the current density is achieving limit value. Therefore, the mass transfer of electro active ions to the electrode surface by diffusion becomes the time-dependent step which determines the rate of electrochemical reaction at the electrode.

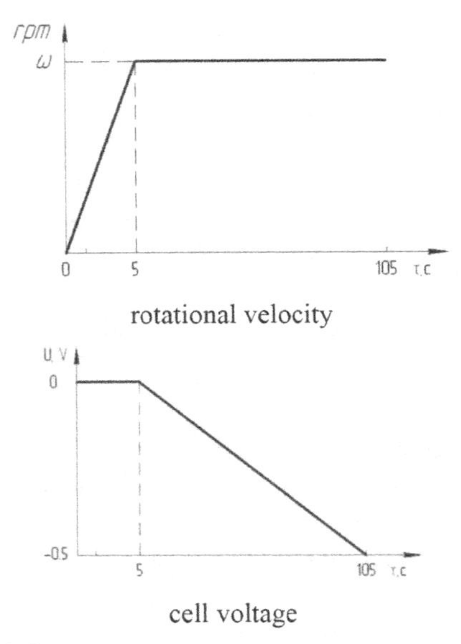

FIGURE 4.36 The modeling parameters.

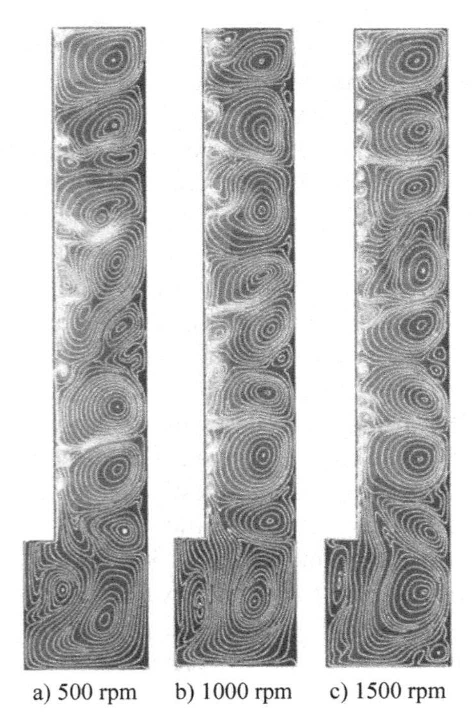

a) 500 rpm b) 1000 rpm c) 1500 rpm

FIGURE 4.37 Developed turbulence in the operating conditions of the electrochemical cell.

The rate of ECD process reaches a certain limiting value because rate of diffusion is limited. It should be pointed out that when the diffusion-controlled stage of deposition process has been achieved the oscillation of current density appears at all operating modes of electrochemical cell (Figs. 4.38–4.40). It is the author's opinion that these oscillations occur by reason of turbulent flow near the electrode surface and because the regions with high or low electrolyte motion speed are continuous appearing near the electrode surface and are being replaced by each other. In consequence of which the stagnant zones with relatively small ion concentration is being arisen from time to time near electrode surface. It means that the diffusion layer, which have a larger thickness, arisen too. In its turn this increases the time, which wanted to transfer the electro active ions to the electrode surface by means of diffusion, and, respectively, decreases the current density. The mass transfer of ions to the cathode surface by convection is increased when a zone with a high velocity of electrolyte reaches the electrode surface. As a result of this the thickness of diffusion layer and, consequently, the time which is required to delivering the Cu^{2+} ions to the cathode surfaces by means of diffusion, are decreasing and the local current density is temporarily increasing.

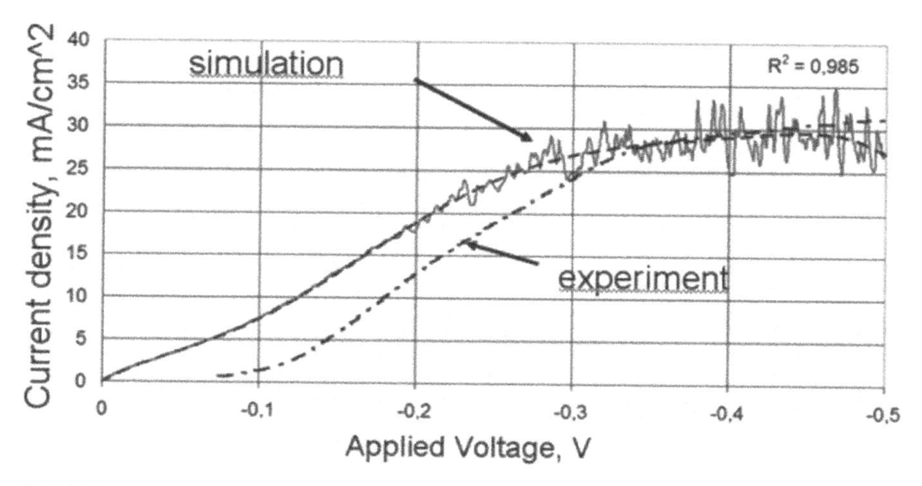

FIGURE 4.38 Current density at 500 rpm of RCE without nanoparticles.

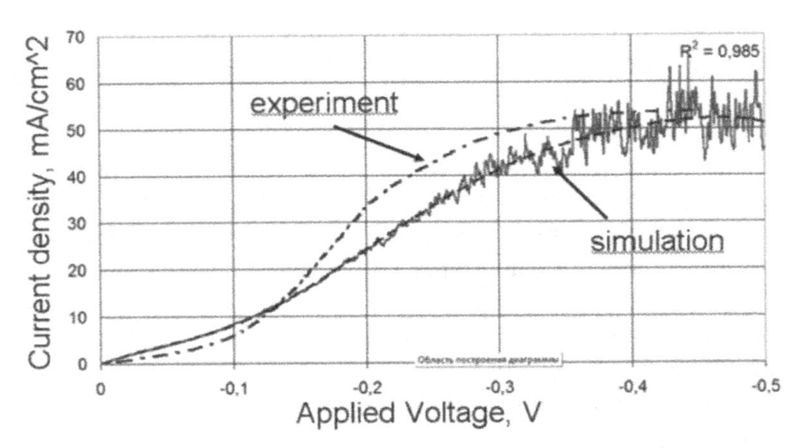

FIGURE 4.39 Current density at 1000 rpm of RCE without nanoparticles.

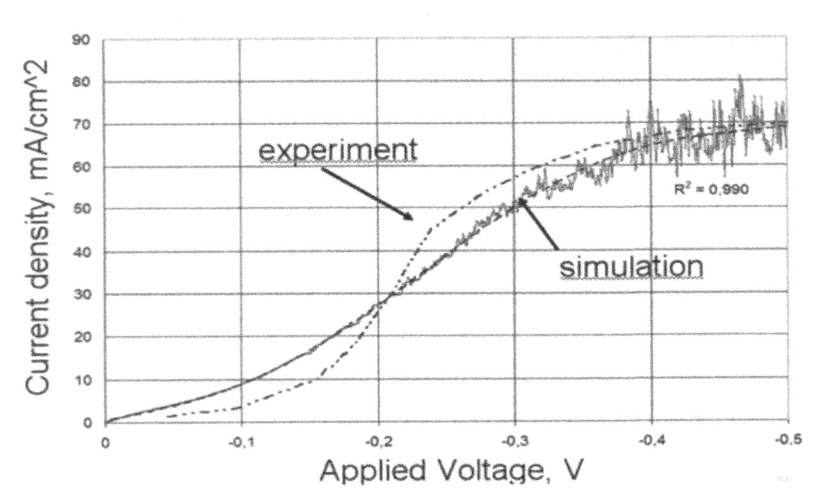

FIGURE 4.40 Current density at 1500 rpm of RCE.

The diffusion layer thicknesses are varied in sufficient wide range, so the diffusion layer thicknesses are changed from 30 to 60, from 20 to 50, and from 15 to 40 μm for rotational velocity of RCE 500, 1000, and 1500 rpm, respectively.

In 1954 Bockris and Reddy[24] determined the empirical relationship for limiting diffusion current density which is commonly used in theoretical electrochemistry:

$$i_L = 0.0791 \frac{nFDC}{2r_i} \left(\frac{\nu}{D} \right)^{0.356} \mathrm{Re}^{0.7}, \tag{4.48}$$

here n, D, and C are, respectively, the charge number, the diffusion coefficient, and the bulk concentration of electroactive ion, F is the Faraday constant. The values of limiting diffusion current density are shown in Table 4.1 for the investigated system configurations. Based on the Brunner's correlation the equation of the thickness of the diffusion layer has the form:

$$\delta = 15{,}562 \cdot D^{0.356} \cdot \nu^{0.344} \cdot \Omega^{-0.7} \cdot r_i^{-0.4}, \tag{4.49}$$

The analysis of eq 4.43 shows that the thickness of the diffusion layer is depended only on geometry of the electrochemical cell, kinematic viscosity of the electrolyte, diffusion coefficient of species and electrode rotational velocity and is not depended on the time during which the process is carried.

TABLE 4.1 Current Density.

Denotation	Limited current density, mA/cm^2 (error, %)		
RCE rotating velocity rpm	500	1000	1500
Experiment[24]	31	50	67
Calculation, eq 4.48	39.3 (26.9%)	63.9 (27.8%)	84.9 (6.7%)
Simulation	29 (6.4%)	51.6 (3.2%)	67.1 (0.15%)

TABLE 4.2 The Diffusion Layer Thickness.

Denotation	Thickness of diffusion layer, µm		
RCE rotating velocity, rpm	500	1000	1500
Calculation, eq 4.49	32,369	20,069	14,977
Simulation	30–60	20–50	15–40
Reynolds number	4896	9792	14,688

Results of mathematical modeling are proved that the diffusion layer is unsteady (Table 4.2). Consequently, the paradigm of steady diffusion layer [142] is not fully sufficed to the RCE when turbulent flow is stated. In order to advance the understanding the nature of electrochemical process it is necessary to modeling the coupled process which takes into account the hydrodynamic of electrolyte flow.

Usually,[141] the influence of electrokinetic forces on mass transfer of electroactive ions are not considered into mathematical modeling of conventional electrochemical processes like a cathodic metal reduction because these processes are carried out in the solution of strong electrolytes with high electric conductivity. However, this assumption is not appropriate when a modeling of ECD process is carried out, since the electrokinetic forces have a significant influence on mass transfer of nanoparticles. Because the double electrical layer (DEL) is formed around nanoparticle suspended in electrolyte solution. The structure of DEL is significantly varied depending on composition and concentration of electrolyte, material, shape, and dimension of the nanoparticles.

The DEL structure is usually characterized by the zeta potential, which can be positive, negative, or equal to zero. During ECD process the nanoparticles with positive zeta potential are additionally attracted to the negatively charged cathode due to electrokinetic forces, while the nanoparticles with negative zeta potential are repelled from it. In its turn the electrokinetic forces does not influence on mass transfer of nanoparticles with zero value of zeta-potential.

The analysis of the experimental results,[141,142] revealed that the nanoparticles weight content is slowly reduced with increasing of current density. However, despite this the nanoparticles flow to the cathode surfaces have to increase with increasing of current density. Because according to the Faraday's law the flow of reduced ions increase with increasing of current density. Consequently, the proportional increasing of nanoparticles flow is necessary to keeping the weight content with the increasing of current density. This is confirmed by the Figure 4.41, which depicted the adducted dependences of nanoparticles flow as function of applied current density and initial concentrations of nanoparticles.

Figure 4.41 reveals that the nanoparticles' flows are linearly increased with increasing of current density and consequently voltage of electrochemical cell. This means that the influences of electrokinetic forces on mass transfer of nanoparticles must be accounted in the mathematical simulation and that the nanoparticles used in the basic experimental work,[141,142] should have a positive zeta potential, because they flow are linearly increases with increasing of the cell voltage.

In the work[141] authors revealed that zeta-potential of nanoparticles is the main factor, which determines the nanoparticles weight content in composite coating. They discovered that the α-Al_2O_3 and γ-Al_2O_3 nanoparticles have, respectively, positive and negative zeta-potential in experimentation conditions. They also found that only α-Al_2O_3 nanoparticles are effective for codeposition with Cu at the same experimental condition.

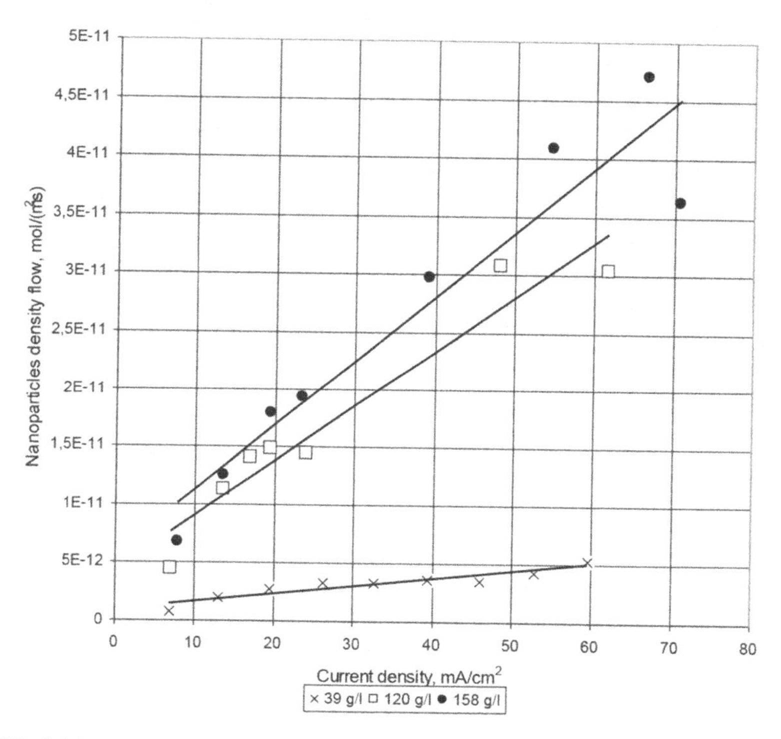

FIGURE 4.41 The adducted dependences of nanoparticles flow depending on applied current density and initial concentrations of nanoparticles.

In the basic experimental work,[142] authors point out that they for the first time ever obtained the MMEC of copper with γ-Al_2O_3 nanoparticles. However, based on the data of experimental work, the γ-Al_2O_3 nanoparticles should have a negative zeta potential for electrolyte concentration used in the work. The surfactants which could change the sign of the zeta potential of nanoparticles in were not used. Therefore the electrokinetic forces should prevent the codeposition of γ-Al_2O_3 nanoparticles and the formation of the CEP with the γ-Al_2O_3 nanoparticles could not occur.

However, as established by the authors used nanoparticles γ-Al_2O_3 were a powder mixture of γ-Al_2O_3 nanoparticles with the low presence of α-Al_2O_3 nanoparticles. And thus, the nanoparticles used in Ref. 141 were a two-component mixture. The actual ratio between phases has not been determined in the work of the authors. However, according to the authors of this book the concentration of each phase can be approximately determined from the difference between the densities of the powder mixture of nanoparticles measured in Ref. 141 and reference values of density of α- and γ-Al_2O_3 nanoparticles. It was determined the approximate ratio between γ-Al_2O_3 and α-Al_2O_3 phases on the basis of the calculation, which is not given in this chapter. It was found that the γ-Al_2O_3 nanoparticles comprise 88.3 of weight percent of powder mixture, while the α-Al_2O_3 comprise only 11.7% of weight percent of powder mixture. Based on the foregoing, it is assumed that instead γ-Al_2O_3 with an initial concentration of nanoparticles in electrolyte bath the α-Al_2O_3 nanoparticles are taken into consideration in this work with concentration of 11.7% from initial.

The resulting dependencies of weight content on applied current density are depicted in Figure 4.42. The good agreement with experimental data[141,142] has been obtained.

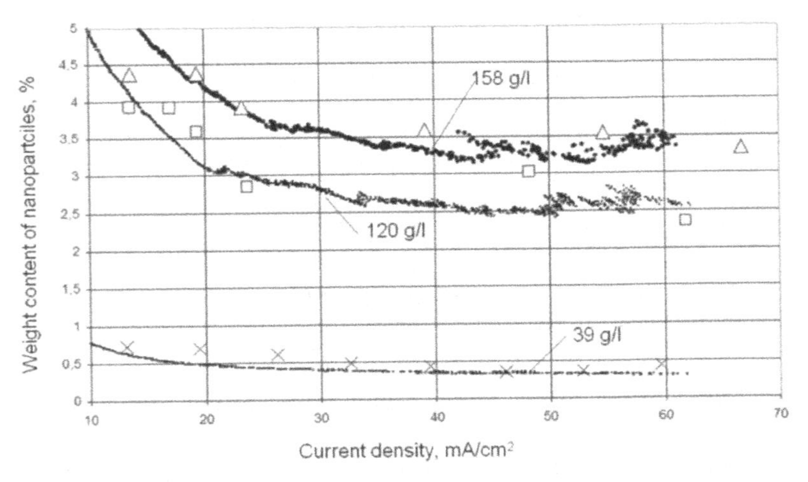

FIGURE 4.42 The dependencies of weight content on applied current density.

4.2.6 *CONCLUDING REMARKS*

The results of mathematical modeling of copper and alumina particles ECD on RCE with consideration of electrolyte turbulent flow are presented. A good correlation with the published experimental date[141,142] has been found. For the first time ever it is found that near the RCE surface the unsteady diffusion layer is formed by the reasons of electrolyte turbulent flow.

A new mathematical model of electrochemical codeposition of nano-composite coatings on the rotating cylindrical electrode, which is taking into account the turbulent flow of the electrolyte, is developed.

The results of hydrodynamic mathematical modeling of copper and alumina particles ECD on RCE with consideration of electrolyte turbulent flow are presented. A good correlation with the published experimental date have been found.

For first time ever it is found that when the current density achieve their limit value the nonstationary diffusion layer with the variable thickness is formed near the electrode surface by the reasons of electrolyte turbulent flow.

The composition of the MMEC is inhomogeneous both on the length and on the thickness of the composite.

It is found that the unsteady diffusion layer is formed close to the rotating electrode surface and the electrokinetic forces are the driving forces of the process. The good correlations with experimental data are received.

4.3 SIMULATION OF FORMATION OF SUPERFICIAL NANOHETEROSTRUCTURES

Nanoheterostructures are objects of sizes from one to several hundred nanometers consisting of layers of different semiconductor materials or having inclusions of semiconductor materials in their structure. There are many varieties nanoheterostructures: nanowhiskers (NWs), thin nanofilms, quantum dots, superlattices, etc. At the present time, heterostructures are used to create infrared LEDs for the use in fire sensors, perimeter protection systems, night vision devices, infrared illumination in the video surveillance equipment, automatic control systems, and other equipment.[8,186,187] Controlled synthesis of heterostructures, especially nanoheterostructures of certain composition and morphology, is one of the priorities of modern nanotechnology. The processes of nanoheterostructures are complicated and little studied, therefore computer simulation of the formation of systems based on nanoheterostructures elements is now one of the most important and urgent problems.[46,47,85,88,144,200] In this part of chapter, the researches of the formation of quantum dots and NWs on silicon substrate by method of molecular dynamic simulation are presented.

The solution to this problem was carried out by using the developed problem-oriented software. The program is based on the molecular dynamic simulation of the processes that accompany the formation of nanoheterostructures. The use of the methods of molecular dynamics for solving this problem was dictated by the necessity to trace the simulated process kinetics, as well as to evaluate the structure and physical properties of nanoheterostructures. The software includes several units: preparation of initial data, computing unit, data matching unit, the unit of analysis, and visualization of results. Simulation of the formation of nanoheterostructures was carried out using the molecular dynamic method which describes the motion of a system of atoms by a system of differential equations.

4.3.1 SIMULATION OF NANOWHISKERS FORMATION

NWs are characterized by a transverse size D (up to 100 nm) and length L, on order or more surpassing the transverse size. The mechanical strength of whiskers in 100 or more times surpasses the strength of large specimens of the same material. Often NWs are heterostructures. A pioneer in the development and creation of heterostructures is a Nobel Prize winner, academician of the Russian Academy of Sciences Zh.I. Alferov.

The first studies of the whiskers formation date back to the 1860s. R. Wagner and W. Ellis in 1964 confirmed the importance of a number of impurities for the whiskers formation; they discovered the mechanism of whisker growth according to the vapor–liquid–solid method (VLS).[181] Currently, extensive experimental and theoretical studies of processes of education and growth of NW is carried out at Ioffe Physical Technical Institute scientific group under the direction of professor V.G. Dubrovsky, the set of the experimental data connected with NW formation is received, the theory of NW growth on the VLS mechanism is developed.[37–39]

New methods of NW growth are developed at the Chemistry Department of Moscow State University in the laboratory of inorganic materials science, headed by Academician Yu.D. Tretyakov.

Today extensive researches of processes of NW formation are carried out. However by means of adequate mathematical models it is possible to predict properties of NW that will allow reducing costs of experiment considerably. The question of the need to control the properties of the NWs: diameter, length, shape, surface density, uniformity, composition, etc., by changing the conditions of surface preparation and deposition of material often arises in the process of NW growing up. The solution to this problem is impossible without theoretical studies and modeling of growth processes. In Russia and abroad extensive work on the development of mathematical models describing the processes of formation and growth of NWs are carried out. Thus, the research team from Novosibirsk under the leadership of I.G. Neizvestnykh simulates the growth of silicon NW on the substrate Si (111), using the Monte Carlo method[112] Monte Carlo method yields good results, however, the disadvantage of this method is the large computation time, and it does not allow us to see the dynamics of the growth process. It is unable to compute a system with a large number of atoms by quantum mechanics method at this stage of technological development. Basically, three phenomenological models: Givargizov–Chernov model, Kaschiev model, and basic model of diffusion growth of NW are used to explain patterns of growth processes.[39] These models include a number of phenomenological parameters for which definition carrying out complex and expensive experiments is necessary. For this reason the research problem of formation and NW growth by methods of the mathematical modeling constructed on the direct analysis of atomic structure, is very actual.

4.3.1.1 *PROBLEM FORMULATION*

Let us consider the formation of Au–Si NWs on the surface of silicon substrate activated by gold. The process of formation of NW on the silicon substrate under the gold drop consists of three sequential stages:

- Apply a thin film of catalyst (gold (Au)) on the surface of silicon (Si).
- Heating of the Si surface to a temperature above the eutectic point at which formation of drops of the catalyst is possible.
- "Bombing" of drops of the catalyst by atoms of Si and formation of NW under the gold drops.

Let us consider mathematical models of each of these stages of NW formation process. In the first stage of NW formation atomic structure has the form shown in Figure 4.43. The boundary conditions used in the simulation are also presented in Figure 4.43.

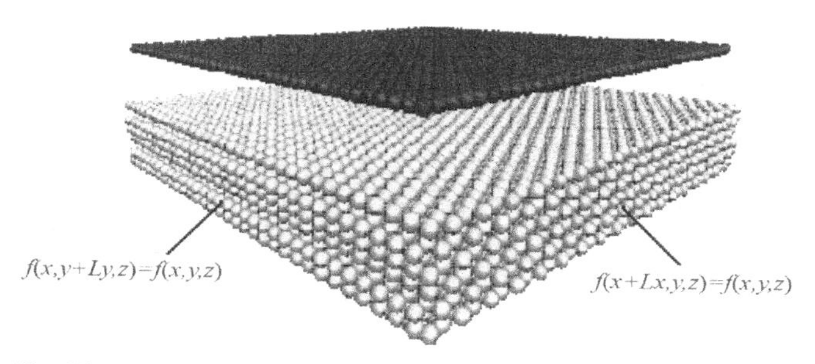

FIGURE 4.43 The model of the first stage of the process (light atoms are the substrate; dark atoms are the gold atoms).

The second stage is the heating of the substrate to a temperature above the eutectic point (~800 K), where formation of liquid solution droplets of the material and catalyst is possible, because heat is continuously supplied to the system. A drop of gold is formed on a silicon substrate as a result of the heating system (Fig. 4.44).

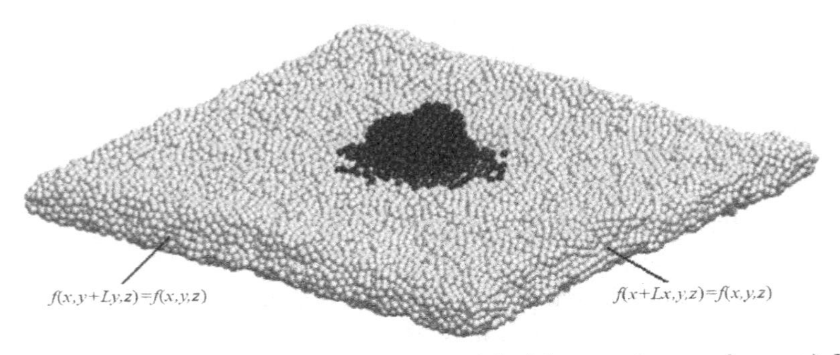

FIGURE 4.44 Drop of gold on the substrate (model of the second stage of process). Light atoms is substrate, dark atoms is Au.

The model of the last (third) stage describes the process of deposition of the silicon atoms on a substrate with gold at a fixed temperature (Fig. 4.45). At experiments at a stage of silicon atoms deposition there is a rebuilding the structure (Fig. 4.46), and NW with diameter $2R$ and height l is formed on a silicon substrate with the longitudinal size l and height h under a catalyst drop (Au).

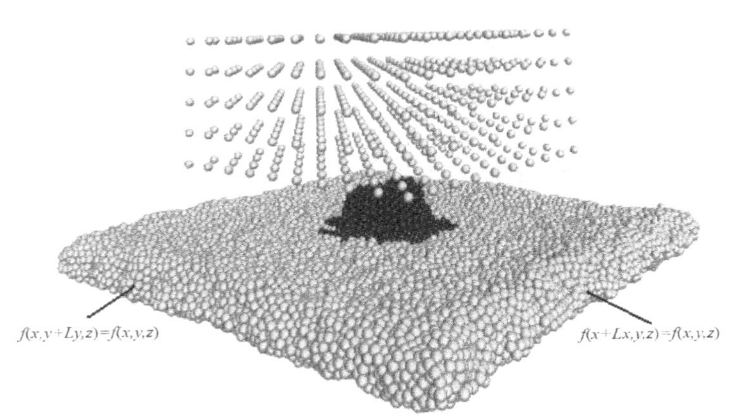

FIGURE 4.45 Silicon deposition (model of the third stage of process). Light atoms are substrate, dark atoms are Au.

The equations describing the considered atomic structure are a system of differential equations of MD (Chapter 1).

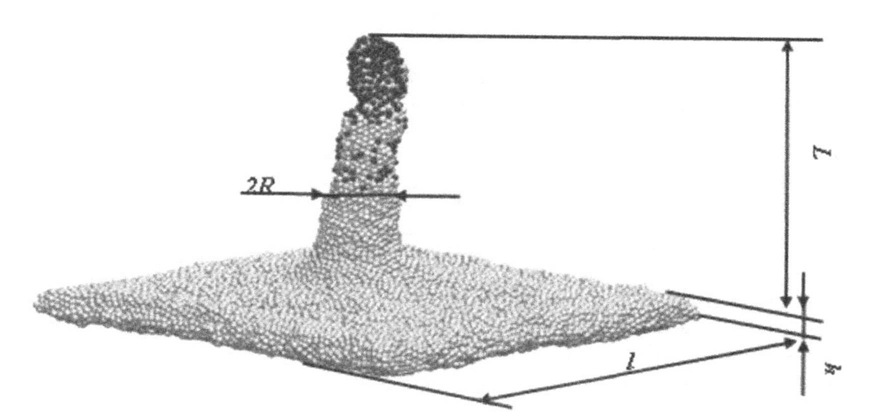

FIGURE 4.46 The image of the simulated system.

Particles of a besieged material get to a drop in two ways: directly from the vapor or molecular beam; by diffusion from the lateral surface of NW.

Diffusive flux from the lateral surface, in turn, consists of:

- particles arriving at the lateral surface of the NW due to diffusive motion along the surface of the substrate,
- atoms penetrate directly to the side walls of the NW.

Having got to the catalytic agent drop, besieged particles at first are dissolved in it, and then crystallize on a surface under a drop; there is a growth on the VLS mechanism.

Let us consider explicitly model of the third stage of NWs formation process. This stage also will be described by the equation of motion of atoms with initial conditions, and the use of a thermostat, but total force acting on each individual atom of the system will be other. In physical models of NWs growth process, as a rule, chemical potentials of the substances included into considered system[37–39,153] are entered. However in a method of molecular dynamics it is impossible to consider chemical interactions directly. It can be done only within quantum-chemical modeling. As mentioned above, currently it is not obviously possible to calculate large systems by means of this method. Therefore, to consider influence of factors of chemical interaction the additional force operating on each atom of system ($\vec{\Phi}_i$), except for atoms of a substrate where $\vec{\Phi}_i = 0$ is entered (Fig. 4.47).

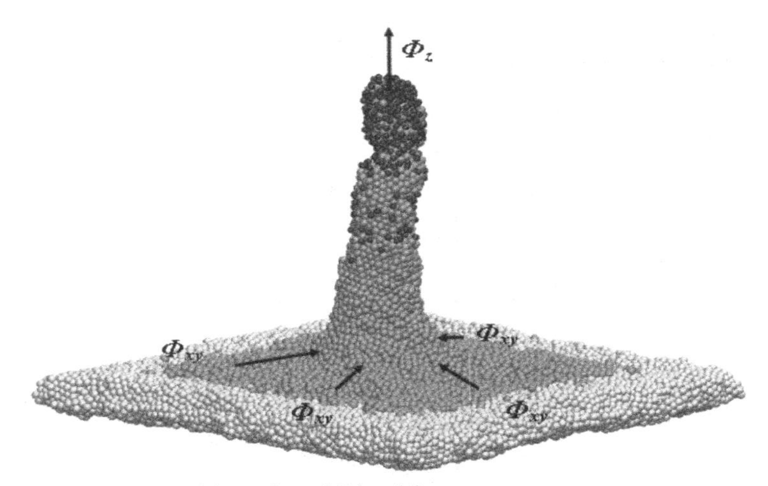

FIGURE 4.47 Diagram illustrating additional forces.

Then the resultant external force acting on the considered atom is calculated as a sum of the derivative of potential function describing the physical interaction of atoms and additional force:

$$\vec{F}_i\left(t,\vec{r}\left(t\right)\right)=-\frac{\partial U\left(\vec{r}\left(t\right)\right)}{\partial \vec{r}_i\left(t\right)}+\vec{\Phi}_i, \tag{4.50}$$

where $U\left(\vec{r}\right)$ is potential function, $\vec{r}_i$ is coordinate vector of atom i, $\vec{\Phi}_i$ is additional force.

The interaction between atoms is a potential,[71] and therefore the first term in eq 4.50 is the gradient of the potential energy of the system. The potential of the system in this case can be written as:

$$U\left(\vec{r}\left(t\right)\right)=U_{vdW}, \tag{4.51}$$

where U_{vdW} is potential of van der Waals interaction.
On the basis of eqs 4.50 and 4.51:

$$F_i=\sum_{i\neq j}\frac{12\varepsilon_{ij}}{r_{\min}}\left(\left(\frac{r_{\min}}{r_{ij}}\right)^{13}-\left(\frac{r_{\min}}{r_{ij}}\right)^{7}\right)+\Phi_i. \tag{4.52}$$

Since the substrate atoms do not participate in the formation of NWs for them $\vec{\Phi}_i$ is zero. Physical characteristics of the substances used in modeling the growth process are presented in Table 4.3.

TABLE 4.3 Physical Characteristics of the Considered Substances.

Characteristics of materials	Si	Au
Lattice type	Diamond 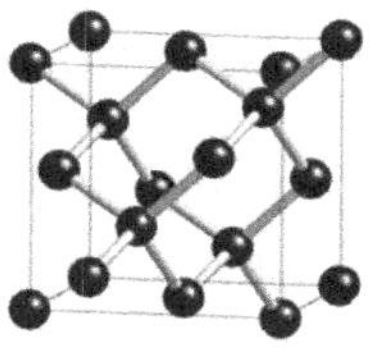	Cubic face-centered 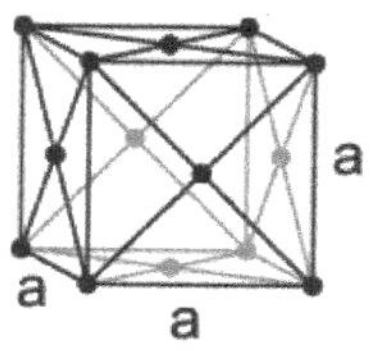
Lattice constant, 10^{-10} m	5.430	4.080
Standard atomic weight, amu	28.0855	196.9665
Melting point, K	1688	1337.58
Boiling point, K	2623	3080

4.3.1.2 CONNECTION OF ENTERED FORCE WITH PHENOMENOLOGICAL MODELS

As mentioned above, in order to reflect the actual physical and chemical processes taking place during the growth of NWs, as well as to increase their growth rate, an additional force acting on each individual atom of the system is entered. For an explanation of the nature of this force it is necessary to apply the main phenomenological models.

At interaction of two phases according to the second law of thermodynamics their condition changes toward equilibrium, which is characterized by the equality of temperatures and pressures of the phases as well as the equality of chemical potentials of each component in the coexisting phases. Motive force of the transfer of a component from one phase to another is the difference between the chemical potentials of the component in the interacting phases. The chemical potential characterizes the ability of the component to the output of this phase (by evaporation, dissolution, crystallization, chemical reaction, etc.). The chemical potential is included into the first law of thermodynamics and represents the energy of adding one particle to the system without performing external work. Transition of a component occurs in the direction of decrease of its chemical potential. Crystallization is used to separate the crystallizing solid phase from the solution by creating the conditions of super saturation required for the component. The achievement of system equilibrium is accompanied by movement of the phase boundary. But this process does not fit within the time frame available for the molecular dynamics simulation at this stage of development of computer technology, therefore the boundary moves through the application of force in direction z (Φ^z) during the simulation. The nature of this force can also be described as the intermolecular interaction.

Diffusion is the process at the molecular level and is determined by the random character of the motion of individual molecules. The rate of diffusion is proportional to the average velocity of the molecules. It is not obviously possible to track completely diffusion process in this temporary framework (about several nanoseconds). Therefore, to increase the diffusion rate in accordance with the model of NWs diffusion-growth and Fick's first law (flux density of matter is proportional to the diffusion coefficient and concentration gradient), an additional diffusion flux to the base of the whisker is entered, that is, force Φ^{xy}, whose direction is perpendicular to z axis, is applied to the deposited silicon atoms. This force, respectively, has nature of intermolecular interaction.

In the model of Givargizov–Chernov30 growth rate of NWs can be written as:

$$\frac{dL}{dt} = K\Delta\mu_v^2 = K\left[\Delta\mu^0 - \frac{2\Omega_s\gamma_{sv}}{k_B TR}\right]^2 = K\left[\Delta\mu^0 - \frac{R_0}{R}\right]^2, \tag{4.53}$$

where K is kinetic coefficient of crystallization; γ_{sv} is the surface energy per unit area on the crystal-vapor boundary; Ω_s is atom volume in a crystal; $\Delta\mu_v$ is an effective difference of chemical potentials in a gaseous and solid phase; $\Delta\mu^0$ is the difference of chemical potentials in a vapor phase and a crystal (vapor supersaturation), R_0 is the Givargizov–Chernov characteristic size.

Integrating the equation of motion we obtain:

$$\frac{d\vec{r}_i}{dt} = \int \frac{\vec{F}_i\left(t,\vec{r}\left(t\right)\right)}{m_i}dt, i = 1,2,\ldots,N\cdot \tag{4.54}$$

As in Givargizov–Chernov model it is considered that growth of a NW is carried out by means of a direct hit of atoms of deposited material in a drop, in a motion equation it is possible to consider only forces component along an axis z. So, for the direction z:

$$\frac{dz_i}{dt} = \int \frac{F_i^z}{m_i}dt, i = 1,2,\ldots,N\cdot \tag{4.55}$$

In turn $\vec{F}_i\left(t,\vec{r}(t)\right)=-\dfrac{\partial U\left(\vec{r}(t)\right)}{\partial \vec{r}_i(t)}+\vec{\Phi}_i$. For the direction z we have: $F^z=-\dfrac{\partial U}{dz}+\Phi^z$.

As atoms of gold are at the top of NW and Φ^z is applied to them for a drop of gold it is possible to write a ratio:

$$\frac{dL}{dt}\sim\frac{dz_{Au}}{dt},$$

then we obtain:

$$K\left[\Delta\mu^0-\frac{2\Omega_s\gamma_{sv}}{k_BTR}\right]^2\approx\int\frac{F_i^z}{m_i}dt=\frac{1}{m_i}\left(\int\frac{\partial U}{\partial z}dt+\Phi_i^z t\right)\Rightarrow$$

$$\Phi_i^z\approx K\left[\Delta\mu^0-\frac{2\Omega_s\gamma_{sv}}{k_BTR}\right]^2\frac{m_i}{t}+\int\frac{\partial U_i}{\partial z}dt. \tag{4.56}$$

The main equation of elementary model of diffusion growth looks as:

$$\frac{\pi R^2}{\Omega_s}\frac{dL}{dt}=\left(\frac{V-V_s}{\Omega_s}-\frac{2Cr_l}{\tau_l}\right)\pi R^2+j_{diff}(L). \tag{4.57}$$

Here R is NW radius, L is NW height, $V=J\Omega_s\cos\alpha$ is the equivalent growth rate taking into account the angle of incidence of a beam α, V_s is the rate of vertical growth of non-activated surface, C is volume concentration of the solution in a drop, r_l is intermolecular distance in the liquid, τ_i is atom life-time in a drop, $j_{diff}(L)$ is atom volume in a solid phase, $j_{diff}(L)$ is a summary diffusion flux on the top of NW, J is flux determined by the temperature, and pressure of the vapor-gas mixture in accordance with the formula $J=C_g\dfrac{P}{\sqrt{2\pi mk_BT}}$. The first term on the right side takes into account the adsorption on the surface of the drop, the second considers the growth of non-activated surface, and the third heeds desorption from the drop and the fourth takes into account the diffusion of atoms from the lateral surface.

The flow to the bottom of NW is:

$$j_{diff}(0)=\frac{V}{\Omega_s}\frac{\varepsilon R}{\langle R\rangle N_w}, \tag{4.58}$$

where $\varepsilon = \frac{V - V_s}{V}$, V is the deposition rate, $\langle R \rangle$ the average radius of drops, N_w is their surface density.

As for the whiskers with a small diameter in the diffusion model the diffusion flux contributes most to the growth process, it is possible not to consider the adsorption and desorption phenomena, and also owing to absence in this time span of growth of not activated surface, for model of diffusion growth we obtain:

$$\int F_i^{xy} dt \approx \frac{m_i \Omega_s}{\pi R^2} j_{diff}(0) = \frac{m_i}{\pi R} \frac{(V - V_s)}{\langle R \rangle N_w} \approx \frac{m_i}{\pi R} \frac{V}{\langle R \rangle N_w} \quad (4.59)$$

or

$$\Phi_i^{xy} \approx \frac{m_i}{\pi R} \frac{V}{\langle R \rangle N_w t} + \int \frac{\partial U_i}{\partial r_{xy}} dt, r_{xy} = \sqrt{x^2 + y^2}. \quad (4.60)$$

4.3.1.3 TIME-FORCE ANALOGY FOR DETERMINATION OF VALUE RANGE OF THE FORCES APPLIED TO THE ATOMS OF THE SYSTEM

The parameter values given in Table 4.4 were used to estimate the input force.[37]

TABLE 4.4 The Main Parameters of the Nanowhiskers Formation Process.

$K, 10^{-9}\ m/s$	$V, 10^{-11}\ m/s$	$\Delta\mu^0$	$\Delta\mu_v$	T, K	$R, 10^{-9}\ m/s$	$N_w, 10^{13}\ m^{-2}$	t, s
1.07	4.2	0.7	0.199	773	35	1	7200

Then

$$\frac{R_0}{R} = 0.5 \Rightarrow \gamma_{sv} = \frac{k_B T R}{4\Omega_s} \Rightarrow \gamma_{sv} = 3\ J/m^2$$

$$\Phi_i^z = 0.63 * 10^{-38} (N) = 0.91 * 10^{-28} \left(kcal \middle/ \left(mol * \mathring{A} \right) \right),$$

$$\Phi_i^{xy} = 9.27 * 10^{-38} (N) = 13.33 * 10^{-28} \left(kcal \middle/ \left(mol * \mathring{A} \right) \right).$$

The received values are conditions of NW growth for the parameters of system provided in Table 4.4. As at present it is not obviously possible to count by means of methods of molecular dynamics process by duration about two hours (namely so much on the average grow up whiskers in practice), for reduce of calculations time and increase in speed of growth process the theory of similarity[173] was used. The introduction of time-force analogy makes it possible to simulate the initial stage of NW growth and evaluate the process kinetics in a relatively short period of time. The forces values calculated above are suitable for modeling NWs growth process; however, in case of such parameters time of growth will be about two hours that in case of a step in one fs will give $7.2*10^{18}$ iterations that is rather problematic for calculations. However, using the proposed time-force analogy, it becomes possible to investigate the initial stage of NW growth with less number of iterations.

The first theorem of similarity is as follows: similar phenomena in one sense or another (complete, approximately, physical, mathematical, etc.) have certain combinations of parameters, called the similarity criteria, which is numerically equal for similar phenomena.

In the second theorem of similarity it stated that: every complete equation of a physical process, written in a certain system of units may be represented by a functional relation between the criteria of similarity derived from parameters participating in process.

Based on the first and second theorems of similarity, proceeding from the main equations of process and phenomenological models, the following criteria of similarity were received:

$$\pi_1 = \frac{FRN}{E}, \pi_2 = \frac{FLN}{E}, \pi_3 = \frac{\Omega_s mV}{F^{xy}\pi R^3 t}, \pi_4 = \frac{mV_L}{Ft}, \tag{4.61}$$

where F is average force acting on each atom of the system; R is radius of NW; N is number of atoms in NW; E is energy of the system; L is NW height; m is atomic weight; F^{xy} is the force acting on the deposited silicon atoms; Ω_s is volume of atom; V is the rate of deposition of the silicon atoms; V_L is the rate of NW growth; t is growth time.

Dependences of growth rate on criteria of similarity are provided in Figure 4.48.

Graphs shows the calculated data for different values of the forces applied to the atoms of the system and the value of the criterion of similarity calculated from the experimental values of growth parameters.

Apparently from the diagrams, the received values are well approximated by linear functions. Value of reliability of approximation is from 0.9153 (for dependence of speed on π_3) to 0.9946 (for dependence of speed on π_4). The linear form of approximating curve shows the correctness of accepted assumptions.

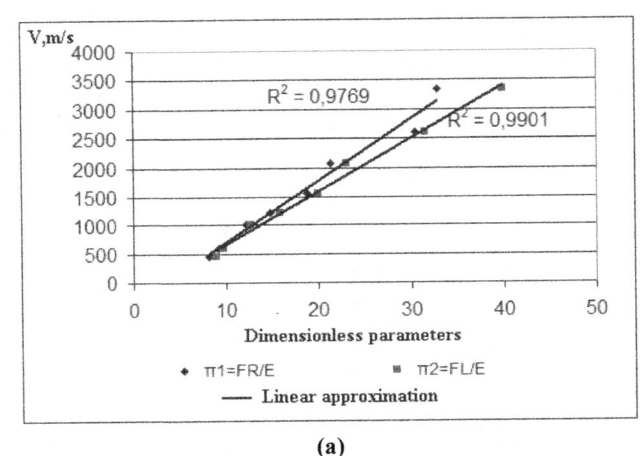

(a)

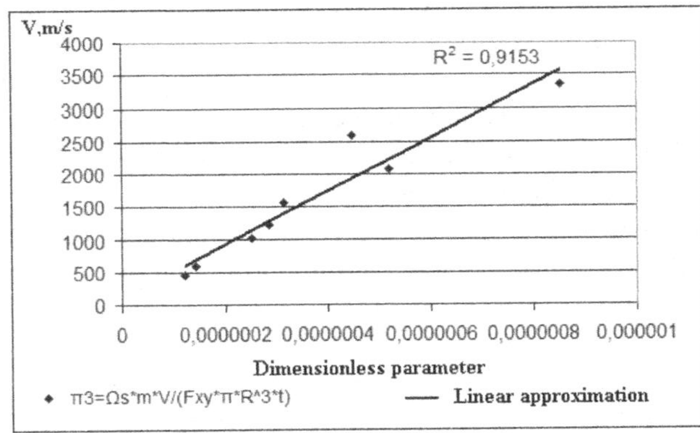

(b)

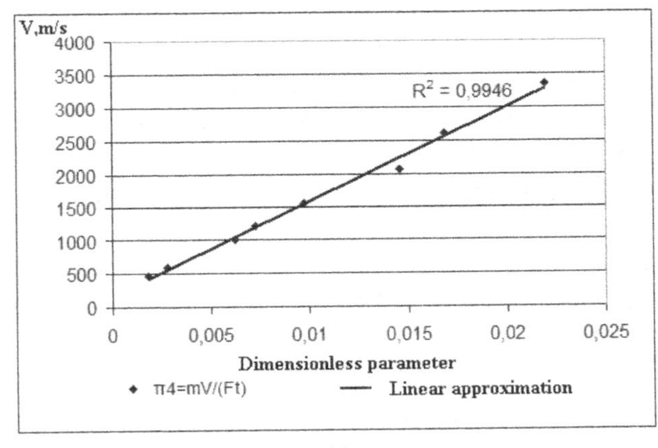

(c)

FIGURE 4.48 Diagrams of dependence of growth rate on criteria of similarity.

The theory of similarity is also used to determine the order of the force applied to the atoms of the system. So the force operating on deposited atoms of silicon, shall be from fraction to several units of kcal/(mol*Å). For example, proceeding from Φ^z and Φ^{xy} values calculated above, and having selected arbitrary two simulated process, it is possible to receive value range of applied force. The value range of forces Φ^z and Φ^{xy} is presented in Table 4.5. It is calculated using a similarity criterion π_4.

The theory of similarity is also used to determine the order of the force applied to the atoms of the system. So the force operating on deposited atoms of silicon, shall be from fraction to several units of kcal/(mol*Å). For example, proceeding from Φ^z and Φ^{xy} values calculated above, and having selected arbitrary two simulated process, it is possible to receive value range of applied force. The value range of forces Φ^z and Φ^{xy} is presented in Table 4.5. It is calculated using a similarity criterion π_4.

TABLE 4.5 Range of Forces Applied to the Atoms of the System.

Φ^{xy} **kcal/(mol*Å)**		Φ^z **kcal/(mol*Å)**	
From	To	From	To
5.73	90.51	0.36	5.68

Proceeding from diagrams of dependence of NW growth rate from similarity criteria given above, it is possible to define dependence of growth rate on system parameters.

Thus: for π_1: $V_L \approx 100.42\, FR/E - 216.85$; for π_2: $V_L \approx 87.963\, FL/\mathrm{E} - 151.75$, for π_3: $V_L \approx 4{*}10^9 * \Omega_S *m* V/(F^{xy}\pi * R^3 * t) + 76.215$, for π_4: $V_L \approx 15.52\, Ft/(Ft - 151184m)$.

4.3.1.4 SOFTWARE PACKAGE

The software package, which diagram is provided in Figure 4.49, was designed to solve the problem of numerical simulation of Si-Au NW growth. Problem-oriented software package is a multistage structure and consists of units and modules. The solution of the problem is consistent from unit to the unit (Fig. 4.49).

In the unit of preparation input files for the computing module are created and initial and boundary conditions are defined. Computing module includes units: a model describing formation of silicon substrate, a model describing formation process of gold catalyst drops on silicon substrate, a model describing process of "bombing" gold drops by silicon atoms and a model describing NW growth.

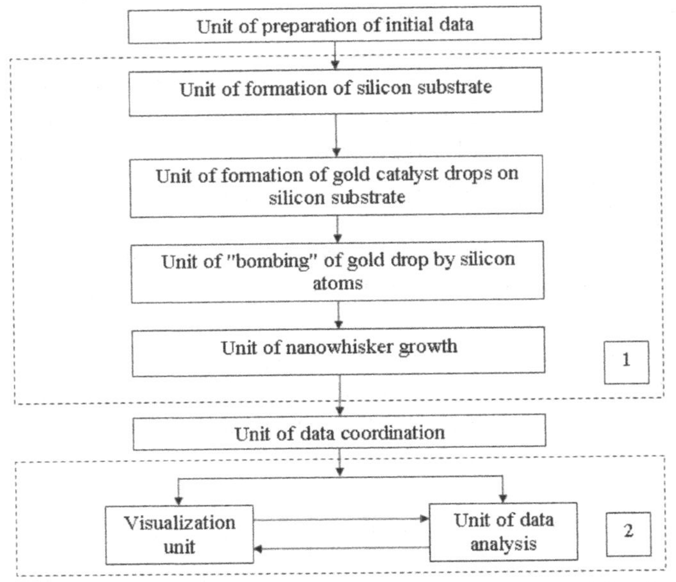

FIGURE 4.49 Structure of software package, 1—calculation module, 2—module of analysis and visualization.

This module is intended for simulation of nanosystem and formation of results. In the module of analysis and visualization analysis of received data is carried out (computation of physical characteristics, creation of dependences, and also visualization of the received results).

4.3.1.5 THE ANALYSIS OF THE CALCULATION RESULTS

Let us consider the formation of Au–Si NWs on the surface of silicon substrate activated by gold. The gold drop appears as growth catalyst. Diameter of the drop was 2 nm. Silicon substrate has following dimensions: length is 6.6 nm, width is 6.6 nm, and height is 2.5 nm. The substrate is not rigidly fixed; it says that the substrate atoms can move freely in any direction.

Convergence of the numerical solution of problem is often depends on a choice of appropriate integration step. Step must be sufficiently small to correctly display the behavior of the system. When using methods of molecular dynamics value of mass of simulated substances influences the value of integration step. The step is selected from limits from 0.5 to 2 fs. In our case, the time integration step was 1 fs. Total time of calculation process was about 2 ns.

The process of NWs forming can be divided into several stages. At the first stage a thin layer of active catalyst (e.g., Au) is applied to the surface of the silicon substrate. Then the whole system is heated to a temperature above the eutectic point at which formation of catalytic agent drops is possible. Further "bombing" of catalytic agent drops by silicon atoms (Si) is carried out. Silicon atoms are dissolved in drops. Owing to Au–Si solution super saturation atoms of silicon under a drop start to crystallize, therefore under the gold drops there is NW formation, that is, growth occurs on the "vapor-liquid-solid" mechanism.

Simulation of the first two stages of NW formation at temperature of 800 K showed that drops of gold are formed on the substrate surface (Fig. 4.50). This is consistent with experimental data obtained by Ref. 39,153 However, there is no formation of the alloy Si-Au, because there is no diffusion of atoms of silicon and gold under the drop (see the cut of system in Fig. 4.51). Gold atoms do not diffuse into the substrate because the binding energy of silicon atoms in crystal lattice of substrate exceeds the depth of potential well of van der Waals interaction energy for gold atoms. Thus from the diagram of energy provided in Figure 4.52, it is visible that the system came to a steady state, that is, there is a minimization of energy of system.

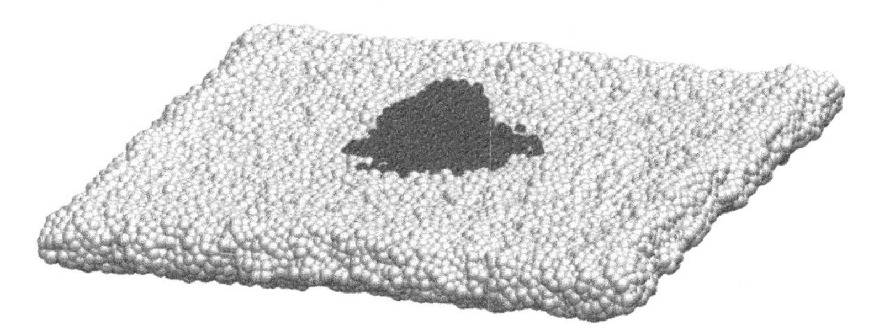

FIGURE 4.50 Drop of catalyst on the substrate surface.

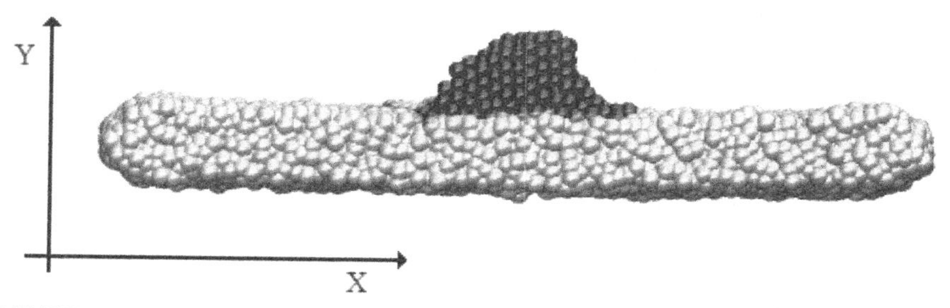

FIGURE 4.51 Drop of catalyst on the substrate surface, cut along the x-axis.

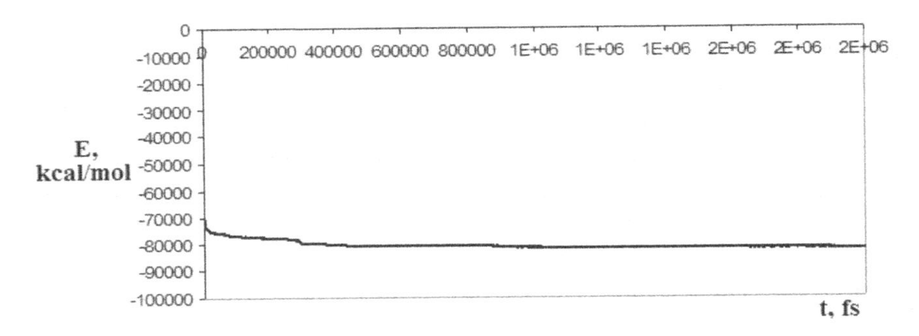

FIGURE 4.52 Total energy of system.

In the model of the third stage of NWs formation process "bombing" a drop of gold by silicon atoms is described. In contrast to the previous stage diffusion of silicon atoms in the drop of gold (Fig. 4.53) is observed. A solution of Au–Si, which is the basis for NWs growth, is formed under the drop due to diffusion. Figure 4.53 and graph in Figure 4.54 show that part of deposited silicon atoms penetrate into the drop, and some remains at the lateral surface. Figure 4.55 allows us to estimate the depth of their penetration. In Figure 4.55, the upper value corresponds to the top of a drop of gold, and the lower corresponds to substrate level.

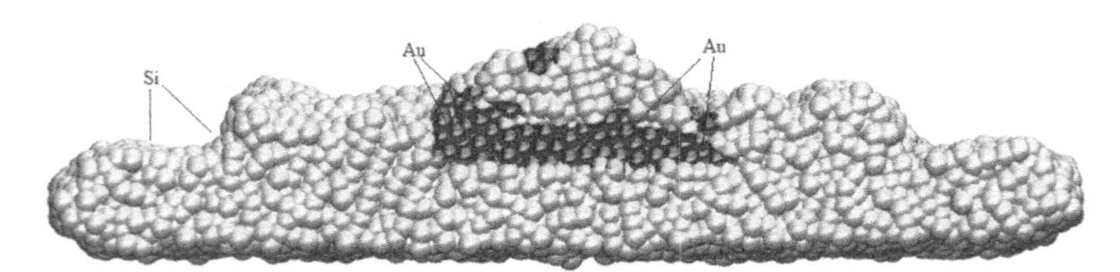

FIGURE 4.53 Diffusion of silicon atoms into the gold drop.

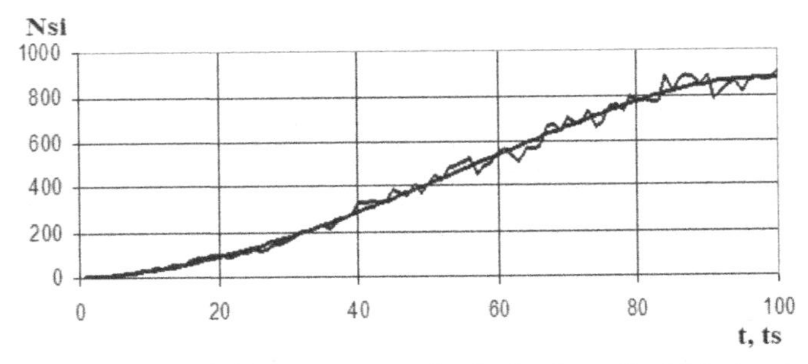

FIGURE 4.54 Changing of silicon atoms number in the drop of gold during deposition.

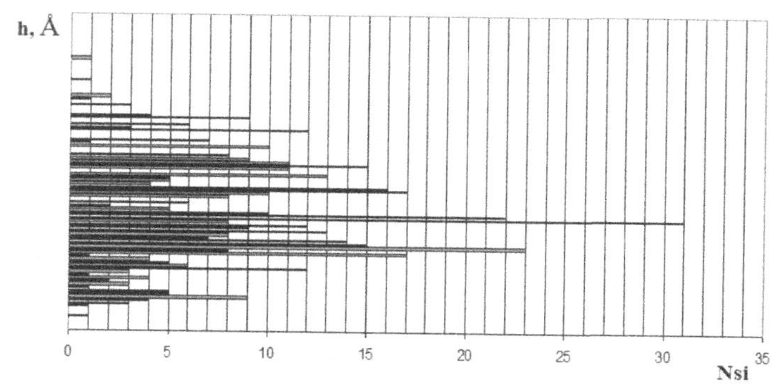

FIGURE 4.55 Penetration of silicon atoms into the gold drop.

The diameter of droplets increases due to diffusion. A similar effect is observed in experiments. In Figure 4.56 it is shown, how the drop radius changes during "bombing." It should also be noted that the diffusion of silicon atoms from the surface of the substrate is not observed, and further "bombing" does not lead to the NWs growth.

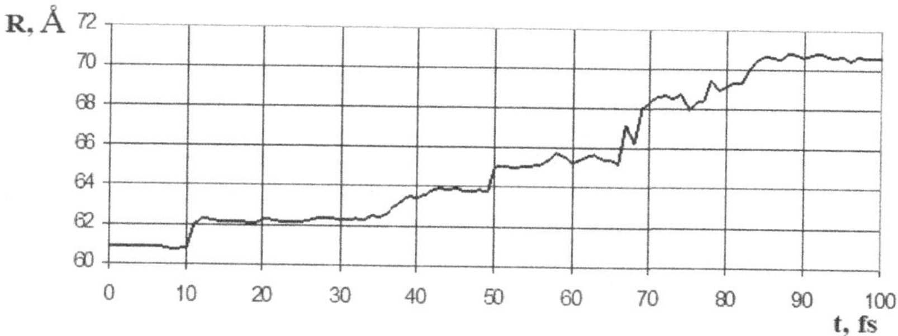

FIGURE 4.56 Changing of the catalyst drop radius during "bombing."

Let us enter force vectors lying in the x-y plane directed toward the central axis of the system, z-axis passes through the center of substrate and droplet (Fig. 4.57). As a result, we obtain the system shown in Figure 4.58. Apparently from a figure, owing to application of forces a slight growth of a NW is observed.

However, it is possible to note that the most part of gold atoms remains on a substrate, and further growth of a NW under such conditions is difficult (Fig. 4.59). Therefore, forces also are applied to the atoms of gold. In this case their direction will match the axis z direction that allows us to increase growth rate of NW and to leave a gold drop on the top of NW.

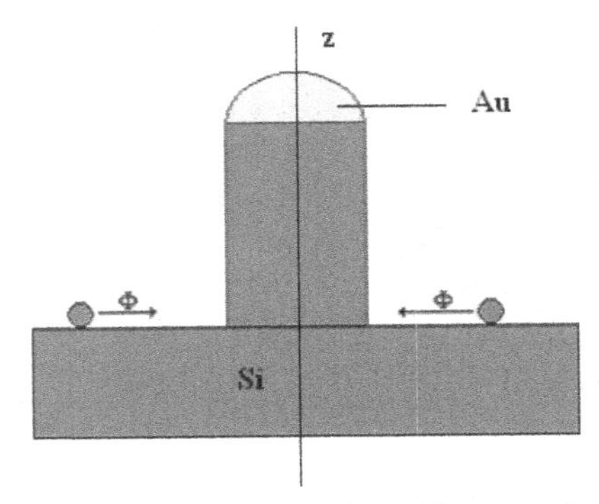

FIGURE 4.57 Diagram illustrating the application of forces to the system in the x-y plane.

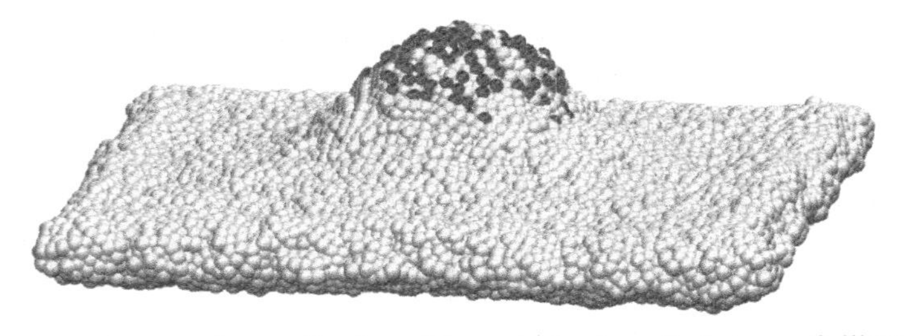

FIGURE 4.58 System after application of forces to the deposited atoms of silicon.

However, it is rather difficult to define exact value of the module of force necessary for growth of NW, just as it is difficult to define kinetic coefficient of crystallization and $\Delta\mu^0$ defining chemical potential in Givargizov–Chernov model.[39] Therefore in case of growth process simulation Φ values are varied. Calculations showed that the nano-whisker growth is observed, for example, at Φ_{xyz} =10 kcal/(mol*Å). In Figure 4.65, the result of use of such value of force is provided.

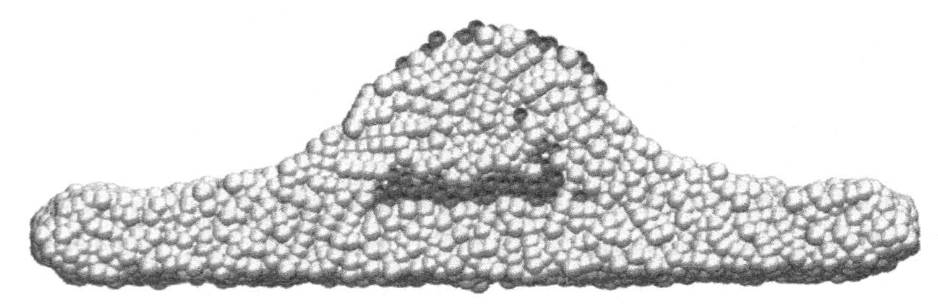

FIGURE 4.59 System after application of forces to the deposited atoms of silicon, a cut along x-axis.

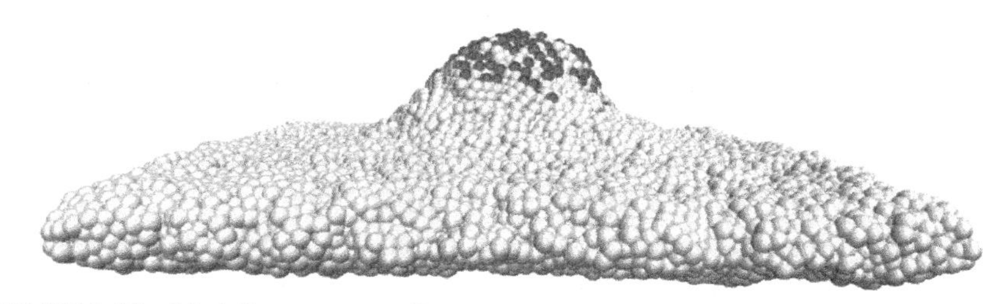

FIGURE 4.60 Modeling system at Φ_{xyz} =10 kcal/(mol*Å).

The height of the NW in this case varies (Fig. 4.61). The graph shows that system reaches a steady state, when the height of NW does not significantly change. Figure 4.67 illustrates dynamics of change of total energy of system. Energy of system also quitted on a certain steady-state value that in turn testifies to stability of the received system (Fig. 4.62).

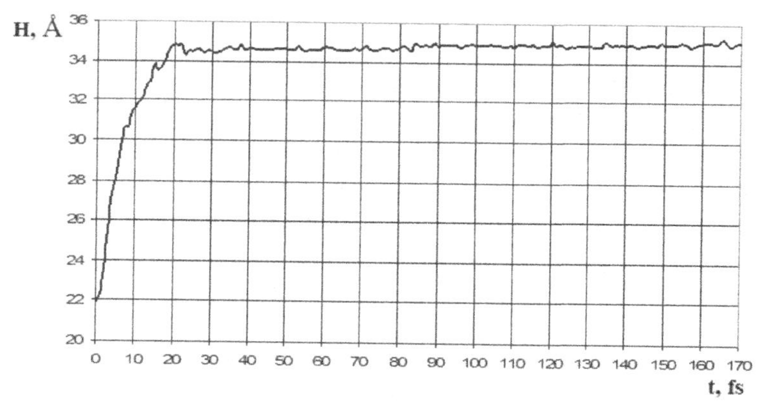

FIGURE 4.61 Change of nanowhisker height in case of Φ_{xyz} =10 kcal/(mol*Å).

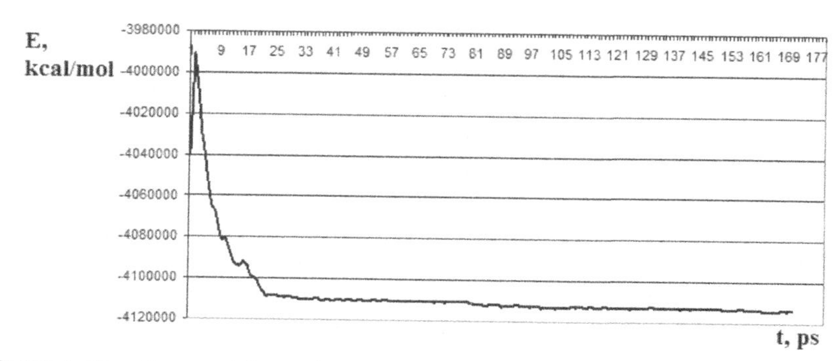

FIGURE 4.62 Change of the total energy of system.

As a result of hit of silicon atoms in a drop, Au–Si solution becomes supersaturated. In turn, the growth of NWs occurs due to crystallization of supersaturated solution under the drop. As a result whisker with lateral size, approximately equal to diameter of a drop, grows under the drop and the drop moves up with a speed equal to NW growth rate.

From the foregoing it can be concluded that the growth of NW is carried out by diffusion of silicon atoms in the "bombing" stage and flux of atoms from the lateral surface of a NW. It is possible to advance the assumption that as the system (Fig. 4.60) came to steady state, further growth of NW is possible in case of "bombing" continuation.

Let us consider the dependence of NWs morphology on the value of force applied to the atoms of system. Undoubtedly, the properties, size, and structure of NWs depend on growth conditions and parameters of simulated system. It should be noted that the steady growth of the NWs is observed only in a certain range of Φ force values[170,171]. For example, at Φ_{xy} =5 kcal/(mol*Å) and Φ_z =10 kcal/(mol*Å) NW growth is not observed (Fig. 4.63). Apparently from a figure part of silicon atoms remains on a lateral surface, part penetrates into a drop, but the diffusion flow of silicon atoms from a lateral surface of NW is absent. It, in turn, suggests that the diffusion flow from a lateral surface contributes significantly to the NW formation process. Thus the contribution of a diffusion flow will be proportional to perimeter of crystal lateral surface.[38] It should also be noted that in case of small diameter of a drop the appropriate diffusion flow leads to the significant increase in NWs height.

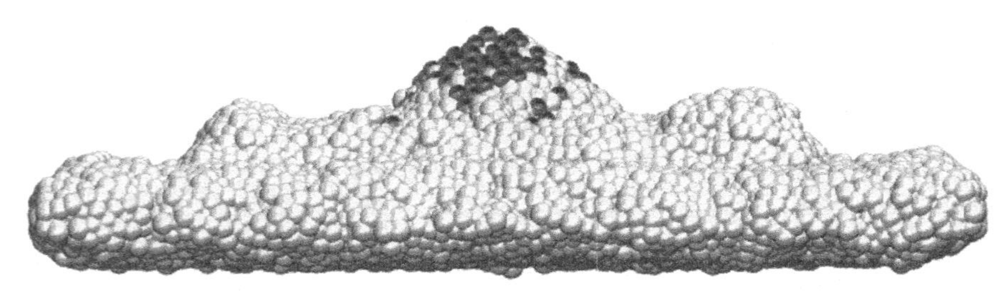

FIGURE 4.63 Simulated system at Φ_{xy} =5 kcal/(mol*Å) and Φ_z =10 kcal/(mol*Å).

In case of small diameter of a drop and small values of diffusion flow of silicon atoms from a lateral surface significant growth of NW is not observed. Further process patterns are provided in case of different values Φ_z, where Φ_{xy} =0 (Figs. 4.64–4.66).

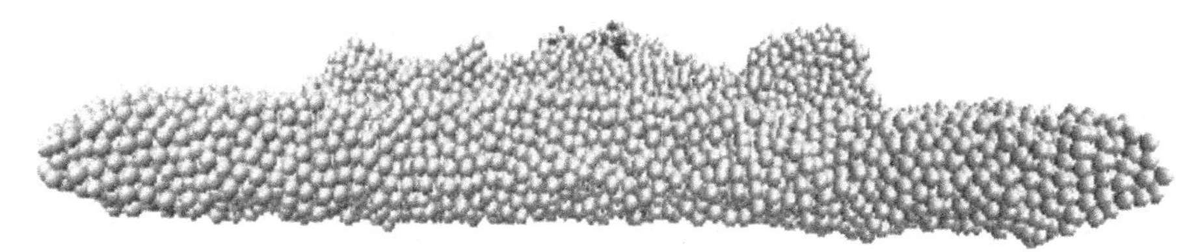

FIGURE 4.64 Simulated system in case of Φ_z=12 kcal/(mol*Å).

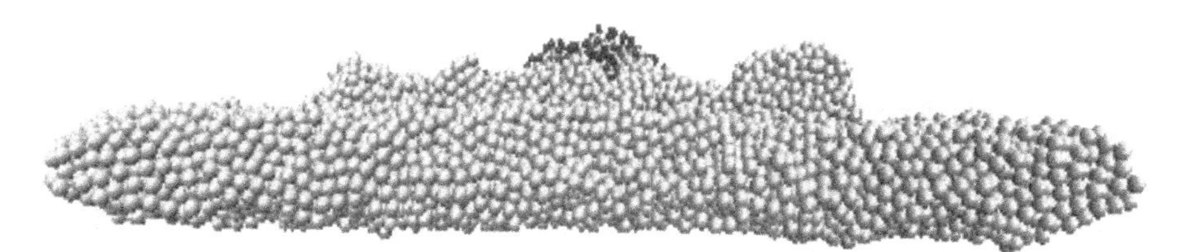

FIGURE 4.65 Simulated system in case of Φ_z=15 kcal/(mol*Å).

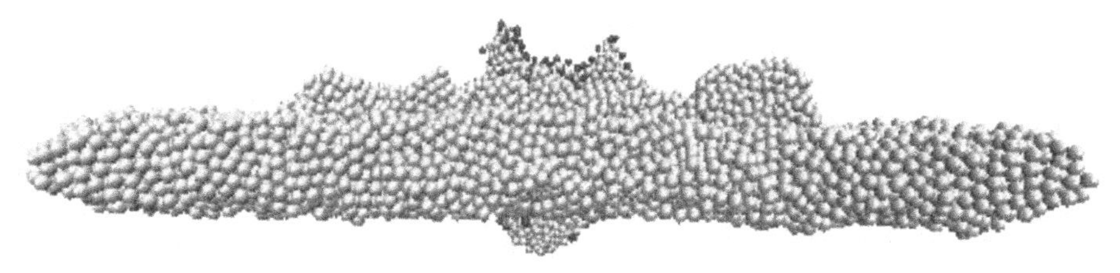

FIGURE 4.66 Simulated system in case of Φ_z =25 kcal/(mol*Å).

After analyzing the figures, it can be said that NW growth in the absence of lateral diffusion is not observed. It once again confirms the experimental data and theoretical calculations that in case of small sizes of catalyst drop the diffusion flow from lateral surface makes the significant contribution to process of NW formation and growth. Concerning a Figure 4.66 it can be said that the application of forces to the gold atoms exceeding depth of a potential hole of the van der Waals interaction, leads to drop corrupting and not physically reasonable results. Figure 4.67 confirms the aforesaid. In the case of Φ_z =12 kcal/(mol*Å) and Φ_z =15 kcal/(mol*Å) system is stable and the graph has a flat look, the energy is minimized and set at a certain level. The dashed line in Figure 4.67 describes the dynamics of change of total energy of system during the simulation at Φ_z =25 kcal/(mol*Å). Apparently from the diagram, energy in case of such value of force enclosed to the gold atoms, changes spasmodically, proceeding from it, it is possible to say that solution the problem of NW formation in case of such value of Φ_z disperses.

Thus, it can be concluded that the potential well depth of van der Waals interaction is the upper limit for selecting values of forces applied to the atoms of catalyst.

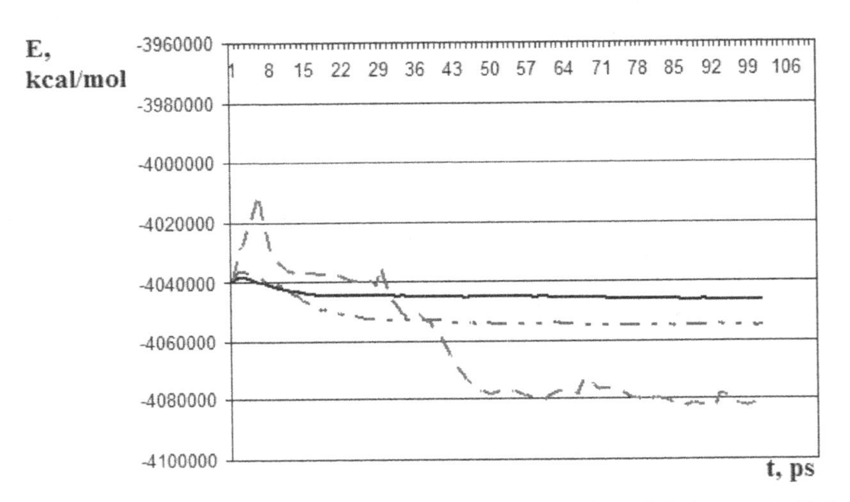

FIGURE 4.67 Change of system total energy for different values of Φ_z (▬ Φ_z=12 *kcal/mol**Å). ▬ ▪ ▪ Φ_z=15 *kcal/mol**Å) ▬ ▬ Φ_z=25 *kcal/mol**Å)

Let us consider other cases of forces application to the atoms of system. So in case of Φ_{xy} =10 kcal/(mol*Å) and Φ_z =10 kcal/(mol*Å) picture presented in Figure 4.68 is observed: after short growth NW starts to fall over on the side.

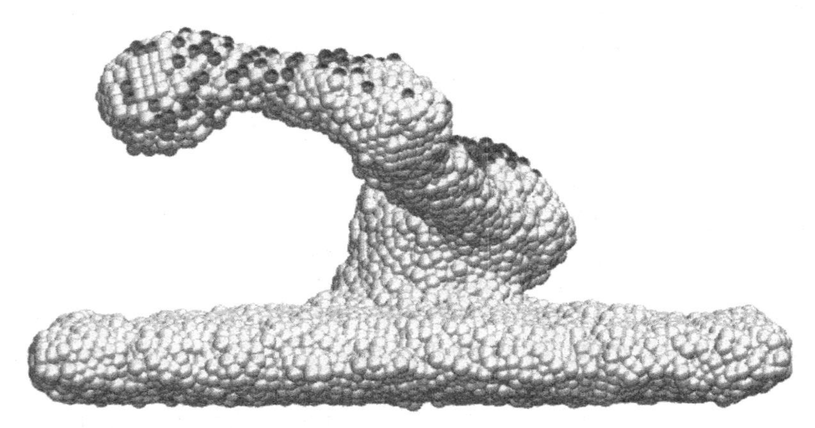

FIGURE 4.68 Nanowhisker falling to the side.

Figure 4.69 presents the state of the simulated system in case of Φ_{xy} =25 kcal/(mol*Å). In this case, after short growth whisker shatters into pieces. It is possible to advance the assumption that nanocrystal destruction during the simulation occurred owing to two reasons. First, the system does not have the force along z-axis and applied to the atoms of gold. Second, value of force enclosed to atoms of deposited silicon was greater than depth of potential well of van der Waals interaction for gold atoms. It in turn led to the destruction of a drop of Si-Au molten.

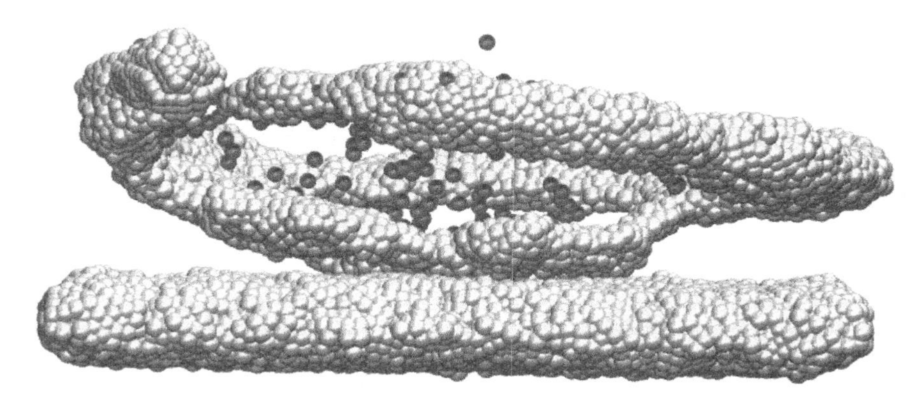

FIGURE 4.69 Nanowhisker destruction.

All this confirms that the steady growth of the NWs occurs only within a certain range of values of applied forces. Let us consider, as value of force applied to the atoms of gold affects process of NWs formation. Let us analyze two relaxed systems. In this case, the relaxation is understood as switch-off of action of forces Φ_{xy} and Φ_z on atoms of simulated system after a certain temporal interval. This time interval is defined as follows: after the bombing at the certain time step change in the height of NW becomes insignificant and from this time relaxation is launched. As a result of relaxation the system becomes steady; height and diameter of nanocrystal are stabilized. In Figure 4.70, appearance of system after relaxation is provided.

For this system the diagram of dependence of NW height from diameter is provided in Figure 4.71. For a system where Φ_z =15 kcal/(mol*Å) this dependence has the form shown in Figure 4.72. Apparently nature of these dependences is not always identical.

As a result of carrying out a relaxation of system the height of a NW is stabilized, it is confirmed by Figures 4.73 and 4.74, but value of initially applied forces Φ_z considerably influence system behavior and in case of relaxation. The stabilization of NW diameter can be seen from the plots provided by Figures 4.75 and 4.76. However, unlike nature of height change profiles of curves of NW diameter change on time in both cases are similar.

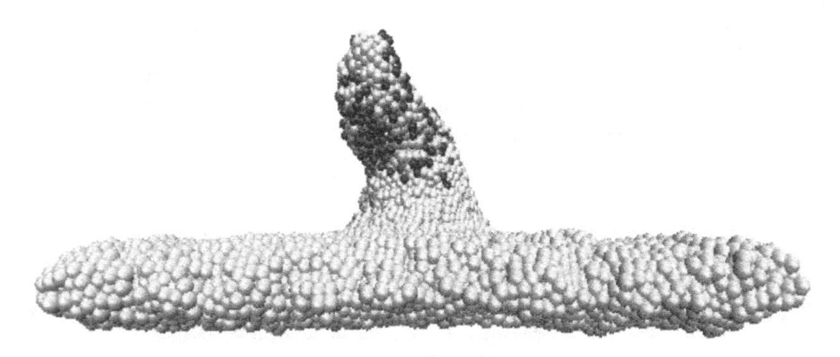

FIGURE 4.70 Appearance of system after relaxation.

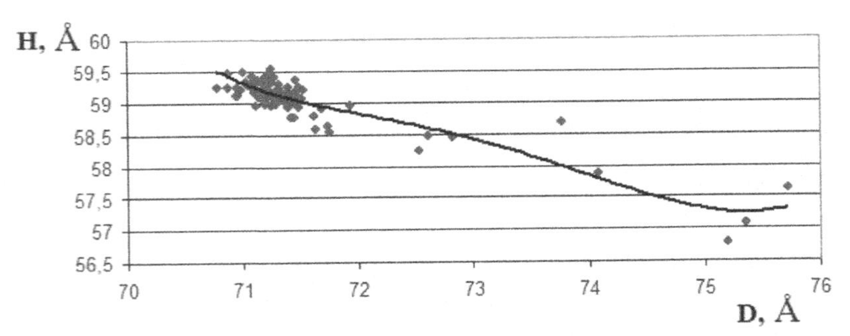

FIGURE 4.71 Dependence of height of a nanowhisker on diameter in case of Φ_z =10 kcal/(mol*Å).

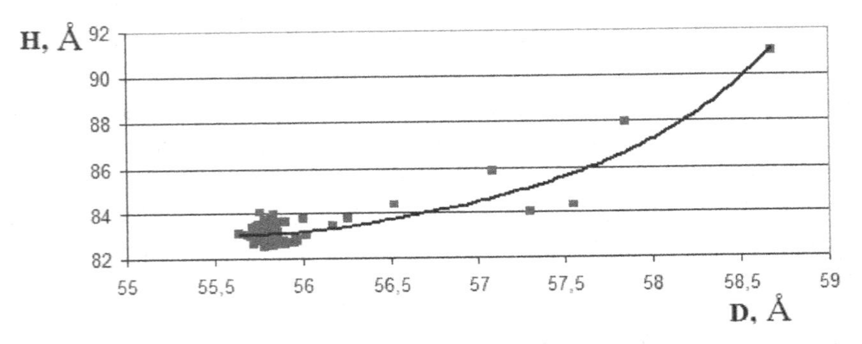

FIGURE 4.72 Dependence of height of a nanowhisker on diameter in case of Φ_z =15 kcal/(mol*Å).

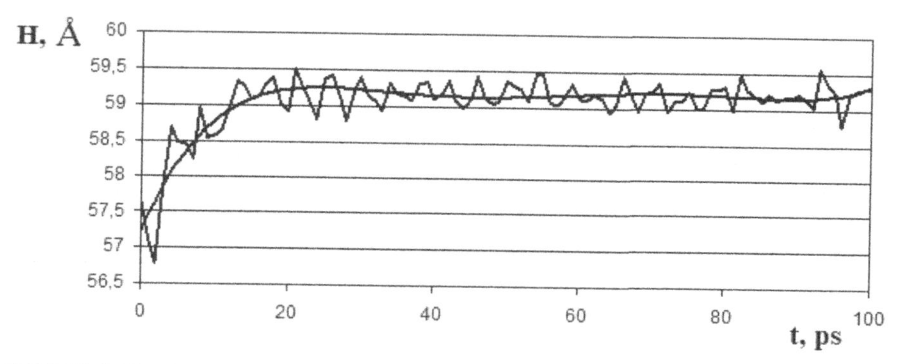

FIGURE 4.73 Change of nanowhisker height on time in case of Φ_z =10 kcal/(mol*Å).

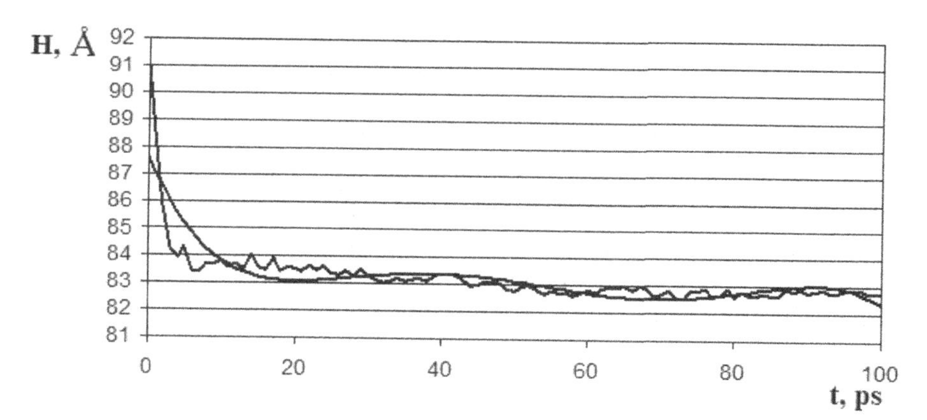

FIGURE 4.74 Change of nanowhisker height on time in case of Φ_z =15 kcal/(mol*Å).

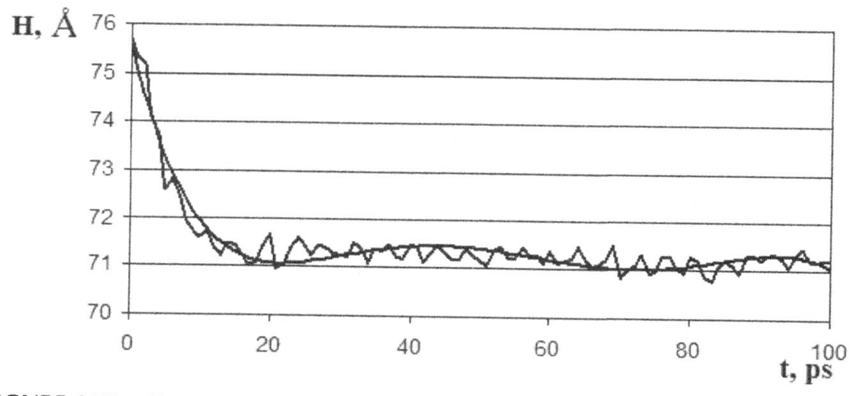

FIGURE 4.75 Change of nanowhisker diameter on time in case of Φ_z=10 kcal/(mol*Å).

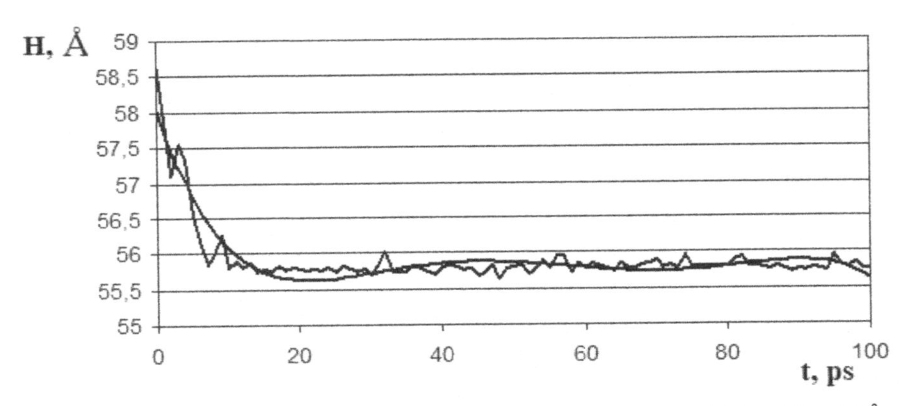

FIGURE 4.76 Change of nanowhisker diameter on time in case of Φ_z=15 kcal/(mol*Å).

NWs formed under appropriate thermodynamic conditions are stable. So, if after carrying out "bombing" by atoms of silicon to remove from enclosed additional forces and to carry out a system relaxation, it is possible to see that the system passes to stable conditions. "Switch-off" of in addition entered forces is caused by reduction of flow of atoms in a drop and respectively reduction of a super saturation of Au–Si solution and as a result growth stop in view of "bombing" extinction. In Figures 4.77 and 4.78 the system status before and after such relaxation is provided.

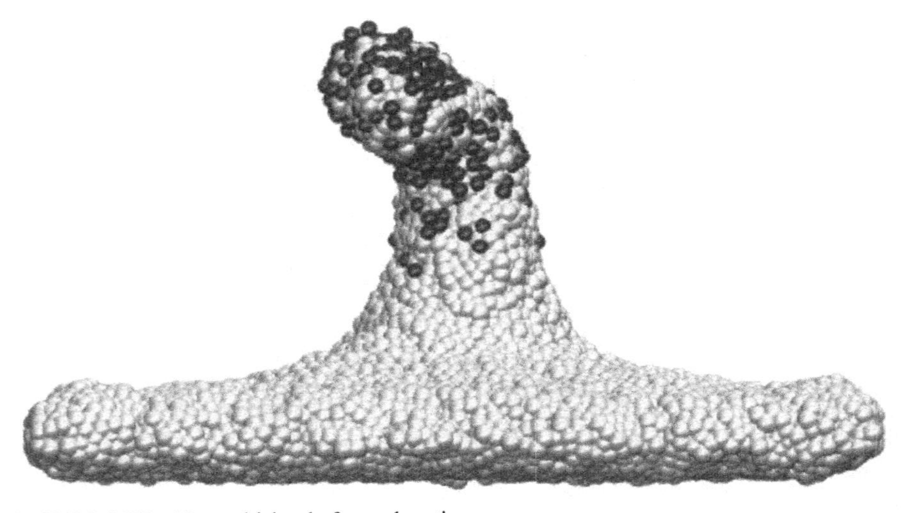

FIGURE 4.77 Nanowhisker before relaxation.

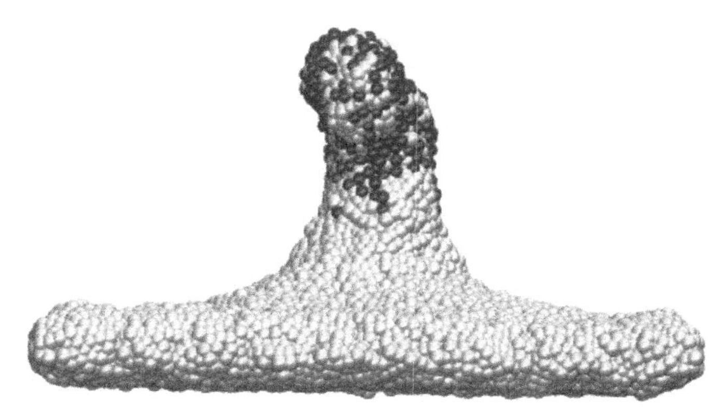

FIGURE 4.78 Nanowhisker after relaxation.

After the system relaxation NW height was 7 nm in case of initial radius of catalyst drop equal to 2 nm. As a result of relaxation height and radius of NW slightly change (Figs. 4.79 and 4.80).

Apparently from figures the smallest changes of height and diameter of NW during a relaxation occur in case of Φ_{xy} = 20 kcal/(mol*Å), Φ_z = 20 kcal/(mol*Å). The largest fluctuations in the values of NW parameters (diameter and height) are observed at Φ_{xy} = 8 kcal/(mol*Å), Φ_z = 20 kcal/(mol*Å), that can testify to poor stability of system before relaxation.

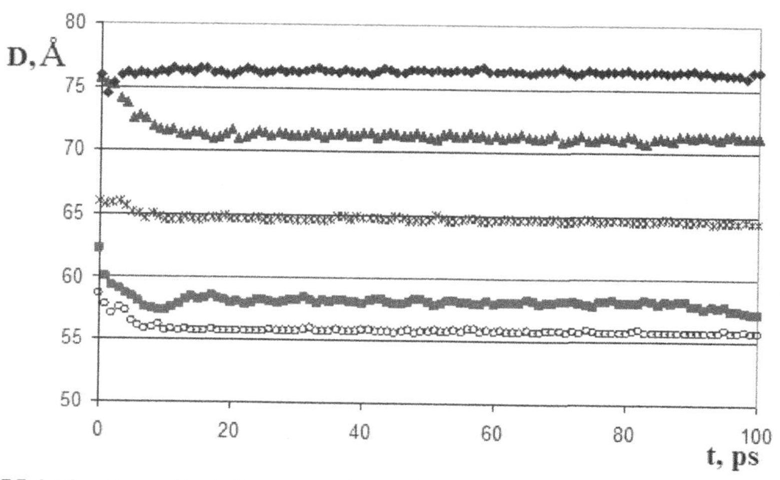

FIGURE 4.79 Generalized diagram of changes in nanowhiskers diameter during relaxation at different values of Φ (■—Φ_{xy} = 8 kcal/(mol*Å), Φ_z = 20 kcal/(mol*Å), ▲—Φ_{xy} = 10 kcal/(mol*Å), Φ_z = 10 kcal/(mol*Å), ○—Φ_{xy} = 10 kcal/(mol*Å), Φ_z = 20 kcal/(mol*Å), *—Φ_{xy} = 20 kcal/(mol*Å), Φ_z = 20 kcal/(mol*Å), ◆—Φ_{xy} = 8 kcal/(mol*Å), Φ_z = 10 kcal/(mol*Å)).

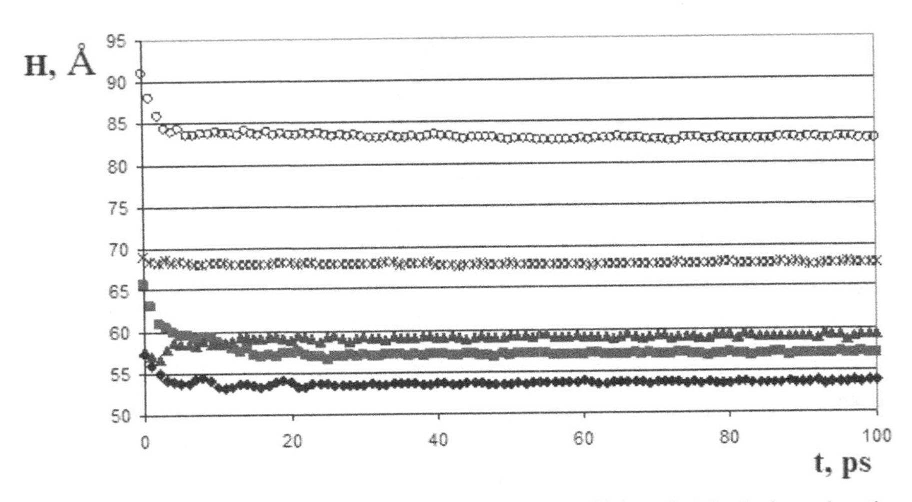

FIGURE 4.80 Generalized diagram of changes in nanowhiskers height during relaxation at different values of Φ (■—Φ_{xy} = 8 kcal/(mol*Å), Φ_z = 20 kcal/(mol*Å), ▲—Φ_{xy} = 10 kcal/(mol*Å), Φ_z = 10 kcal/(mol*Å), ○—Φ_{xy} = 10 kcal/(mol*Å), Φ_z = 20 kcal/(mol*Å), *—Φ_{xy} = 20 kcal/(mol*Å), Φ_z = 20 kcal/(mol*Å), ◆—Φ_{xy} = 8 kcal/(mol*Å), Φ_z = 10 kcal/(mol*Å)).

It is necessary to mark that the compound of gold and silicon is formed at the top of NW owing to diffusion (Fig. 4.81); it is consistent with the experimental data.

During the simulation the effect of NWs curling was observed in some cases. Figure 4.82 graphically illustrates this process.

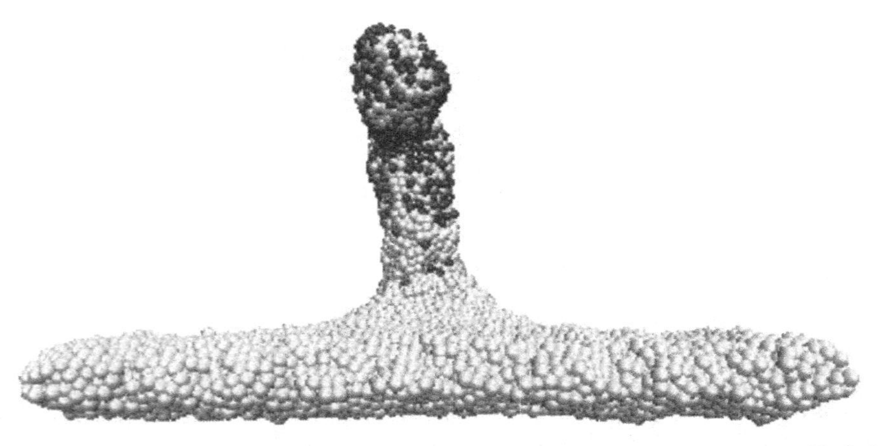

FIGURE 4.81 A mixture of silicon and gold on the top of whisker (light atoms—Si, dark atoms—Au).

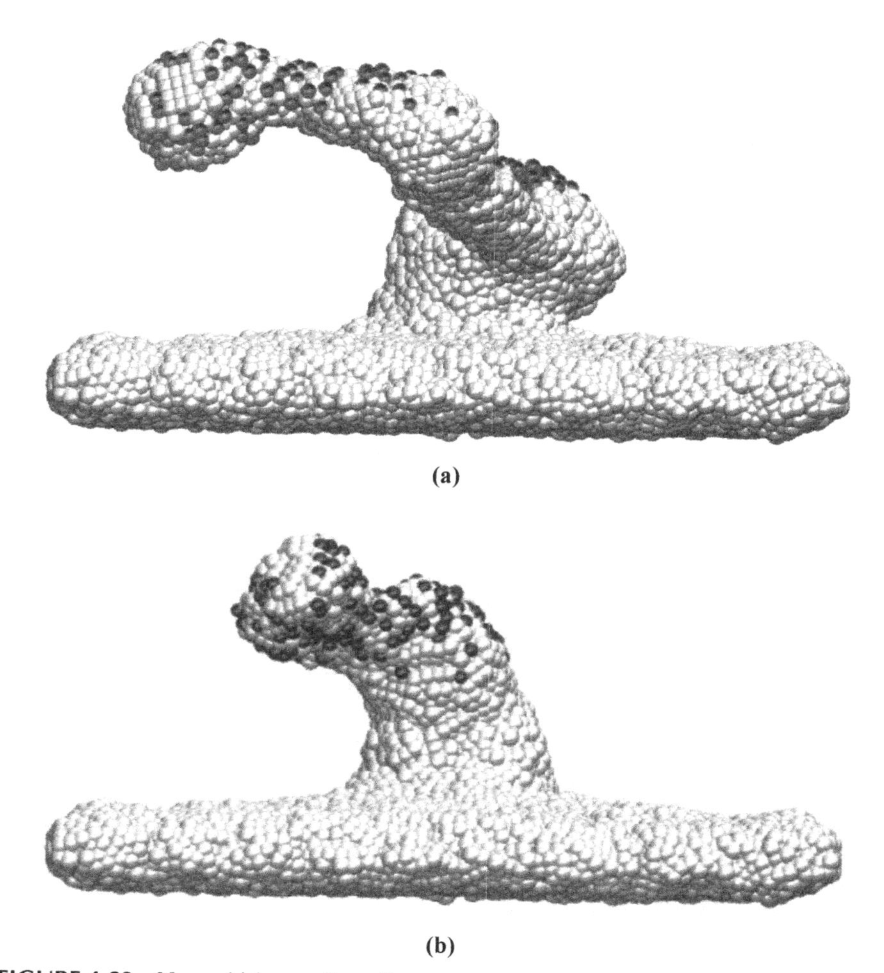

(a)

(b)

FIGURE 4.82 Nanowhisker curling effect.

The emergence of this phenomenon can be explained by rebuild the structure of NW that is in qualitative agreement with experimental data. It is possible to assume that in this case there is a formation of so-called dense packing. Figure 4.83 shows the variation of internal structure of NW when curling. The figure shows that the structure of the NW is rebuilt. Figure 4.84 also illustrates the rebuilding of the internal structure of the crystal. The selected area is similar to the dense packing. Figure 4.85 shows a face-centered crystal lattice, one corner is cut off to show the (111) plane. The planes (111) are close-packed layers of hard spheres. The trajectories of individual atoms in NW curling process are shown in Figures 4.86 and 4.87.

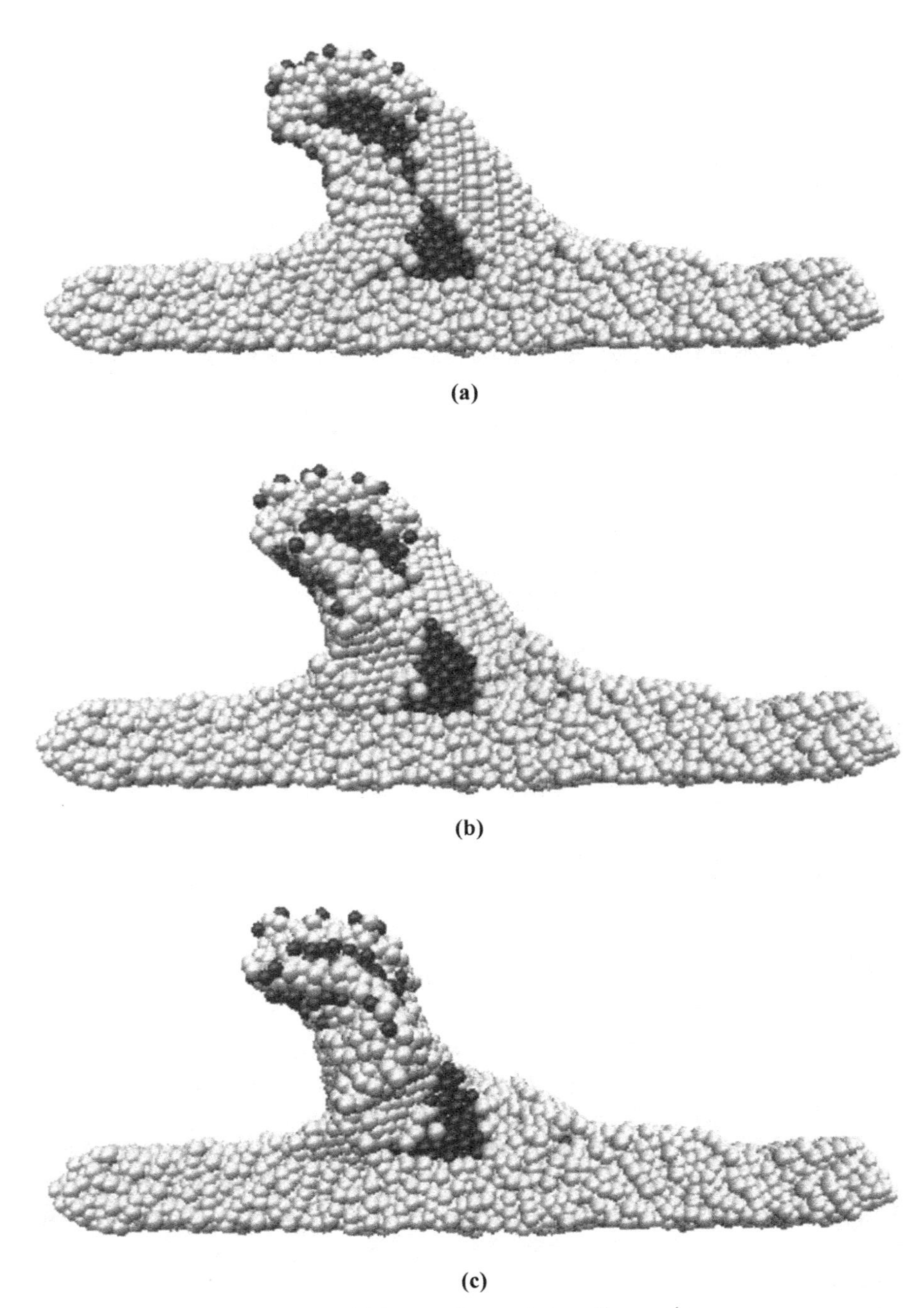

FIGURE 4.83 Stages of nanowhiskers curling, cut along the x-axis.

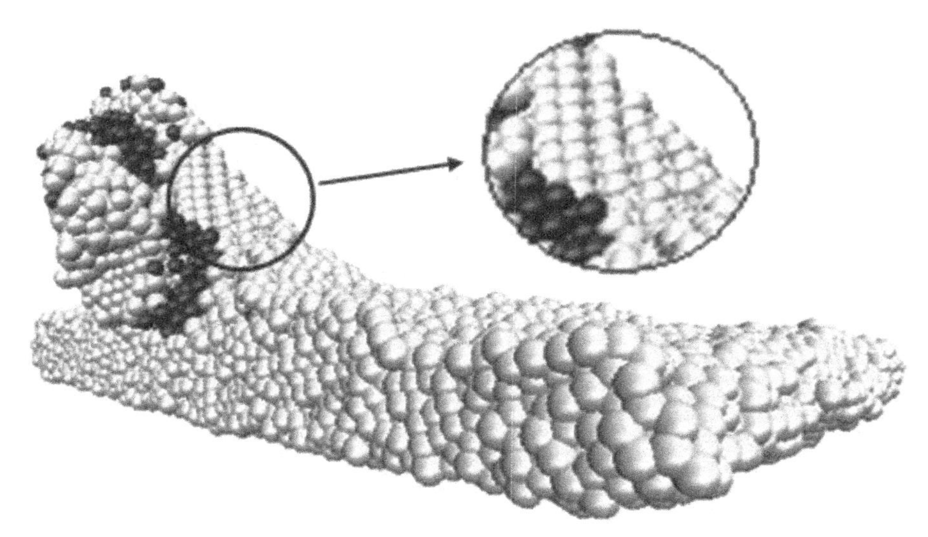

FIGURE 4.84 Rebuilding the crystal structure.

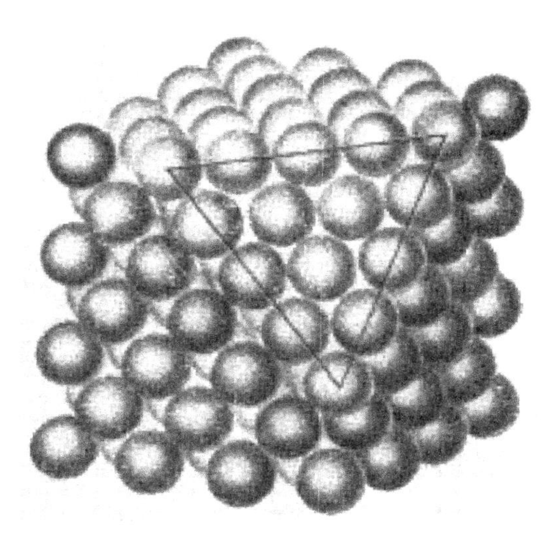

FIGURE 4.85 The dense packing.

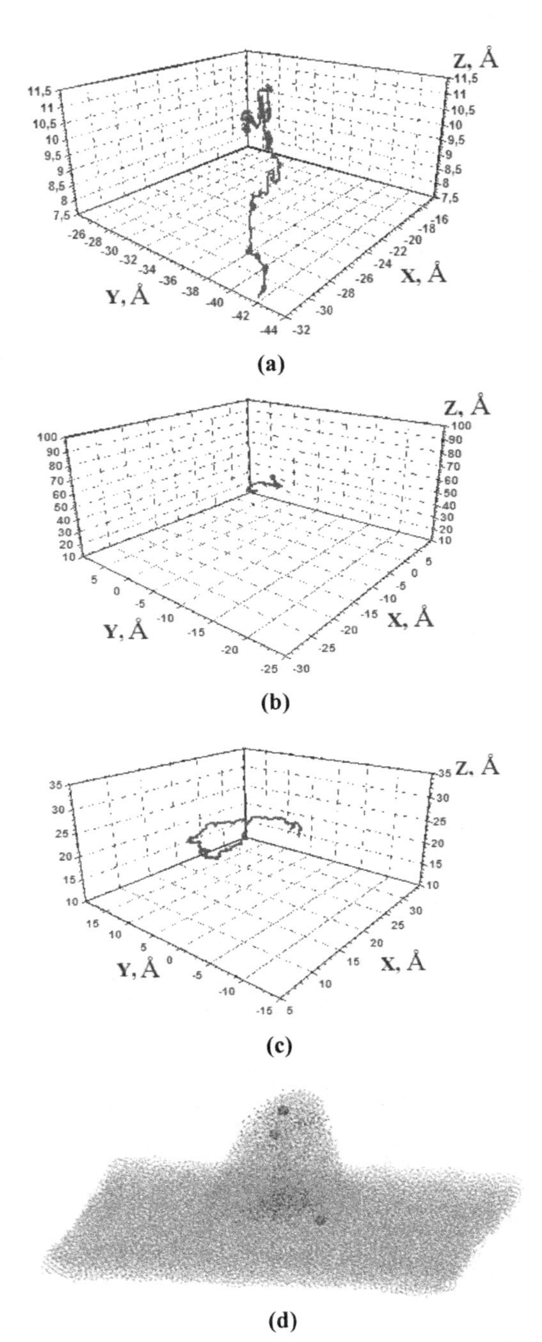

FIGURE 4.86 Movement of atoms in case of nanowhisker curling ((a), (b), and (c)—trajectory of atoms motion and (d) illustrates selected atoms).

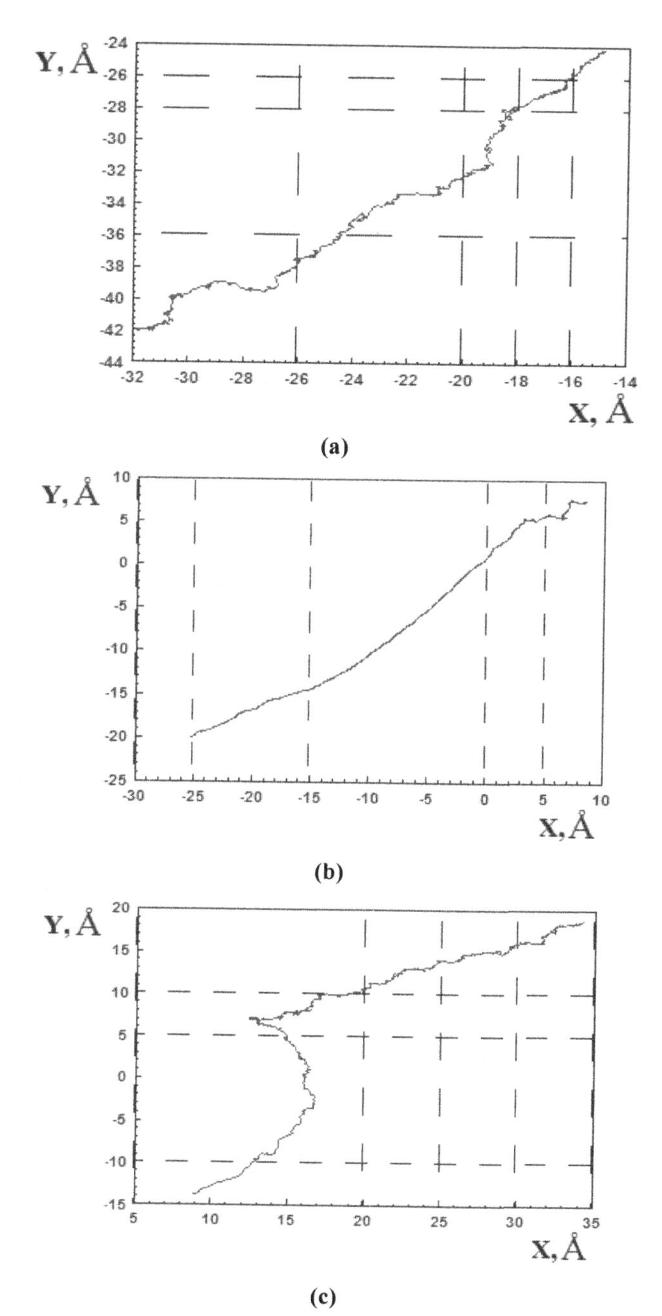

FIGURE 4.87 Movement of atoms in case of nanowhisker curling projection to the x-y plane.

In some cases after curling the melt of Au–Si falls from peak of NW resulting in so-called silicon nanotips formed on the substrate.[39] Figure 4.88 shows the silicon nanotips obtained during the simulation.

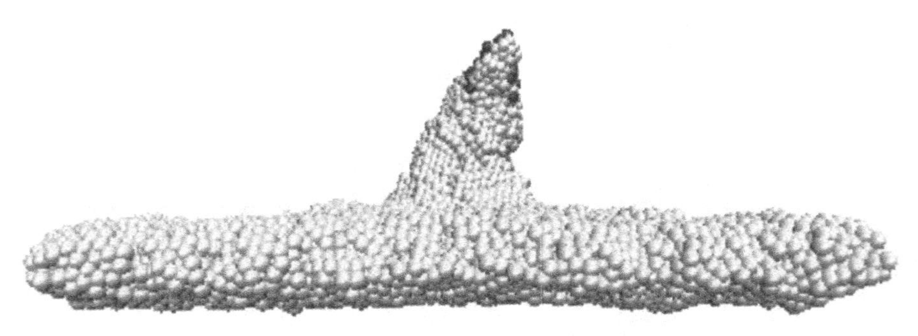

FIGURE 4.88 Nanotips received as a result of simulation.

Often at high growth temperatures NWs at first have a constant radius equal to the initial radius of the catalyst drop and then after reaching a certain critical height begin to sharp. It occurs as follows: when the height of NW becomes comparable with value of diffusion length of adatom on the lateral surface, atoms of silicon from the crystal lateral surface evaporate earlier, than manage to get to the drop.[38] As a consequence, the diffusive flux into the drop decreases that leads to decrease in solution concentration in the drop, reducing the size of the drop and as a consequence—the sharpening of whisker. Nature of NW diameter change depending on the height for the different values of forces enclosed to atoms of system is provided by the graphs in Figure 4.89. The lower value on the ordinate axis corresponds to the substrate, the top—to the height of NW. Based on the given dependences it is possible to note that sharpening of NWs is observed in most cases. Research showed that the size and shape of NWs depend on the value of chosen force. Change of diameter of a drop and NW is called change of quantity of atoms of silicon in a drop and the amount of gold in a drop is constant. However, at sufficiently high temperatures the migration of gold from the drop can occur. Thus, the calculations confirm the experimental studies which show that the steady growth of whiskers is observed only within a certain range of thermodynamic conditions.

Let us consider comparison of simulation results of NWs formation with experimental data. A plot of dependence of height of whiskers on their diameter is presented in Ref. 37,146 During the simulation of growth process the results provided in Figure 4.90 were received.

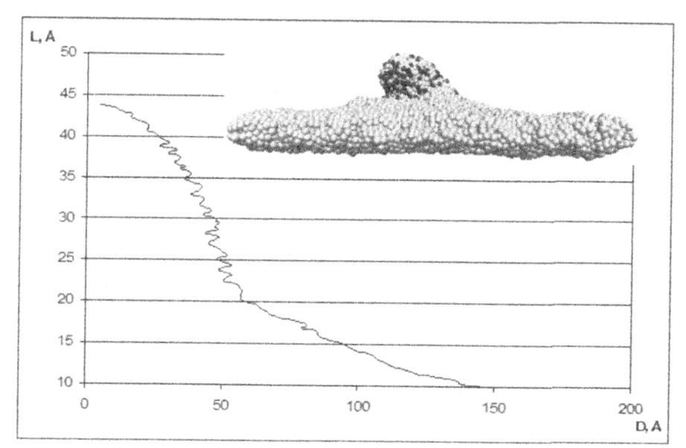

a) Φ_{xy}= 8 kcal/(mol* Å), Φ_z= 5 kcal/(mol* Å).

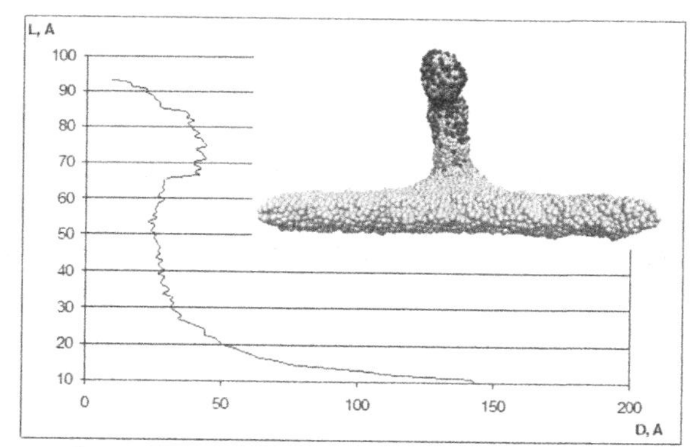

b) Φ_{xy}= 9 kcal/(mol* Å), Φ_z= 15 kcal/(mol* Å).

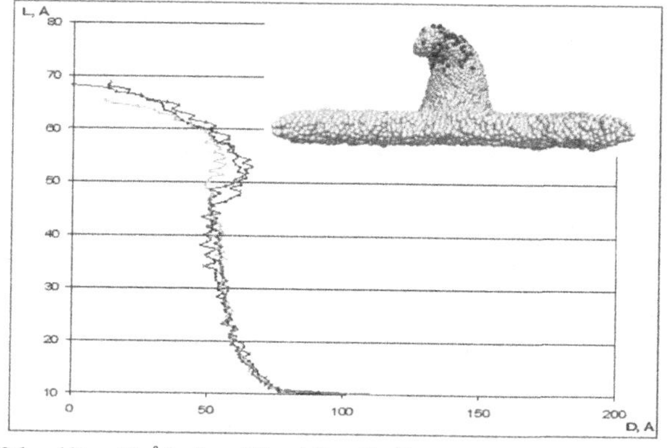

c) blue—Φ_{xy}= 20 kcal/(mol* Å), Φ_z= 5 kcal/(mol* Å).

FIGURE 4.89 Changing the diameter of the nanowhisker on height: green—Φ_{xy}= 20 kcal/(mol* Å), Φ_z= 10 kcal/(mol* Å); red—Φ_{xy}= 20 kcal/(mol* Å), Φ_z= 15 kcal/(mol* Å).

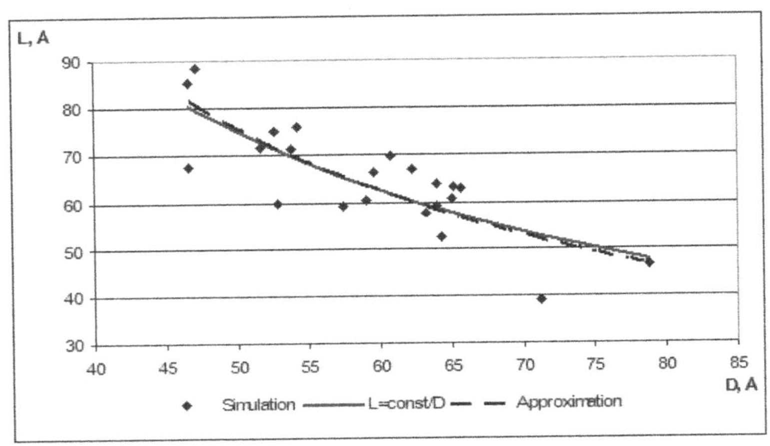

FIGURE 4.90 Dependence of height nanowhisker on diameter received by results of simulation.

Apparently from figures character of curves of dependence of NW height on its diameter is similar, in both cases dependence of NW height on diameter is decreasing. In Figure 4.95, the dark dashed line corresponds to the optimal approximation of simulated values, the bright line corresponds to the curve $L = const/D$. In this range of values of L and D these functions are almost identical. Table 4.6 shows the average deviation of approximating function from theoretical curve $L = const/D$. It confirms adequacy and correctness of offered mathematical model.

TABLE 4.6 Comparing of Results of Simulation with Model from Paper[146].

The relative error in %	Absolute error, Å
0.67	0.43

It should be noted that the migration of gold from the drop can occur in silicon NWs at high temperatures. Similar conclusions can be made from the results of simulation. This phenomenon is also observed in simulation of processes of whiskers growth by Monte Carlo methods[112] and in our calculations (Fig. 4.91). These data also show a good coordination of simulation results with experiments and other methods of computer simulation.

Let us consider internal crystallographic structure of a NW. From experiments it is known that often NW contain regions with various crystallographic structure, and in most cases takes place random periodic changes of structure.[36] To assess the internal structure of the NWs obtained during the simulation, radial pair distribution function $g(r)$, which measures independent correlations between particles, was constructed. That is, $g(r)$ is a probability of that in dr volume element in neighborhood of r the particle is found in case of simultaneous existence in origin of coordinates of other particle[71]

$$g(r) = \frac{Vn(r)}{4\pi r^2 \Delta r N},$$

where $n(r)$ is average number of particles located at distances $r,\ldots,r + \Delta r$ from given particle; V is volume of considered system; N is total number of particles. Schedule of radial pair distribution function is presented in Figure 4.92.

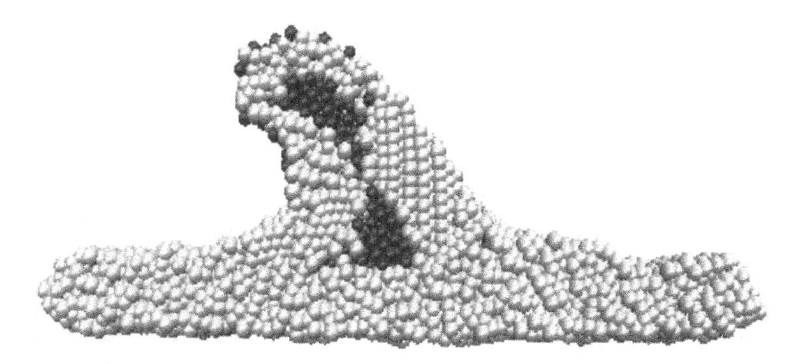

FIGURE 4.91 Gold atoms inside the NW, cut along x-axis.

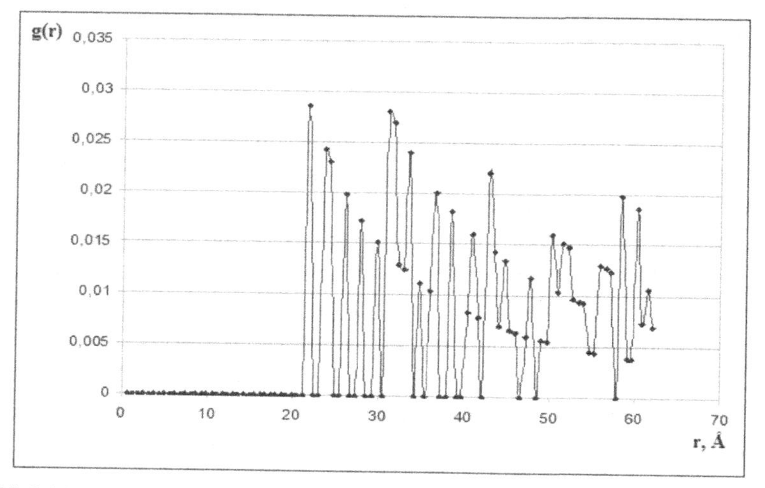

FIGURE 4.92 Radial pair distribution function.

Comparing the graphs 4.92 with this function it is possible to conclude that crystallographic structure of simulated NWs is not uniform. It is fully consistent with experimental data[37]

4.3.2 SIMULATION OF QUANTUM DOTS FORMATION

4.3.2.1 TASK FORMULATION

The process of the formation of nanostructures on the substrate surface is greatly influenced by lattice mismatch of deposited material and substrate. There are different mechanisms of growth depending on lattice mismatch. It is known from experimental and theoretical investigations[39] that for the formation of three-dimensional islands it is necessary that the parameter of the mismatch of the lattices in the "deposited material/substrate" system be sufficiently large ($\varepsilon_0 > 2\%$). Moreover, the larger size of the mismatch of lattices the earlier the formation of coherent islands occurs.

For example, the formation of quantum dots is observed in the InAs/GaAs system, where $\varepsilon_0 = 7\%$. In this case, the growth follows the Stranski–Krastanov mode, initially an elastically strained wetting layer, having the same lattice parameter as the substrate material, is formed on the surface. As soon as the wetting layer attains a certain critical thickness, misfit dislocations begin to form. After the formation of such dislocations, the epitaxial layer grows with the lattice constant of the deposited material. For this mechanism, the critical wetting layer thickness exceeds one monolayer.

When $\varepsilon_0 < 2\%$, quantum dots are not formed, and the growth of heterostructures proceeds by the layer-by-layer growth mechanism. At very large values of the mismatch parameter the growth follows the Volmer–Weber mechanism. Then the critical wetting layer thickness is less than one monolayer, and the formation of three-dimensional islands occurs directly on the substrate surface. An example of such a system is the InAs/Si system, where $\varepsilon_0 = 10.6\%$. In a system of silicon and chromium the lattice mismatch parameter is high; therefore here we may also speak of the formation of quantum dots. The increase in the chromium layer thickness up to 0.6–1.5 nm can lead to a change in the morphology of silicide islands. It is noted[58] that since the subsequent deposition of silicon is carried out at a temperature of 1023 K, an increase in the substrate temperature can lead to a change in the size distribution of islands, their coagulation, and coalescence with subsequent crystallization due to the increase in the intensity of diffusion processes.

The process of formation of metal-silicon quantum dots is as follows:

1. Silicon substrate is heated to a certain temperature T_1 (Fig. 4.93).

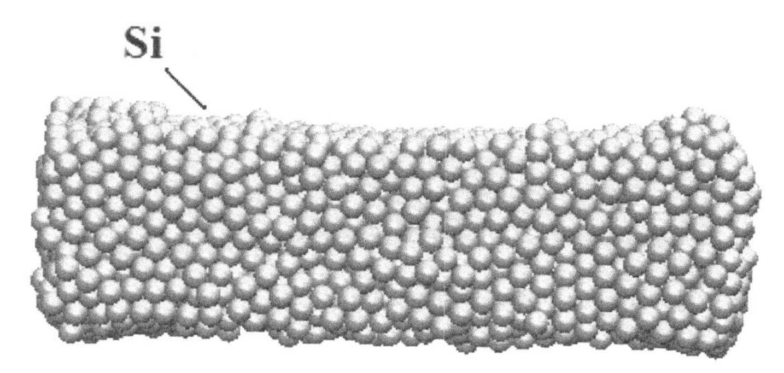

FIGURE 4.93 First stage of quantum dots formation (substrate heating).

2. Then, at the same temperature T_1 atoms of metal (Me) are deposited on a substrate surface. The picture illustrating this stage of process is shown in Figure 4.94.

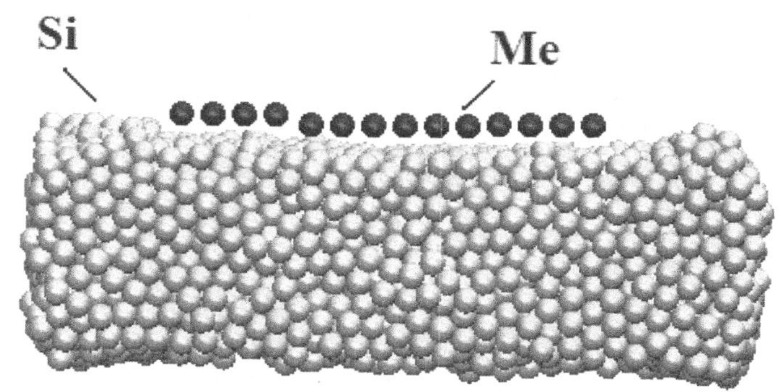

FIGURE 4.94 Metal deposition.

3. Further system annealing is carried out at temperature T_2 (Fig. 4.95). There is a formation of islands on a substrate surface. From the experimental and theoretical studies, it is known that for the formation of three-dimensional islands it is necessary to the lattices mismatch parameter in system deposited material/substrate was sufficiently large ($\varepsilon_0 > 2\%$). Silicon has a lattice spacing of 5.4307 Å, and lattice spacing of most of metals significantly differ from it. For example, its value for chromium is 2.8850 Å, for iron is 2.866 Å, that is, ($\varepsilon_0 > 2\%$) for most of metal-silicon systems. Therefore, for these systems there is a formation of quantum dots.

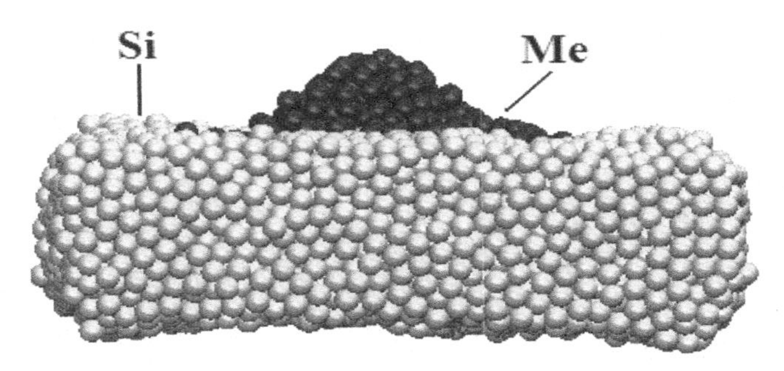

FIGURE 4.95 System annealing at temperature T_2.

4. Silicon atoms are deposited on the obtained system, and as a result there is a formation of Me-Si quantum dots embedded in silicon. This process is shown in Figure 4.96.

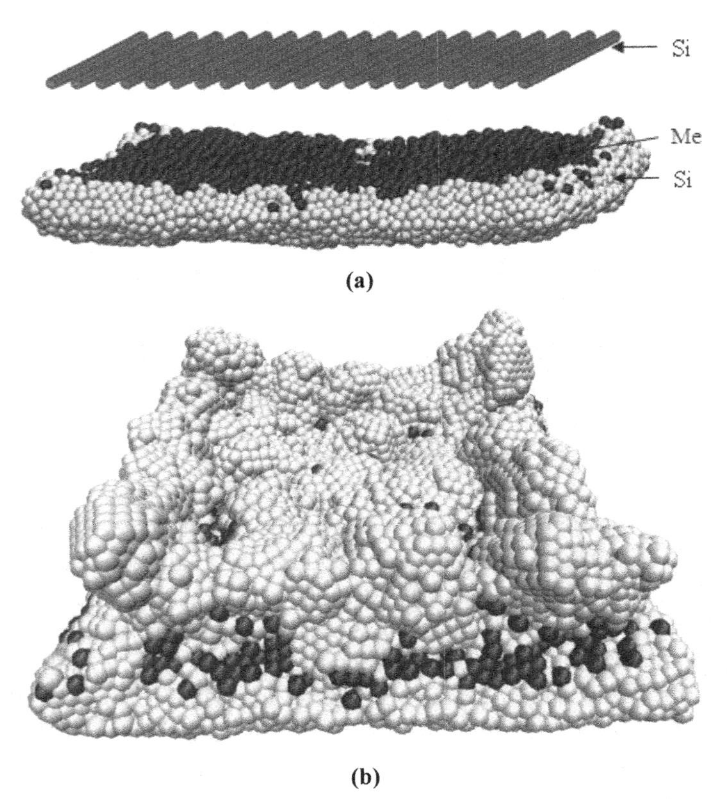

FIGURE 4.96 Silicon deposition (a) beginning of a stage and (b) end of a stage.

Simulation of the above stages of quantum dots formation was carried out using molecular dynamics method, which describes the motion of a system of N atoms in the form of a system of differential equations (Chapter 1).

In the study of heterostructures based on silicon and chromium, it was noted that after the deposition of chromium atoms on the surface of Si (111) substrate and further deposition of silicon, the $CrSi_2$ crystallites are formed. Therefore, starting with the second phase (metal deposition) check the formation of molecules Me_nSi_m is carried out. If atoms are located at distance less than or equal to the equilibrium bond length b_0 (Fig. 4.97), formation of bonds carried out, recalculation of the angles, make the necessary changes to configuration files. When the atoms that make up a molecule are removed to a distance equal to or greater than r^* (the critical bond length), bonds are dropped. Then, the molecular-dynamic calculation continues. On this basis, the total interaction potential of the molecular system can be written as:

$$E_{total} = E_{vdW} + E_{bond} + E_{angle}, \tag{4.62}$$

where E_{vdW} is a potential of van der Waals interaction, E_{bond} is a potential of covalent bonds, which describes the change in bond length, E_{angle} is the angular potential between a pair of covalent bonds. In the first stage (modeling a substrate heating up to) E_{bond} and E_{angle} are zero, since at this stage there are no chemical bonds.

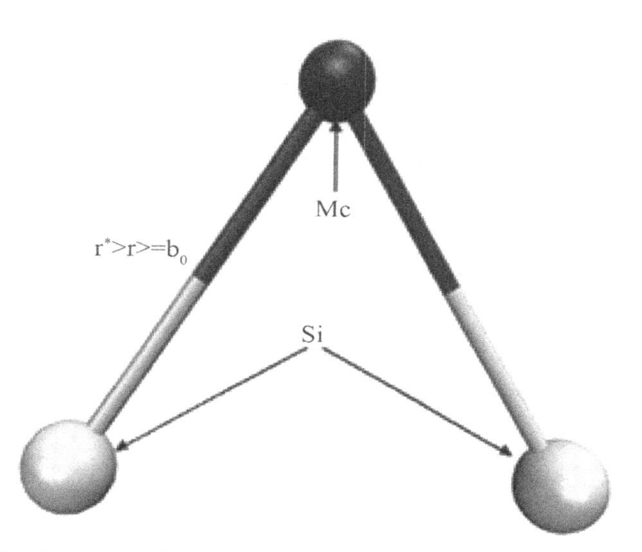

FIGURE 4.97 Bonds formation (for the molecules of the form $MeSi_2$).

The potential of van der Waals interaction is calculated between pairs of atoms in system. This type of interaction can be described by different potentials, for example, by Lenard-Jones potential:

$$E_{vdW}^{ij} = 4\varepsilon_{ij}\left[\left(\frac{\sigma_{ij}}{r_{ij}}\right)^{12} - \left(\frac{\sigma_{ij}}{r_{ij}}\right)^{6}\right], \tag{4.63}$$

where ε_{ij} is the depth of the potential well, σ_{ij} is the finite distance at which the inter-particle potential is zero, r_{ij} is the distance between the particles.

The potential of covalent bonds interaction are described as:

$$E_{bond}^{ij} = k_{bond}^{ij}\left(b^{ij} - b_0^{ij}\right)^2, \tag{4.64}$$

where k_{bond}^{ij} is a constant of stretching (compression) of bond, b_{ij} is a current bond length, b_0^{ij} is equilibrium bond length.

Angular potential between pair of the covalent bonds with common atom in top can be written as:

$$E_{angle}^{ijk} = k_{angle}^{ijk}\left(\theta^{ijk} - \theta_0^{ijk}\right)^2, \tag{4.65}$$

where k_{angle}^{ijk} is a constant of angular interaction, θ_0^{ijk} is an equilibrium angle in degrees corresponding to the minimum of energy,θ^{ijk} is a current value of angle.

The task of initial and boundary conditions is necessary for the solution of considered problem. The initial conditions include the coordinates and velocities of the atoms. The initial coordinates are based on the structure of the modeled system. The values of the velocity field at the initial time for the metal atoms and silicon were chosen according to the Maxwell distribution.[97]

The Maxwell velocity distribution for the vector velocity $\vec{V_i} = \left[V_x, V_y, V_z\right]$ is the product of the distributions for each of the three directions:

$$f_V\left(V_x, V_y, V_z\right) = f_V\left(V_x\right) f_V\left(V_y\right) f_V\left(V_z\right), \tag{4.66}$$

where the distribution for a single direction is

$$f_V\left(V_i\right) = \sqrt{\frac{m}{2\pi kT}} \exp\left(-\frac{mV_i^2}{2kT}\right), \tag{4.67}$$

where m is an atomic mass, k is the Boltzmann constant, T is the temperature of the system.

The Born–von Karman boundary condition[71] is chosen. According to the Born–von Karman boundary condition it is required function to be periodic in each coordinate. In three dimensions:

$$f(x+L,y,z)=f(x,y,z), f(x,y+L,z)=f(x,y,z), f(x,y,z+L)=f(x,y,z). \quad (4.68)$$

4.3.2.2 THE ANALYSIS OF THE CALCULATION RESULTS

Consider the results of the simulation of quantum dots formation as an example the formation of quantum dots based on chromium in silicon. The solution to this problem was carried out by using the developed problem-based software. Molecular dynamics simulation of the processes accompanying quantum dots formation is at the core of the software. Using molecular dynamics techniques for solving this problem is caused by the need to trace the simulated process kinetics, as well as to evaluate the structure and physical properties of quantum dots. The software includes several units: preparation of initial data, computing unit, data matching unit, the unit of analysis, and visualization of results.

The task of the unit of initial data preparation is to set the initial parameters of the simulated system, such as pressure, temperature, quantity, kind, and properties of atoms used in the simulation, and also the formation of the initial and boundary conditions. The computing unit is designed for direct molecular dynamics simulation of the metal quantum dots formation in silicon. Results are created in this unit. Processing and data analysis, received in computing unit, are made in the unit of analysis and visualization of results. Properties and structure of the simulated systems are researched in this unit.

Table 4.7 shows the values for the $CrSi_2$ equilibrium bond length, obtained by different methods. Proceeding from data analysis in Table 4.6, value of 2.3 Å was selected for equilibrium bond length in $CrSi_2$ molecule. Other parameters of interaction used in case of simulation are specified in Table 4.8.

TABLE 4.7 The Equilibrium Bond Length Cr-Si in the Molecule $CrSi_2$.

	NAMD	ABINIT	HyperChem	Theory	Paper[13]
Bond length b_0, Å	2.040	2.377	2.099 (*ab initio*), 2.214 (semi empirical)	2.425	2.361

TABLE 4.8 Parameters of Interaction Potentials.

Type of interaction	Parameter	Value
Van der Waals interaction for silicon atoms	ε, kcal/mol	53.5000
	$r_{min} = 2^{1/6}\sigma$, Å	2.3500
Van der Waals interaction for chromium atoms	ε, kcal/mol	11.8125
	$r_{min} = 2^{1/6}\sigma$, Å	2.5000
Van der Waals interaction for silicon and chromium atoms	ε, kcal/mol	25.1390
	$r_{min} = 2^{1/6}\sigma$, Å	2.4250
Angular potential between pair of covalent bonds silicon—chromium	θ, degree	60.5080

After the deposition of chromium atoms and short-term annealing three-dimensional islands are formed on the substrate surface. It corresponds to the experimental data. The size and number of islands depends on the thickness of the deposited chromium. For example, Figure 4.98 shows a system with two (a) and five (b) islands.

The process of nanostructures formation on the substrate surface is greatly influenced by lattice mismatch of deposited material and substrate. There are different mechanisms of growth depending on lattice mismatch. From the experimental and theoretical research[39] it is known that for the formation of three-dimensional islands it is necessary that the parameter mismatch of the lattices in system "deposited material/substrate" was sufficiently large ($\varepsilon_0 > 2\%$).The big size of a mismatch of lattices causes faster formation of coherent islands.

For example, the formation of quantum dots is observed in the InAs/GaAs system, where $\varepsilon_0 = 7\%$. In this case, growth occurs by the Stranski–Krastanov mode, initially elastically strained wetting layer, having the same lattice parameter as the substrate material, formed on the surface. Upon reaching a certain critical thickness of the wetting layer misfit dislocations begin to form. After the formation of such dislocations epitaxial layer grows with the lattice constant of the deposited material. For this mechanism, the critical wetting layer thickness exceeds one monolayer.

When $\varepsilon_0 = 2\%$, quantum dots are not formed, and the growth of heterostructures is carried out on the layered mechanism (layer-by-layer growth). For very large values of the parameter mismatch on the mechanism of growth is Volmer–Weber. Then the critical wetting layer thickness is less than one monolayer, and the formation of three-dimensional islands are directly on the substrate surface. An example of such a system is the system InAs/Si, where ε_0 = 10,6%. In a system of silicon and chromium lattice parameter mismatch is large, so here it is also possible to talk about the formation of quantum dots.

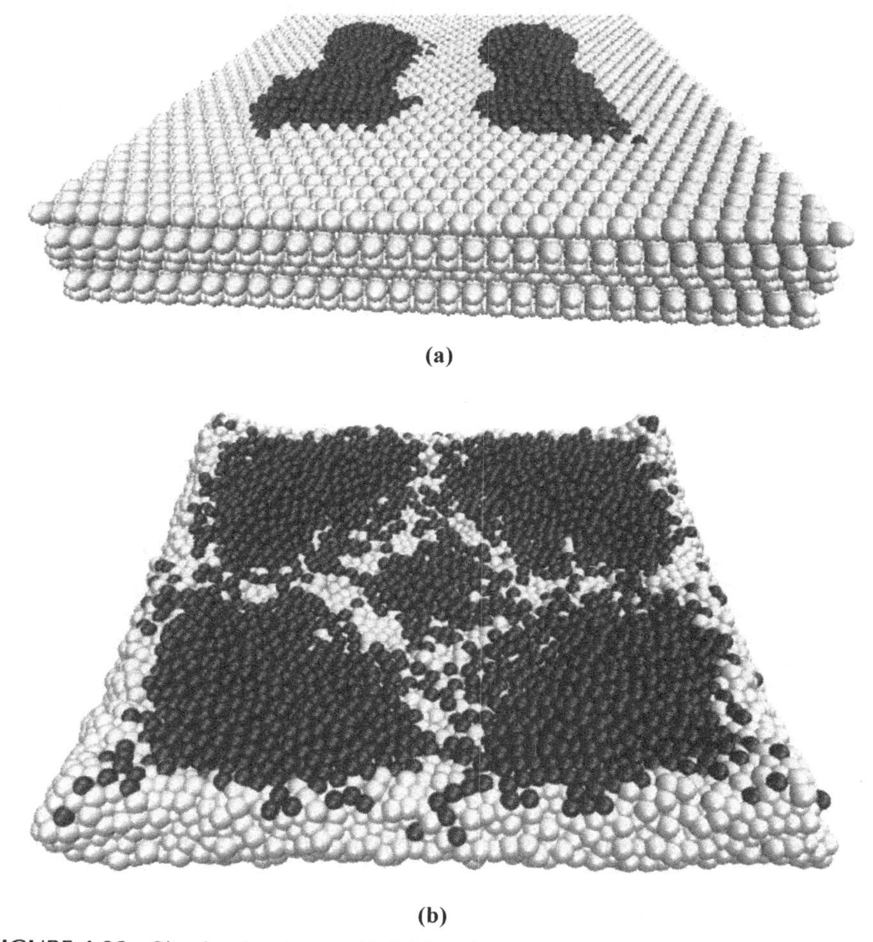

FIGURE 4.98 Simulated systems with 2 (a) and 5 (b) islands (dark atoms are chromium, bright atoms are silicon).

In papers[26,27] it is established that after chromium deposition (0.07–0.12 nm) on Si (111) 7 × 7 substrate at temperature 773 K there is a formation of three-dimensional islands of $CrSi_2$ in height of 0.5–3 nm and the lateral size of 30–80 nm. Similar results were obtained in the simulation. So the height of islands received as a result of simulation is 0.67–1.07 nm, and the lateral sizes are from 33.35 to 55.58 nm. These results indicate correspondence between dimensional characteristics of simulated islands and experimental data.

The increase of chromium layer thickness up to 0.6–1.5 nm can lead to change of morphology of silicides islands. In[57,59] noted that since the subsequent deposition of silicon is carried out at a temperature of 1023 K, increasing the substrate temperature can lead to a change in the distribution of islands by size, their coalescence and coagulation followed by crystallization by increasing the intensity of diffusion processes.

It was found that with increasing thickness of chromium deposited on the surface of the silicon substrate there is coagulation of the islands. Figure 4.99 presents the change in the morphology of the islands, depending on the amount of the deposited chromium (light atoms is silicon substrate, the dark is chromium atoms).

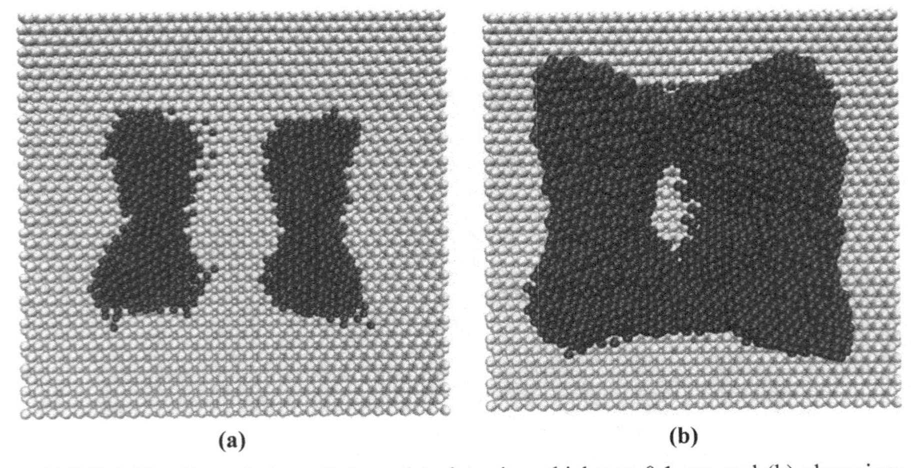

FIGURE 4.99 Coagulation of atoms (a) chromium thickness 0.1 nm and (b) chromium thickness 0.4 nm.

The space between the islands is filled with chromium atoms with further deposition of chromium on the substrate surface and islands. It leads to increase in the lateral sizes of islands (Fig. 4.100).

The morphology of the islands depends on the method of chromium deposition on the substrate surface. The shape and number of islands depends on that as chromium is deposited (at a time or in batches).

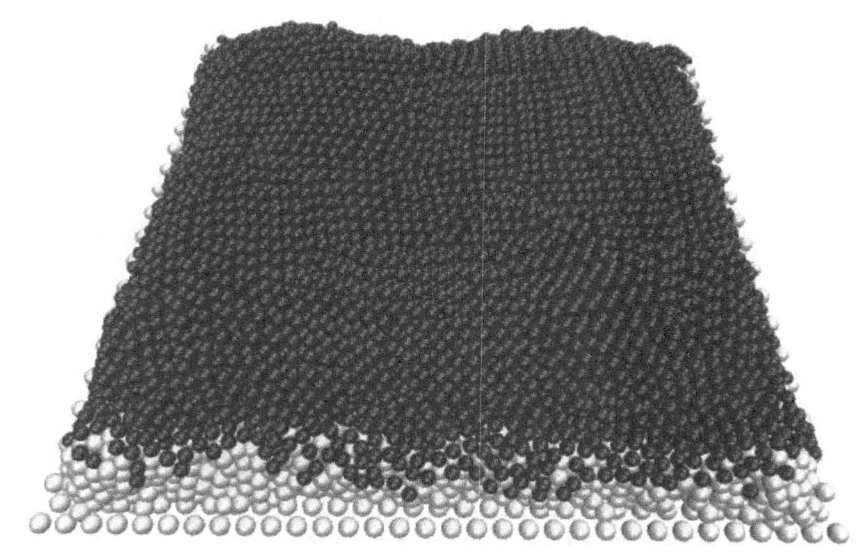

FIGURE 4.100 Increase in the lateral sizes of islands.

Figure 4.101 shows the system obtained at different deposition methods: layerwise, at a time and mixed mode (part with a layerwise method, part at a time). Atoms of chromium are represented by dark color, atoms of silicon by light color.

System shown in Figure 4.101a was obtained by layerwise deposition of seven monolayers of chromium. Four large and one small island are clearly distinguished. The picture presented in Figure 4.101b corresponds to a one-time deposition of seven monolayers of; Figure 4.101c corresponds to chromium deposition of mixed-mode (four monolayers of chromium deposited by layerwise method, followed by three monolayers at a time). It was noted that the heterostructure with deposition of chromium at a time is an almost uniform layer of chromium on a silicon substrate without formation of small islands.

Diffusion makes a great impact on morphology and properties of simulated systems. Chromium atoms diffuse into the substrate, but diffusion is observed only in the surface layers. It agrees well with experimental data.[57–59] The presence of diffusion in the surface layers is caused by the structure and properties of the elements participating in simulation. However, most of the deposited atoms diffuse into the substrate. For example, in case of deposition of 8000 silicon atoms 7733 atoms diffuse in substrate, that is, nearly 97%. It can be assumed that these atoms form $CrSi_2$ molecules. From the experiments it is known that $CrSi_2$ is actively formed during the formation of quantum dots. The process of bonds adding is shown in Figure 4.102.

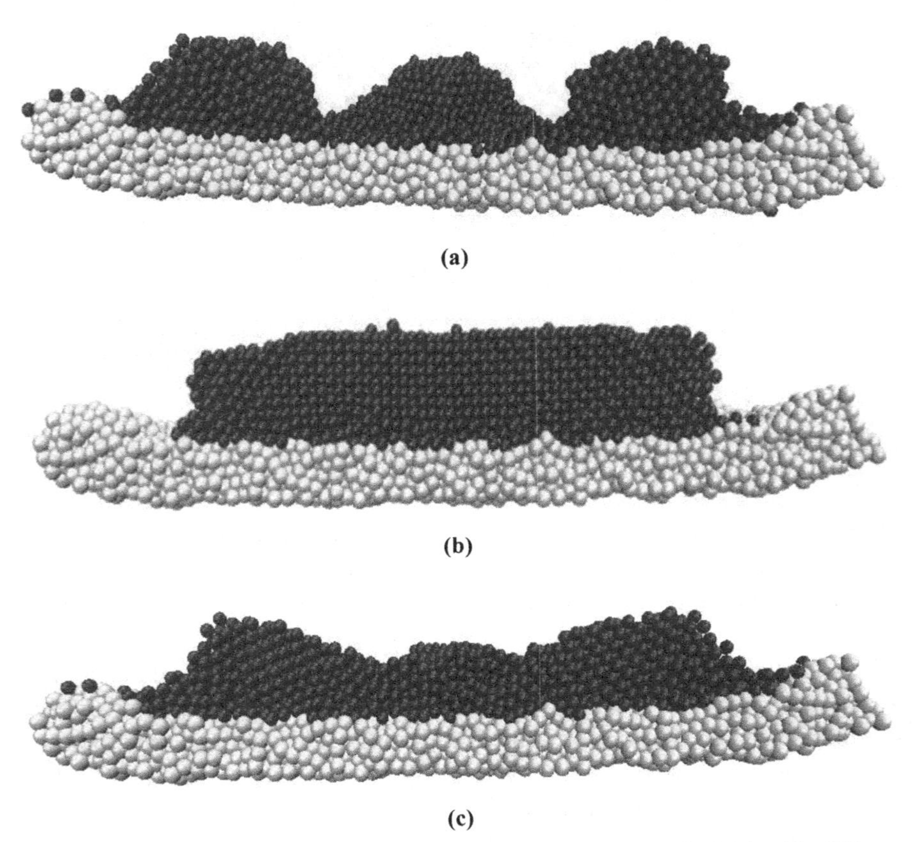

FIGURE 4.101 Diagonal section of the simulated structures obtained with different methods of chromium deposition.

Molecules of $CrSi_2$ are absent at the initial moment of time. By the time of $t = 10^3$ fs the formation of separate molecules uniformly distributed over the surface of the substrate $CrSi_2$ is observed. By the time of $t = 10^5$ fs the formation of $CrSi_2$ islands is observed.

It is known that chromium disilicide belongs to space group $P6_222$. However, is difficult to verify whether such structure is formed as a result of the simulation. Therefore, the radial pair distribution functions of silicon and chromium atoms were constructed to evaluate the crystallographic structure of the resulting chromium disilicide. Figure 4.103 shows plots of the radial pair distribution functions for the $CrSi_2$ unit cell and for the system received as a result of simulation. Proceeding from good compliance of profiles of the distribution curves, it is possible to conclude that the simulation results qualitatively correspond to experimental data, and that $P6_222$ structure of $CrSi_2$ is formed in the simulation.

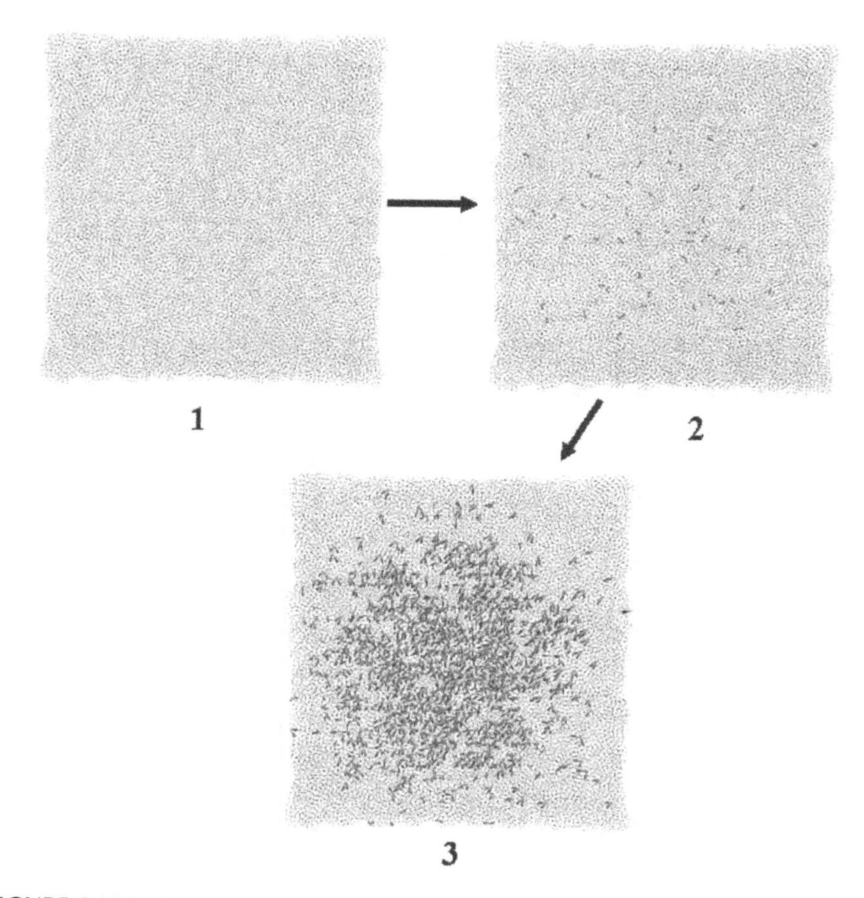

FIGURE 4.102 $CrSi_2$ formation process on the substrate surface (top view). The generated molecules of $CrSi_2$ are shown by dark color. (1) $t = 0$; (2) $t = 10^3$; and (3) $t = 10^5$ fs.

Quantity of formed CrSi2 depends on the thickness of the deposited chromium and formation time. Figure 4.104 shows the graphs of dependence of bonds number from time for various simulated systems. It is evident that these dependences are non-linear, despite uniformity of deposition of chromium atoms. The presented graphs also show a non-linear dependence of the number of bonds on the number of chromium atoms deposited on the substrate surface (plots for 1, 4, and 7 chromium monolayers are provided).

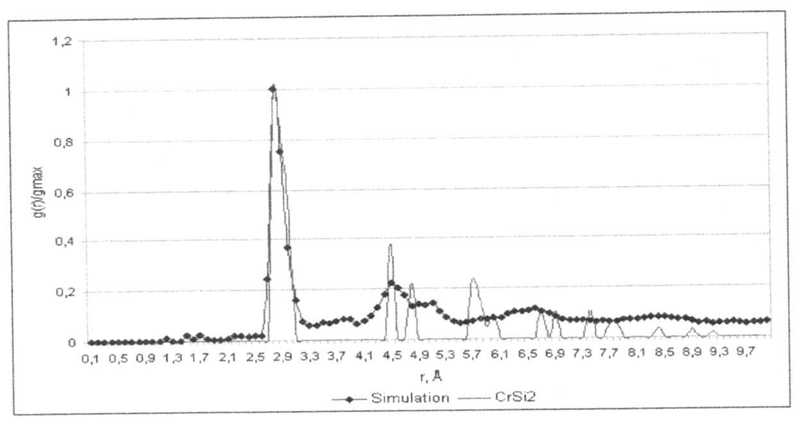

FIGURE 4.103 Radial pair distribution functions (for the $CrSi_2$ unit cell and for the system received as a result of simulation).

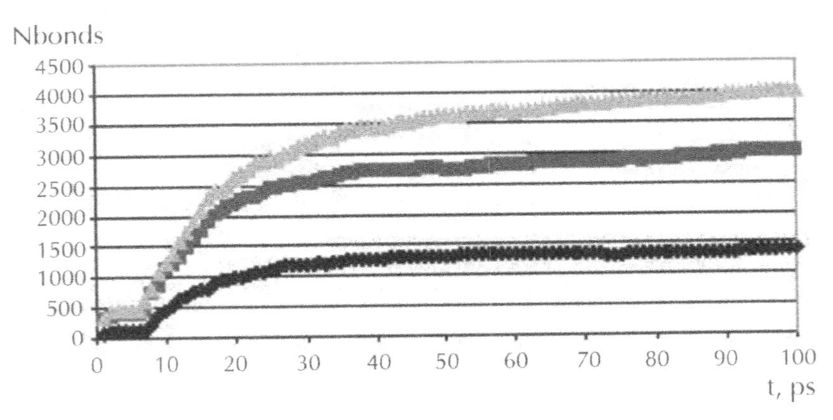

FIGURE 4.104 Change the bonds number with different thickness of deposited material —1 monolayer, —4 monolayers, —7 monolayers.

Let us consider, whether the thickness of a deposited layer influences process of diffusion of atoms of a chromium in a substrate. Systems with amount of chromium from 1 to 7 monolayers were considered. Figure 4.105a shows the dependence of penetration depth of silicon atoms in the substrate on their number under different number of deposited material. It should be noted that, as well as in the previous cases, diffusion here is observed only in the top layers, without reaching the basis of silicon substrate. Figure 4.110b illustrates the influence of the chromium layer thickness on the number of atoms diffusing into the substrate. It is visible that this dependence is not linear. It testifies that with increase in total quantity of deposited chromium not all atoms of chromium are built in a silicon substrate that leads to formation of three-dimensional islands on a substrate surface.

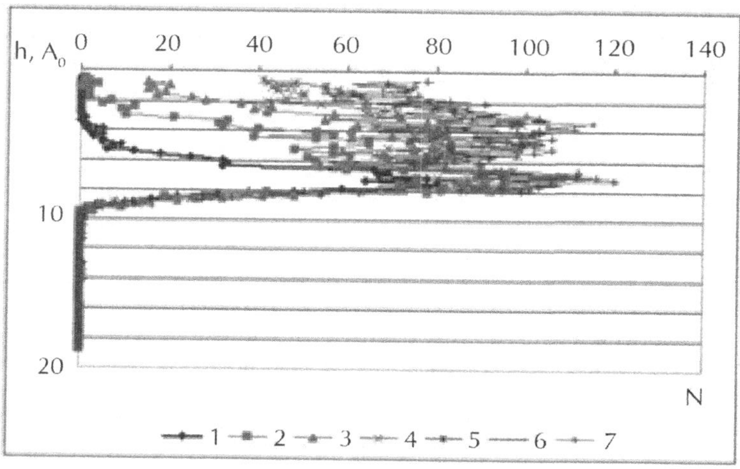

(a) N—amount of diffused atoms; h—penetration depth.

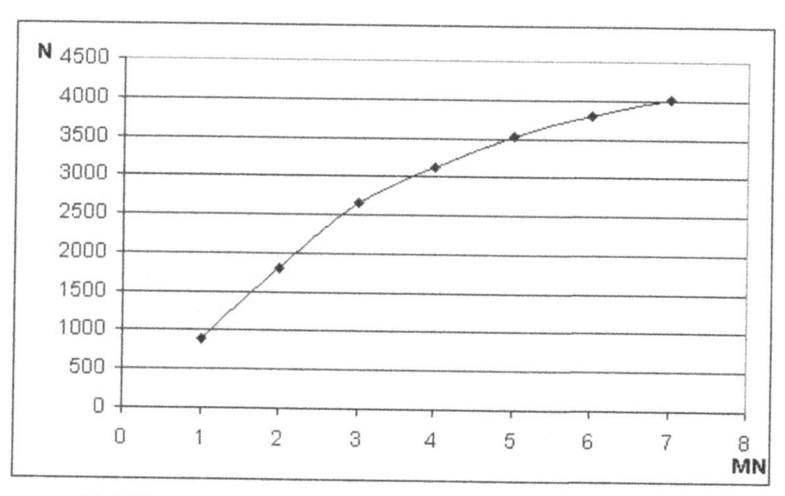

(b) MN—quantity of monolayers, N—amount of diffused atoms.

FIGURE 4.105 Influence of diffused atoms amount on a process of its diffusion in the substrate.

As it was noted above, annealing of system and deposition of silicon are carried out at temperature of 1023 K. As a result of annealing, formation of bonds occurs more actively, the quantity of $CrSi_2$ increases. It is confirmed by experimental work. The number of bonds in the simulated system increases by an average of 10–15% at modeling short-term annealing (annealing duration of 10 ps). For example, for a system with seven monolayers of chromium quantity of formed bonds increases from 1036 (at the stage of chromium deposition) up to 1172 (during the annealing), that is, by 13.13%. Active formation of bonds occurs during annealing, and at the initial stage of silicon deposition (Fig. 4.106).

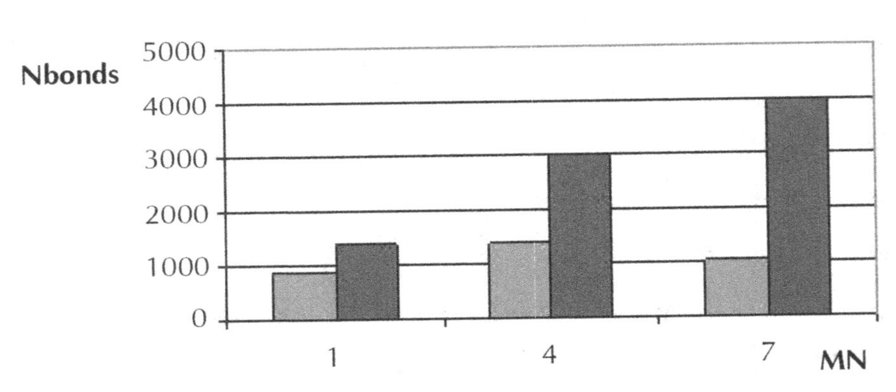

FIGURE 4.106 Effect of annealing stages and silicon deposition on the system condition (Nbonds—number of bonds and MN—amount of monolayers).

Morphology and properties of heterostructures depend on the thickness of the deposited chromium. Moreover, the number of chromium atoms deposited on the substrate surface affects the structure of the emerging islands. For small thicknesses of chromium (1–2 monolayers) almost all the chromium atoms embedded in the silicon crystal lattice, thus forming $CrSi_2$. However with increase in layer thickness up to 7 monolayers not all atoms are built in and even after the deposition of silicon a certain amount of unbound chromium atoms remain inside the islands. Chart presented in Figure 4.107 shows the ratio of the number of atoms involved in the formation of chromium disilicide to the total amount of chromium deposited at different chromium layer thickness. It is seen that the deposition of seven monolayers of chromium more than two thirds (68%) of them remain unrelated to the silicon atoms inside the islands following the overgrowth of silicon.

In case of further deposition of silicon there is a formation of layered heterostructures. The properties of the resulting quantum dots are often dependent on the islands dimensions, the time of deposition and annealing.

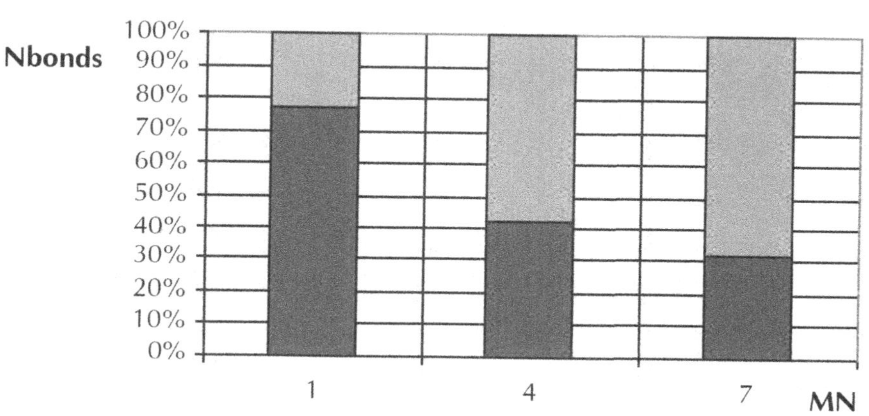

FIGURE 4.107 Influence of layer chromium thickness on islands structure (Nbonds—number of bonds and MN—amount of monolayers).

In work[58] the assumption is put forward that growth of silicon occurs layerwise between islands in the form of two-dimensional blocks of the various sizes, and over each island there is an accretion of silicon blocks with the maximum quantity of defects, as leads to formation holes (recesses) formed in the silicon film. Structures with cavities and holes are also formed during the simulation at the stage of silicon deposition on the islands surface. Figure 4.108 shows the patterns obtained (a) from the simulation and (b) in a real experiment.

In some works relating to the investigation of the properties and morphology of the heterostructures[23,31,60,61] it is noted that after the chromium disilicide islands overgrowth with silicon picture corresponding to an atomically pure silicon surface Si (111) 7 × 7 kept for different thicknesses of chromium. It testifies about the epitaxial growth of silicon and the possible embedding of nanoscale hetero epitaxial islands $CrSi_2$ in the crystal lattice of silicon and transformation them into nano crystallites mostly spherical in shape during the integration process. The picture obtained after the deposition of silicon is also consistent with atomically pure surface, but in some cases, chromium disilicide nanocrystallites which not built into the crystal lattice of silicon come to the surface (Fig. 4.109). The quantity of $CrSi_2$ raised on a surface or located in the surface layer is extremely small. It testifies that nanocrystallites of CrSi2 remain on certain depth in an epitaxial layer of silicon that confirmed by experimental data.[57–59]

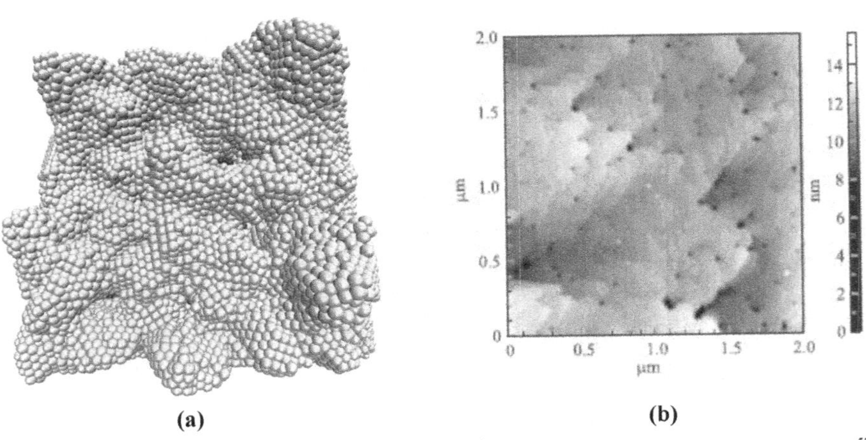

FIGURE 4.108 Heterostructures after silicon deposition (a) simulation and (b) experiment[57]).

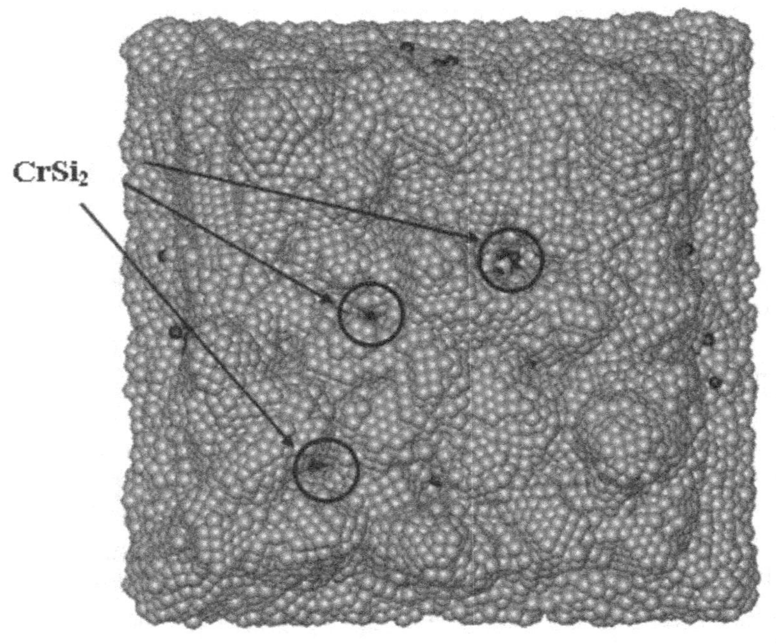

FIGURE 4.109 Simulation results (selected segments—non-embedded nanocrystallites of $CrSi_2$).

It follows that for large thicknesses of chromium not all chromium atoms are embedded in a silicon lattice that leads to formation of three-dimensional islands on a substrate surface. At the stage of silicon deposition silicon atoms, which penetrate into the surface layers of islands, form a molecules of chromium disilicide. Therefore it should be noted such moment as uniformity of chromium disilicide molecules distribution on islands surfaces, that is, their density per surface area unit. Figure 4.110 shows a plot of changes in the density of molecules $CrSi_2$.

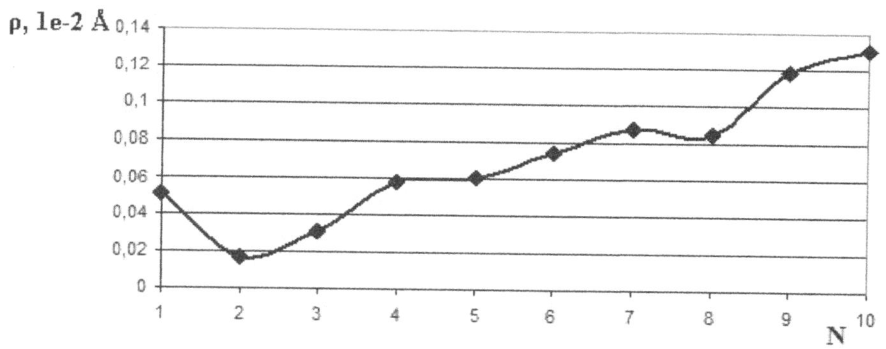

FIGURE 4.110 Assessment of uniformity of $CrSi_2$ molecules distribution on a surface of islands (ρ—density of $CrSi_2$ molecules and N—number of divisions).

In this case, a number of divisions mean the following. The surface of the island is shaped like a spherical segment with height h and diameter d. This surface is divided into several parts (in this case ten) by planes parallel to the substrate surface (Fig. 4.111). Schedule in Figure 4.110 illustrates the density of $CrSi_2$ molecules depending on number of divisions. Character of a curve in Figure 4.110 testifies to non-uniform distribution of chromium disilicide molecules on the island surface.

Let us consider how the quantity of chemical bonds, which formed at a stage of sedimentation of silicon, changes depending on the method of chromium deposition. As it was already noted, the method of chromium deposition affects the structure of samples received as a result of simulation. However, the quantity of the chemical bonds formed at stages of annealing and silicon deposition practically does not depend on it. It is fully confirmed by the graphs in Figure 4.112. In Figure 4.112, mixed deposition is figured by dark color (four monolayers layerwise and three at a time), layerwise sedimentation is shown by light color (seven layers layerwise).

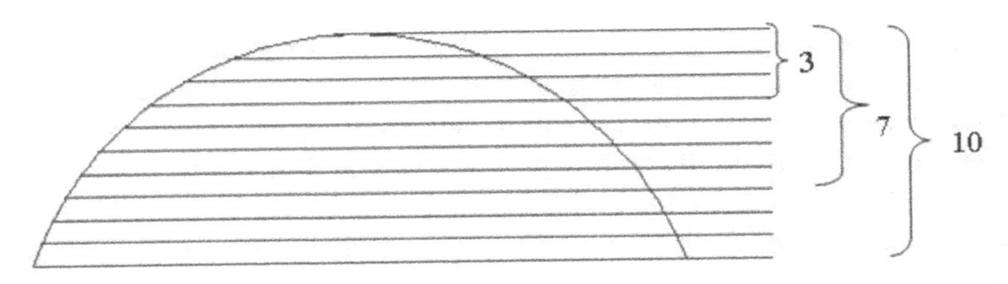

FIGURE 4.111 Scheme of the island surface division (projection to the y-z plane).

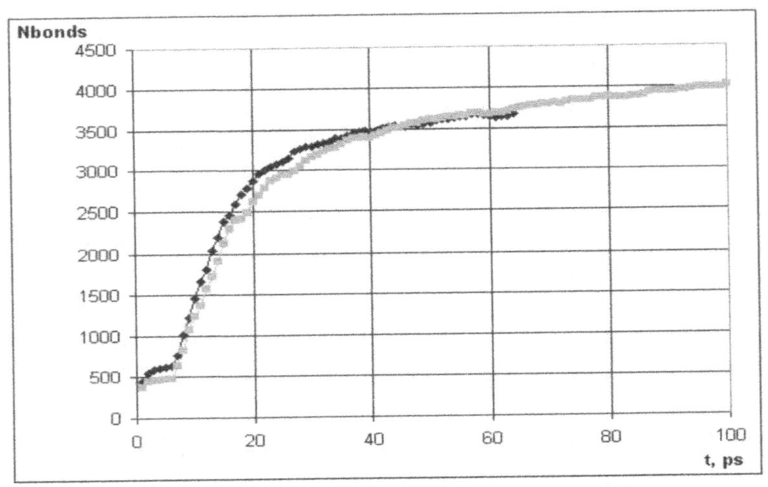

FIGURE 4.112 Change of number of chemical bonds formed at the stage of silicon deposition on time (—◆—mixed method and —■—layerwise deposition).

Experimental studies of multilayered heterostructures based on silicon and chromium show that during the quantum dots formation there is the formation of $CrSi_2$ nanocrystallites, and to neglect this fact would be not correct. Therefore, to assess the impact of accounting processes of formation and rupture of chemical bonds graphs of the kinetic energy of the system taking into account chemical bonds and the system without taking chemical bonds into account were considered. Figure 4.113 shows the graphs of the change of kinetic energies on time for described systems.

The provided diagrams explicitly testify that bonds formation leads to reduction of kinetic energy of system which in turn passes to energy of bonds and angular interactions. However, both in the first and in the second case the system comes to stable conditions where change of kinetic energy is minimal. It should also be noted that the accuracy of the set of bonds potential parameters influences the dependence of formed bonds number on quantity of deposited chromium. Diagrams given below (Fig. 4.114) correspond to value of b_0 = 2.2 Å. The pattern of similar dependence for value of b_0 = 2.3 Å is given in Figure 4.115. The thickness of four monolayers of chromium is optimal (in which the maximum number of bonds formed) for value b_0= 2.2 Å. For b_0 = 2.3 Å maximum number of bonds is formed by lowering the seven monolayers of chromium. However, for any value of b_0 graphics of bonds quantity change on time have non-linear character. Represented plots correspond to the stage of silicon deposition on the surface of islands and substrate.

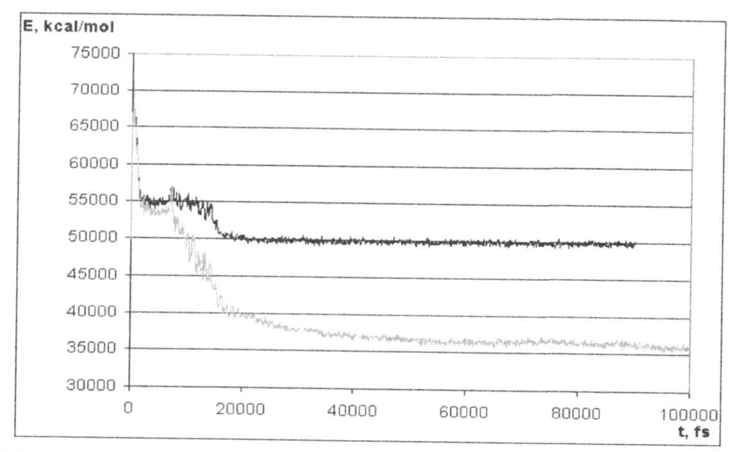

FIGURE 4.113 Change of kinetic energy on time (——system with chemical bonds and ——system without chemical bonds).

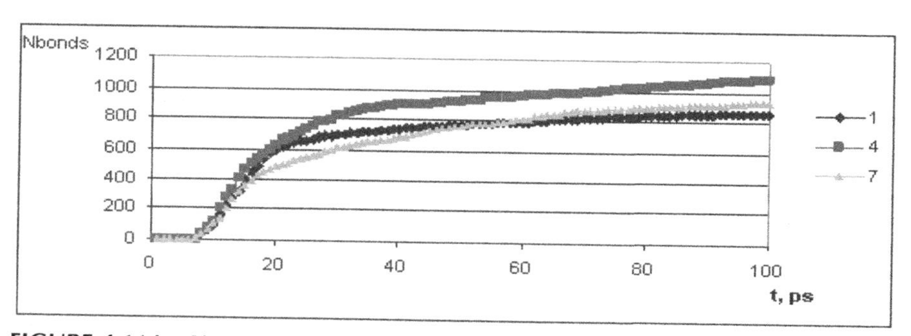

FIGURE 4.114 Change of the bonds number at different amount of deposited chromium for b_0 = 2.2 Å.

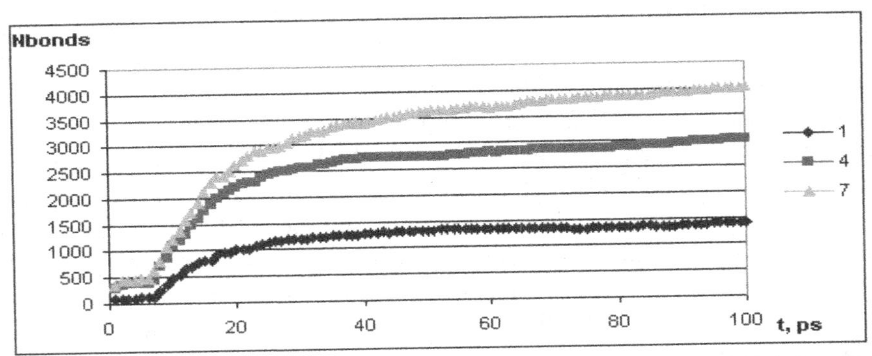

FIGURE 4.115 Change of the bonds number at different amount of deposited chromium for $b_0 = 2.3$ Å.

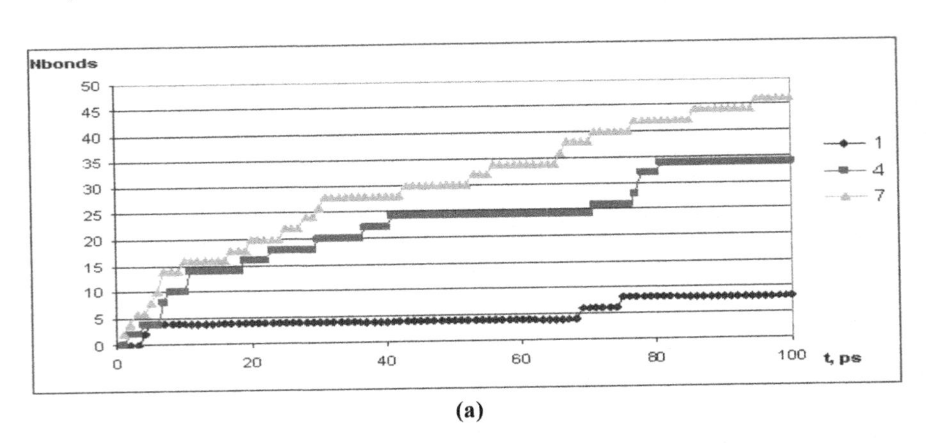

(a)

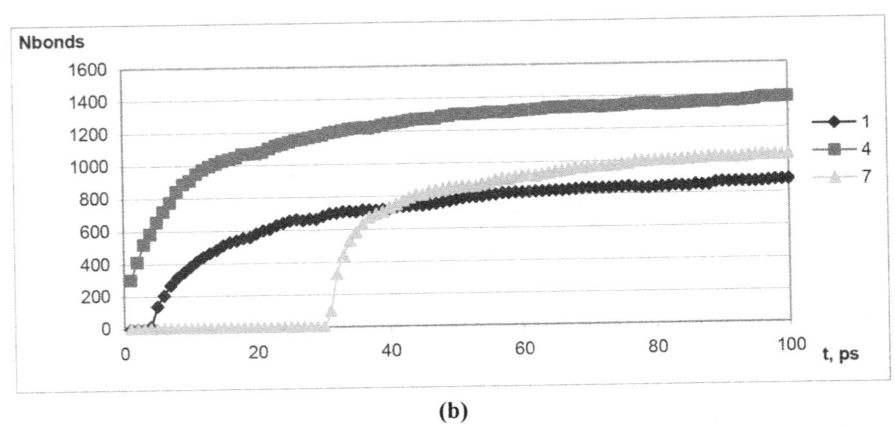

(b)

FIGURE 4.116 Change on time of the formed bonds number at the stage of chromium deposition (a) for $b_0 = 2.2$ Å and (b) for $b_0 = 2.3$ Å.

The effect of the equilibrium bond length b_0 is even more noticeable during deposition of chromium, where the number of formed bonds differs by two to three orders of magnitude (Fig. 4.116). It is seen that even character of the curves received at this stage of simulation in case of different values of b_0 is various. It once again confirms the fact that the exact determination of the parameters of interaction potentials is crucial at the modeling stage.

KEYWORDS

- **numerical simulation**
- **nanoparticle interaction**
- **nanocomposite coatings**
- **nanowhiskers formation**
- **quantum dots formation**

CHAPTER 5

NUMERICAL SIMULATION OF NANOSYSTEM PROPERTIES

CONTENTS

ABSTRACT

This chapter is dedicated to the numerical calculation of nanosystem properties. It is demonstrated that the liquid structure considerably changes under different current conditions and changes with time. The dependence of water viscosity on nanochannel diameter is obtained. The dynamics of complex nanosystem consisting of a nanotube and fullerene and different types of gases as applied to the processes of gas accumulation, storage, and isolation in the given system is considered. The dynamics of fullerene motion is studied, thermodynamic parameters defining the processes indicated are determined, and limiting characteristics of the given system on gas absorption are evaluated. The task for calculating the elastic properties of metal nanoparticles is given.

5.1 CALCULATION OF NANOSTREAMS PROPERTIES

Nanostreams are widely applied in such scientific disciplines as biology, chemistry, physics, and engineering. The forces expressed in liquids on nano level play a very important role in the nature of processes studied. Due to the peculiar molecular structure of the medium various phenomena not observed on the macro level are possible.[42,78] In particular, it was investigated that the formation of lamellar structure of liquid streams is mainly determined by molecule shapes. Thus, symmetric molecules are self-organized into pseudo-solid structure with clear layers; however, asymmetric molecules remain in irregular liquid state.[65,75,76,89] Another peculiarity of nanostreams is the slippage along the channel surface that significantly decreases the stream hydraulic resistance. In this case, the usual condition that the rate on the solid boundary equals zero becomes senseless. In theory this is expressed by introducing a so-called slippage length.[21,40, 107,109,130,131,182,184,193,197–199,201]

The investigation presented is dedicated to the modeling of water structuration in nanochannel, and also discusses the phenomenon of viscous stream slippage along the solid surface. The interaction of the system "liquid-channel" is modeled by the method of molecular dynamics (MD). The liquid motion is given with the help of hydrostatic pressure in the region modeled. The analog of water penetration through the membranes and shell of live cell is used as the main mechanism.

Solid surfaces of the channel are formed by multi-layer carbon nanotubes; this allows ignoring the surface roughness.

5.1.1 PROBLEM DEFINITION

The channel of single or multi-layer carbon tube filled with water molecules was considered as the region modeled (Fig. 5.1). Additionally, the hypothesis on the normal conditions of the surroundings (temperature of molecular system $T = 300$ K) was accepted. The numerical modeling was carried out by the methods of MD.

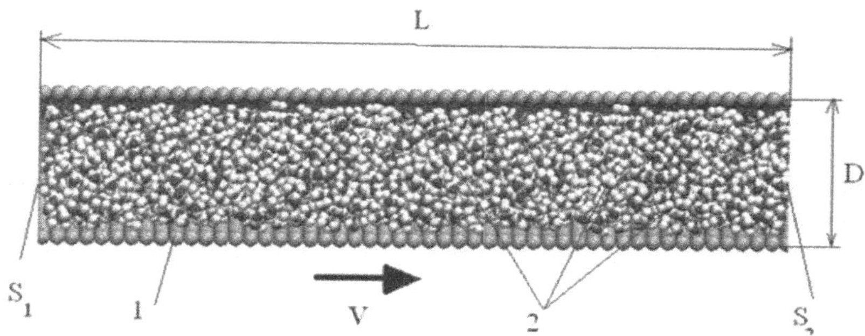

FIGURE 5.1 A carbon tube filled with water molecules.

The peculiarity of water composition is the availability of hydrogen bonds. The mechanism of their formation and destruction will be discussed later. At this stage, it should be pointed out that in numerical algorithms used the availability of hydrogen bonds is given in a fuzzy way. The energy contribution of this type of interactions is taken into account when calculating Van der Waals forces with the help of Lenard-Jones potential

$$\Phi^{ij}_{vdW} = 4\varepsilon_{ij}\left[\left(\frac{\sigma_{ij}}{r_{ij}}\right)^{12} - \left(\frac{\sigma_{ij}}{r_{ij}}\right)^{6}\right], \tag{5.1}$$

where ε_{ij}, σ_{ij}—potential parameters (depth of potential pit and equilibrium radius) that depend on the type of interacting atoms.

To model the liquid medium stream in the channel we used the approach applied, in particular, when calculating the liquid stream through nanopores and membranes. It is based on the modeling of pressure difference in the region calculated. For this, the force applied along the vector parallel to the liquid stream direction set is given.

Not all the liquid molecules, but only those at the entrance into the nanochannel are subjected to the action of the above force (external for the system). Thus, the indicated regions of the molecular system subjected to the action of external forces become the source of liquid motion.

As a result, the pressure value applied to the system externally is calculated as

$$\Delta P = P_{\hat{a}\tilde{o}\hat{\imath}\,\ddot{a}} - P_{\hat{a}\hat{u}\,\tilde{o}\hat{\imath}\,\ddot{a}} = \frac{n_{bulk} \cdot \leq f_{const}}{S_{bulk}}, \tag{5.2}$$

where n_{bulk}—number of molecules in the liquid layer; f_{const}—force applied to the liquid molecules; S_{bulk}—area of internal channel of the multi-layer tube.

In the calculations the correlations (5.2) were used together with periodic boundary conditions.

5.1.2 CALCULATION RESULTS AND ANALYSIS

At the first modeling stage the initial molecular system was formed (Fig. 5.2). It represents a multi-layer carbon nanotube filled with water.

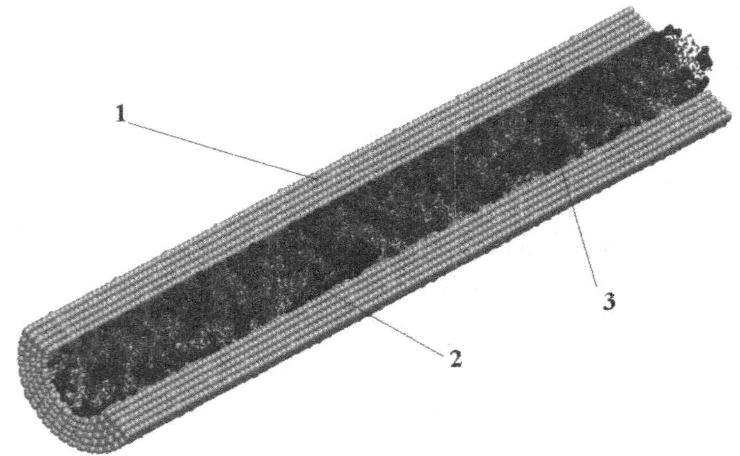

FIGURE 5.2 Initial molecular system in section: 1—multi-layer nanotube; 2—liquid molecules; and 3—cluster structures.

The wall thickness was selected in such a way that next layers did not practically influence the liquid inside the channel, that is, their interaction with the medium studied can be ignored.

The tube faces were fixed and the temperature $T = 300K$ was kept in the walls. Thus, the system of tubes enclosed acted as a damper medium absorbing the excessive heat from the liquid stream. As a result, the water moved at the temperature for which the experimental values of viscosity and density were known.

At the second stage the liquid medium behavior was investigated. The water filling the carbon nanochannel was heated up to the normal temperature $T = 300K$. In the course of numerical experiment, the capillary effect expressed in pulling the liquid inside the channel due to surface tension force was observed. A series of calculations demonstrated that if all numerical investigations were carried out without the aforesaid phenomenon, the cavities were formed inside the moving stream (Fig. 5.3).

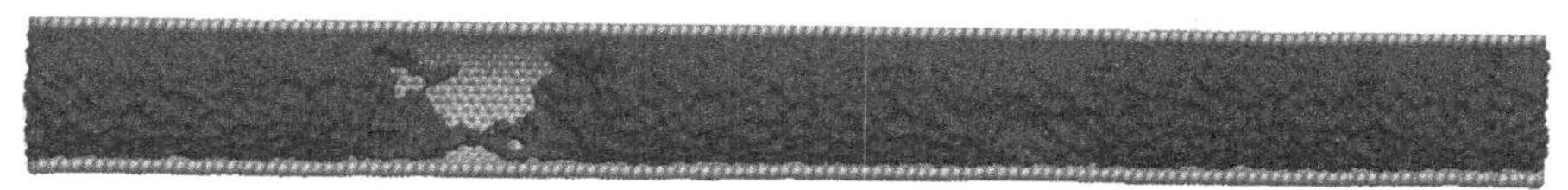

FIGURE 5.3 Formation of cavities in the liquid.

Further using the above methodology, the water stream along the carbon nanotube was modeled. Figure 5.4 demonstrates the calculation results, namely, the stream initial stage. It is seen that the consumption graph is obviously of discrete character, the value of growth or decrease of the consumption is proportional to the atomic mass of water molecule.

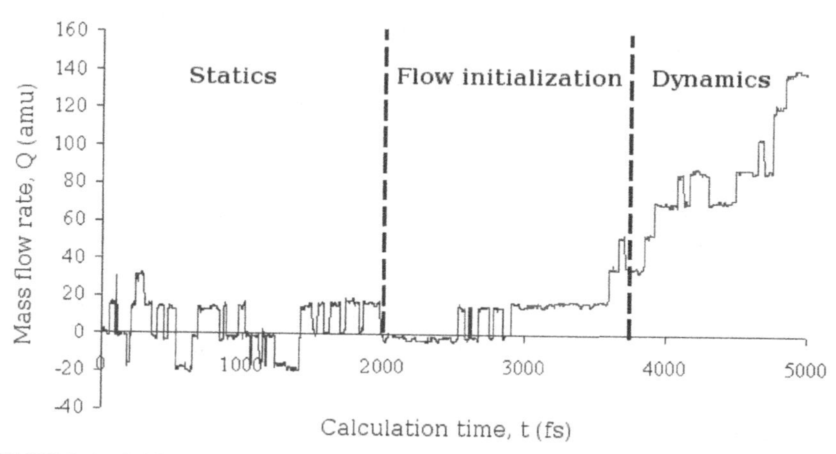

FIGURE 5.4 Initiation of molecular stream inside the carbon nanotube.

The time interval 0…2000 fs corresponds to the static state of the liquid. In this period the system accumulates the energy to start the liquid motion due to the pressure source at the channel entrance. The consumption in the central section is positive as well as negative, since the liquid molecules move chaotically under the action of Brownian movement.

The next time interval 2000…4000 fs does not contain the negative consumption values and moves to the region of water motion in the positive direction. Thus, the pressure impulse transfers between the liquid layers practically instantly, the relaxation time is 10^{-12}c, this confirms the hypothesis on water incompressibility in molecular systems as well.

As discussed earlier, the peculiarity of water chemical composition is the availability of hydrogen bonds between liquid molecules. They are formed under certain conditions and determine "the outstanding" properties of this liquid. In particular, these are relatively high boiling and melting temperatures conditioned by a high water specific heat capacity in solid and liquid states, as well as the strength of "diamond" packing of ice molecules. During the water phase transitions from one state into another, the hydrogen bonds are considerably destructed, or, on the contrary, they are formed.

The calculations showed that when the water moves inside the nanochannel, the hydrogen bonds between certain molecules are constantly formed and destructed. This results in the formation of water cluster structure (Fig. 5.5).

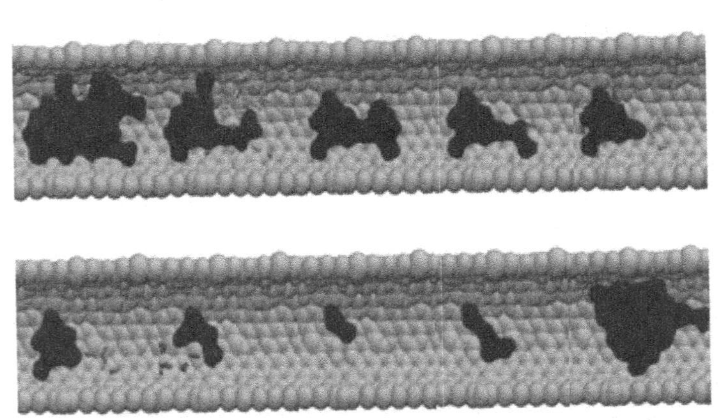

FIGURE 5.5 Evolution of cluster in water from the formation till destruction.

From the Figure 5.5 it is seen that when the structure changes as a static system "nanotube–water" and when the liquid motion is available, the sizes and shapes of water clusters change. These changes are connected with the processes of association (incoming) of water molecules, and dissociation (separation) of molecules from the cluster.

It is interesting to trace the evolution of certain clusters, in particular, the largest at the given moment of time. For this we introduce the notion of cluster nucleus. In accordance with the algorithm developed, one of the new clusters into which the initial one is split in the process of liquid flow and containing the largest number of molecules of the initial cluster is the cluster nucleus.

The time interval during which molecules contained in other clusters, similar in size with the given one, for example, as in Figure 5.5, are attaching to the cluster is taken as the cluster lifetime. In the process of water cluster evolution, they can attach to other clusters or completely destroy to the molecule level (Figs. 5.6 and 5.7).

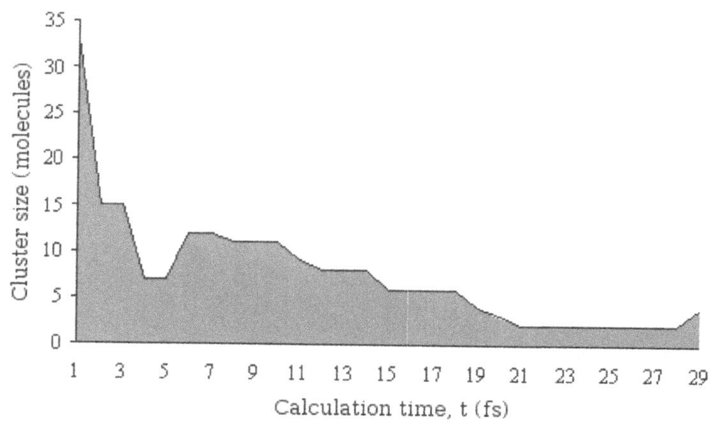

FIGURE 5.6 Change in the independent cluster sizes before its destruction.

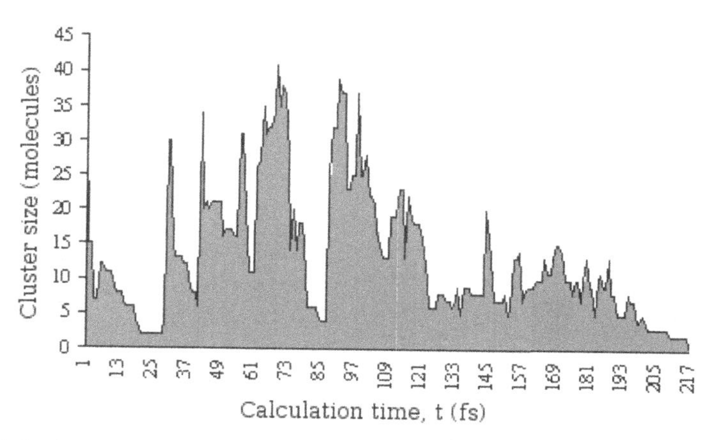

FIGURE 5.7 Process of cluster formation existence with the attachment to other elements of liquid structure.

The typical view of the cluster is given in Figure 5.8. It is seen that it represents a complex structure combining liquid molecules.

By the quantitative evaluation of molecular stream it is revealed that the channel resistance to viscous liquid stream decreases considerably. It is found that hydrogen bonds and clusterization of liquid stream are mainly responsible for this. Let us consider how the water cluster structure changes with the increase of pressure gradient in the liquid stream.

FIGURE 5.8 Example of cluster structure.

The results given below correspond to the system modeled consisting of 31,152 atoms. In the range of pressure gradients investigated the average size of maximum cluster (Fig. 5.9) changes from 35 to 17 molecules of H_2O. That is water structure is refining.

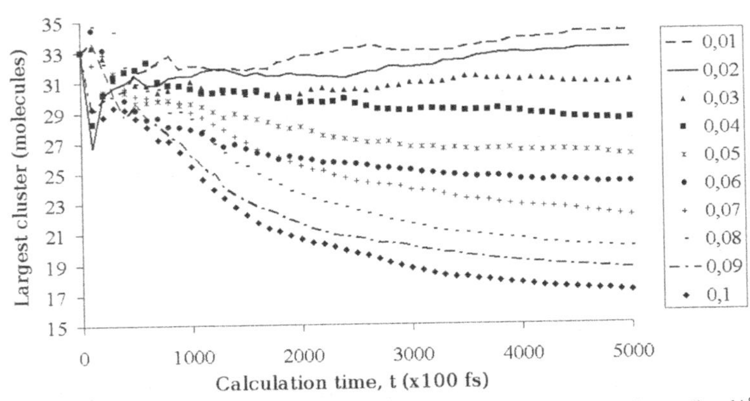

FIGURE 5.9 Size of maximum cluster depending on the force applied f_{const} = (kcal/Å· mol).

Simultaneously the number of clusters in viscous stream increases from 118 up to 135 (Fig. 5.10).

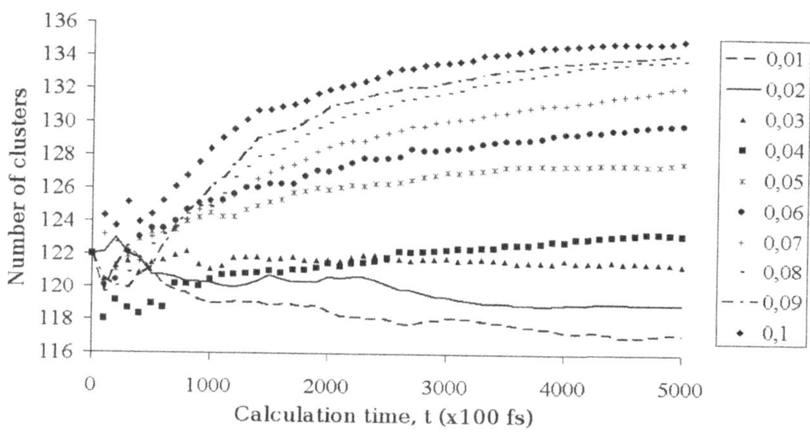

FIGURE 5.10 Number of clusters in the system depending on the force applied f_{const} = (kcal/Å· mol).

The change in the cluster structure is also seen in the change in the share of structured liquid from 77 to 64% (Fig. 5.11).

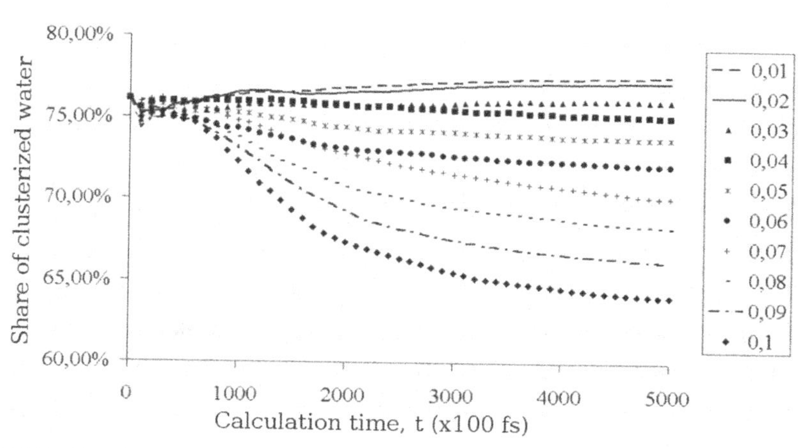

FIGURE 5.11 Share of clustered water depending on the force applied f_{const} = (kcal/Å· mol).

Apart from the water cluster structure in nanochannel the behavior of certain liquid molecules both near the wall and in the stream core was investigated.

The spatial orientation of certain molecules and the liquid medium can be evaluated by the direction of dipole moment. In Figure 5.12 you see the graph of the time change reduced to the number of molecules of the system dipole moment along the axis Oz. It is seen that the time interval 0...400 ps corresponds to the period of liquid stream establishment. In a stationary stream the value of dipole moment is constant.

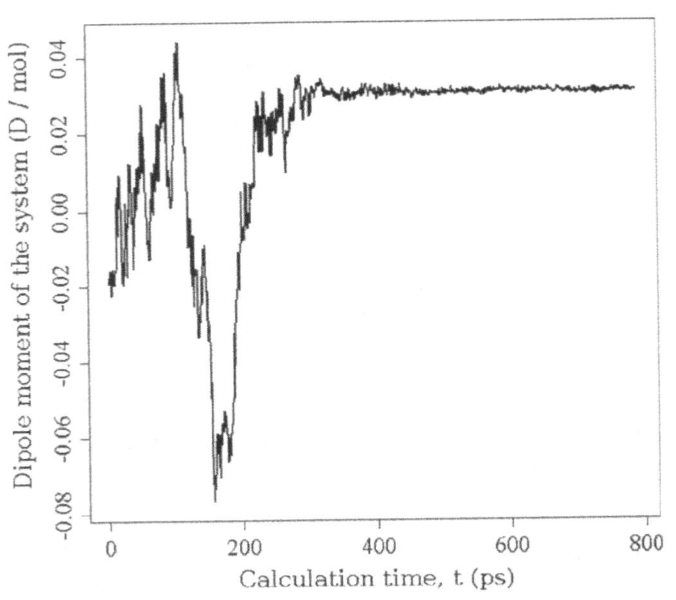

FIGURE 5.12 Dipole moment of the system with the slippage on the wall.

The situation is different if the condition of adhesion to the solid surface is set (Fig. 5.13). Although the stream itself reaches the quasi-stationary mode, "the regularity" in the viscous medium structure is not observed.

In Figure 5.14 you see the graph of the function of coupled atomic distribution of g(r) water molecules relatively to carbon molecules forming the channel solid surface. The values are averaged along the channel length and the trajectory of 1 ns. This dependence reflects the distribution of liquid density in the channel cross-section.

The investigations of liquid medium density in carbon nanotube showed that "the depleted" layer, which does not contain the liquid molecules is formed near the solid surface. On the contrary, liquid molecules form dense packing in the center of the channel.

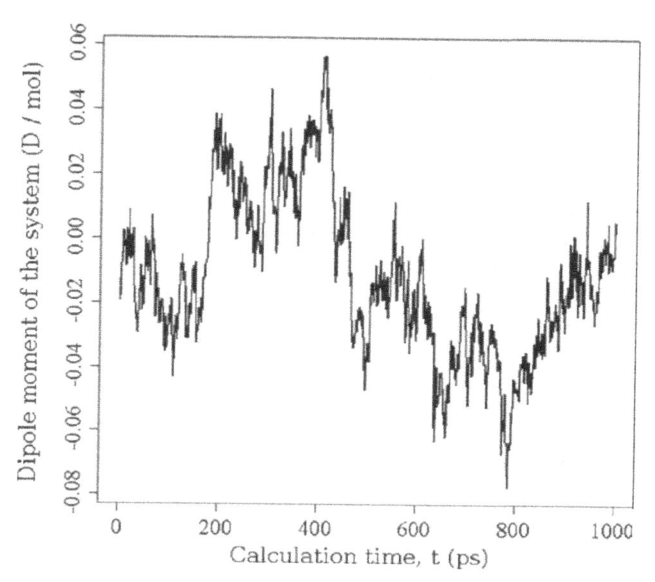

FIGURE 5.13 Dipole moment of the system with adhesion to the solid surface.

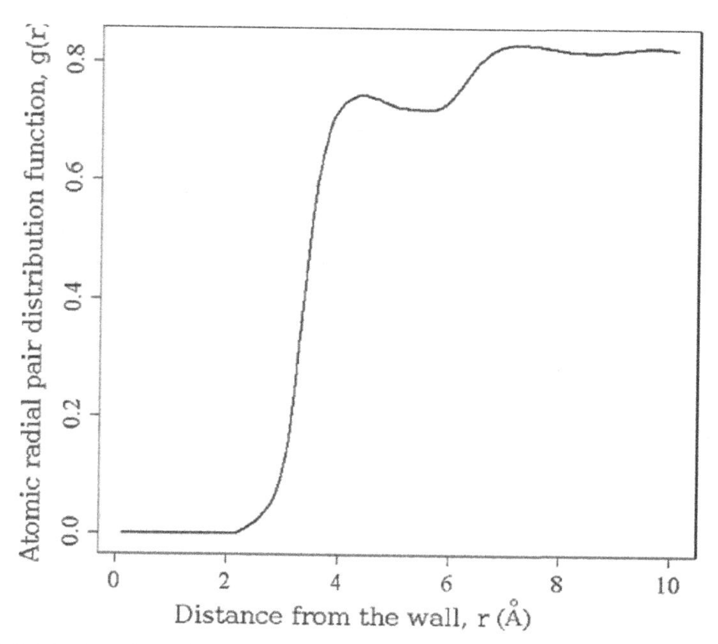

FIGURE 5.14 Function of coupled atomic distribution of g(r) atoms of H_2O relatively to the atoms of carbon, C.

This fact explains the availability of slippage and observed increased consumption characteristics of carbon nanotubes.

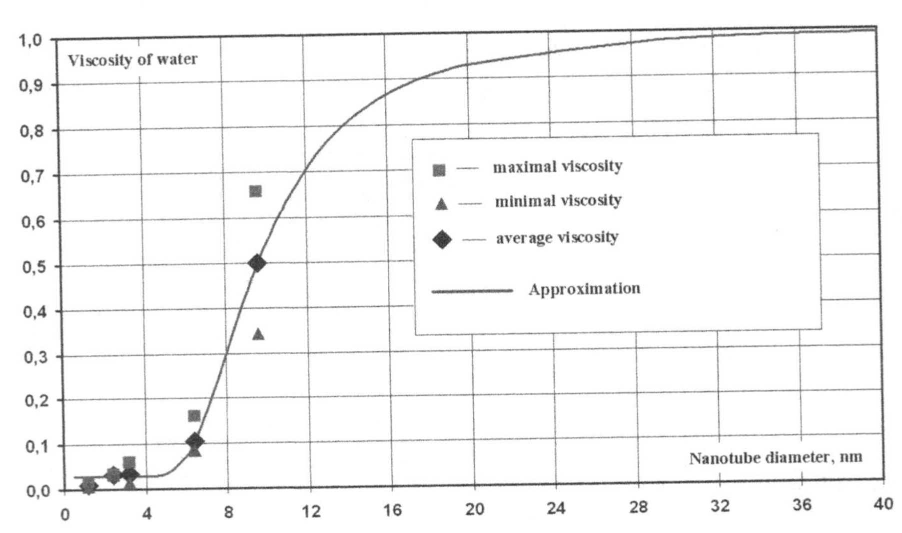

FIGURE 5.15 Changing the viscosity of the water in the flow of inside nanotubes, depending on the diameter of the nanotubes.

Very interesting a graph of viscosity on nanotube diameter (Fig. 5.15), for small diameter nanotubes $d < 5$ nm viscosity of water is very small and is 0.01–0.05 of the viscosity of water at the macro flow. Then, by increasing the nanotube diameter, the viscosity of water increases dramatically and tends to the viscosity of water at the macro flow.

5.1.3 CONCLUDING REMARKS

Thus, in the course of investigation it was found out that viscous properties of the liquid change due to the reconstruction of water cluster structure based on a special type of interactions—hydrogen bonds between water molecules.

It was revealed that the processes of cluster structure formation in the molecular liquid investigated flow in a free state. The start of the motion of viscous stream along the nanochannel is characterized by the destruction of structures indicated.

The size of the maximum cluster, total number of clusters and share of structured liquid depend on Reynolds' number and have the fixed average value in a stationary stream. It was found that "the depleted" layer is formed near the nanochannel wall, defining the liquid slippage along the solid surface.

In the channel with a hydrophobic surface the homogeneous orientation of water molecules is observed both along the channel axis and radially in the process of stream setting. When the smoothness of channel surface is distorted and the adhesion occurs, the liquid layers are mixed, the orientation of molecules is constantly changing, and consumption characteristics of the stream decrease by several orders.

The results obtained highlight the fact that structural peculiarities of the viscous medium and mechanisms of its interaction with the solid surface are the determining factor influencing the properties of molecular streams.

5.2 MODELING OF ABSORPTION PROPERTIES OF CARBON NANOCONTAINERS FOR GAS STORAGE

The discovery of fullerene and nanotube initiated a wide interest to their properties. One of the research areas is the investigation of adsorption properties of nanostructures. Nanocapsules can be such nanostructures. They usually consist of two elements: a nanocontainer, where the gas is stored, and a blocking nanoelement,[161] which allows them absorbing the gas at a low temperature and under a high pressure, and maintain it under the normal conditions. The gas isolation from them has to be controlled.

The works on the investigation and application of nanocontainers (nanocapsules) are being actively developed in various fields: medicine, electronics, and substance storage.

The properties of nanocontainers have been actively discussed and investigated by computational simulation and experiments for a number of years for the storage of hydrogen, methane, and noble gases.[17,69,91–92,113,114,126,136,148–149,176,177,188–190,192,194,195] Nanocontainers can be divided into two main groups: single-use (destructed during the utilization of substances stored in them) and multi-use (opened and closed in accordance with technological regulations). For multi-use nanocontainers "bucky shuttle" carbon nanostructures synthesized by Smith B.W. and Luzzi D.E. in 2000 are usually used. A multi-use nanocontainer is a nanotube with one or several fullerenes in the internal cavity. Figure 5.16 demonstrates a similar structure. "Bucky shuttle" carbon nanostructures are a good prototype for creating the ideal nanocontainer, in which the nanotube internal space is used to store the required substance, and the fullerene—as a blocking nanoparticle preventing the substance stored from emission.

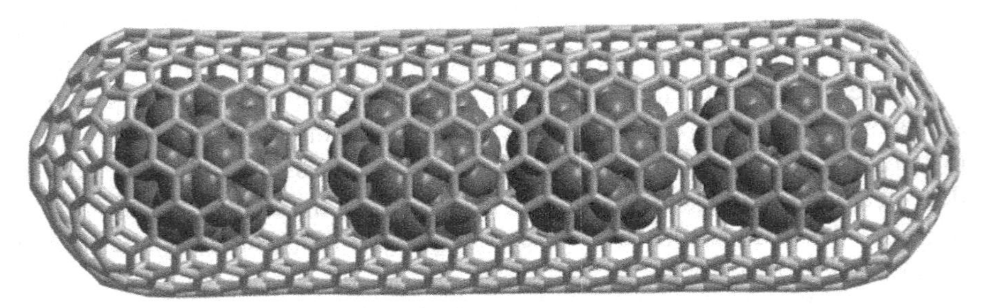

FIGURE 5.16 "Bucky shuttle" single-wall carbon nanostructure: nanotube (10.10) containing fullerenes C_{60}.

It should be pointed out that theoretical aspects of gas storage in a nanocapsule, physical mechanisms of their accumulation and isolation, dynamics of nanocapsule opening and closing have not been studied sufficiently so far.

The processes of adsorption, storage, and desorption of various gases with different absorbers have been purposefully investigated at our group, including self-organizing nanocontainers and those controlled by electrostatic field based on "bucky shuttle" single-wall carbon nanostructures.[161] Figure 5.17 demonstrates schematic images of these nanocontainers.

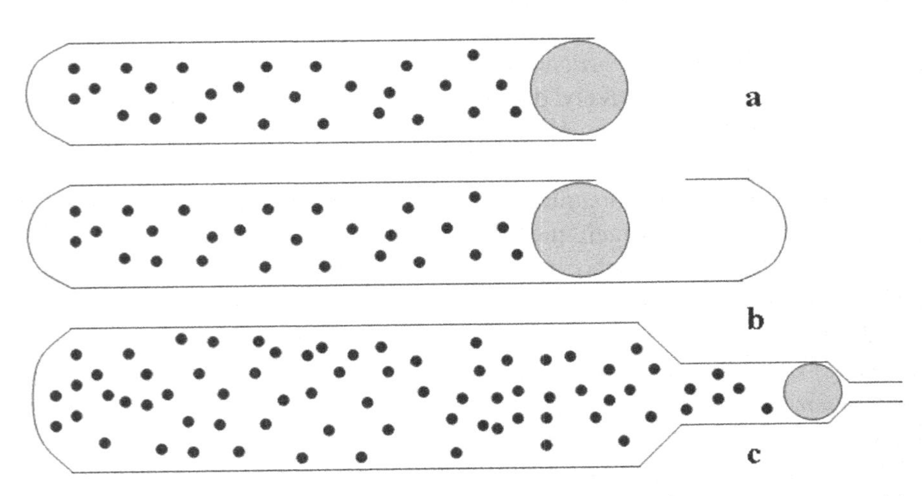

FIGURE 5.17 Nanocontainers based on "Bucky shuttle" carbon nanostructures: (a) nanocontainer in which the fullerene closes the exit from one end; (b) nanocontainer with the structure defect; and (c) nanocontainer with the modified structure for the increased gas content.

In this section the summarized results of the dynamics of filling and opening processes of multi-use nanocontainers based on "Bucky shuttle" carbon nanostructures are given and methods to increase mass proportions contained in nanocontainer are theoretically analyzed, the main being: the pressure increase inside the nanocontainer and increase in its useful volume.

5.2.1 PROBLEM FORMULATION

Modeling of adsorption, storage, and desorption of molecular hydrogen by nanocapsules was carried out with the method of MD. In Figure 5.18 you can see the calculation scheme.

When modeling, five types of interaction existing in the nanosystems were considered: wavelength, bond angle change, torsion angles, flat groups, and Van der Waals interactions. The bound interactions of atoms define the structure of fullerenes, nanotube, and gas molecules. Free interactions of atoms describe the adsorption of gas molecules onto the nanotube and fullerene, as well as the interactions between the nanotube and fullerene. To calculate the charge values of hydrogen and carbon atoms in gas molecules the following potential was applied: B3LYP/6-31G(d).[18,19] The calculations were made with the program Gaussian.[27]

Type of potentials and values of their force constants are given in Ref. 114. For numerical modeling the leap-frog algorithm was used.

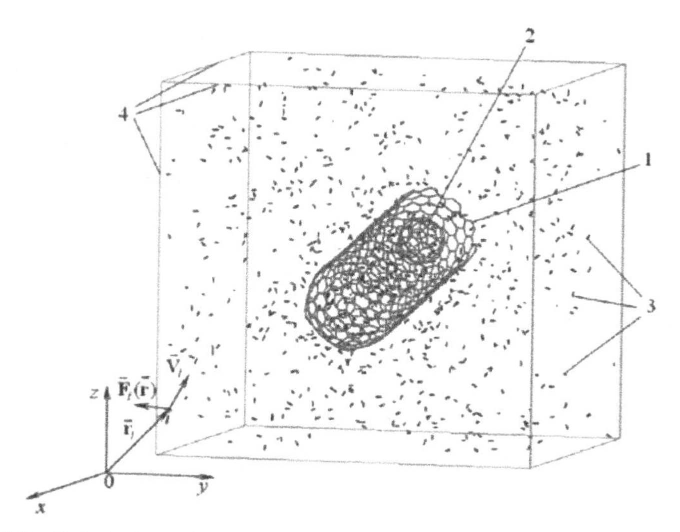

FIGURE 5.18 Calculation scheme: 1—nanotube, 2—fullerene, 3—gas molecules, and 4—rated cell with periodic boundary conditions.

Let us consider the aspects of dynamics of various types of nanocontainers for gas storage.

5.2.2 NANOCONTAINERS OPERATING UNDER THE CHANGE OF THERMODYNAMIC PARAMETERS

Figure 5.19 demonstrates a single-wall armchair nanotube (10.10) and fullerene C_{60}, forming a nanocontainer, as well as hydrogen molecules. The nanotube (10.10) has the internal diameter of 11.8 Å, and external—15.2 Å. The length of the nanotube (10.10) is 32 Å. Fullerene C_{60} best suits to block substances, including gases, in the nanotube (10.10). The distance between the wall of the nanotube (10.10) and fullerene C_{60} does not allow hydrogen molecules pass between them.

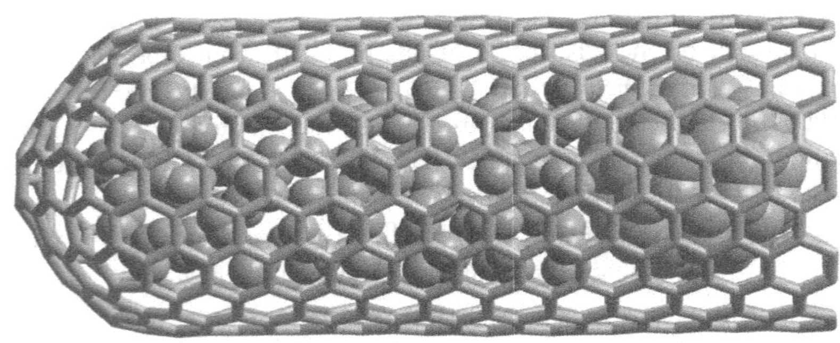

FIGURE 5.19 Nanocontainer consisting of the nanotube (10.10) and fullerene C_{60} containing hydrogen molecules.

Initially the molecular-dynamic modeling of adsorption of hydrogen molecules into the nanotube (10.10) internal space was carried out. The modeling was carried out with the canonical assembly (with constant temperature, volume, and number of particles in the system) at 77 K and under 10 MPa. To investigate the nanotube blocking by the fullerene under the action of capillary forces, the assembly was converted into microcanonical (with constant energy, volume, and number of particles in the system) at the initial temperature 77 K. Initially the fullerene C_{60} was placed near the nanotube edge. During the modeling process C_{60} was attracted to the nanotube shear under the action of capillary forces, however, too close to the nanotube wall. As a result, the fullerene was repelled. The next try was more successful, and the fullerene got inside the nanotube under the action of capillary forces, pressing the hydrogen there. The dependence of C_{60} movement inside the nanotube (10.10) on time is shown in Figure 5.20, where zero distance L corresponds to the nanotube shear. The fullerene fluctuations inside the nanotube are dying. This is explained by the action of capillary forces trying to keep the fullerene inside the nanotube and counteraction of compressed hydrogen pressure.

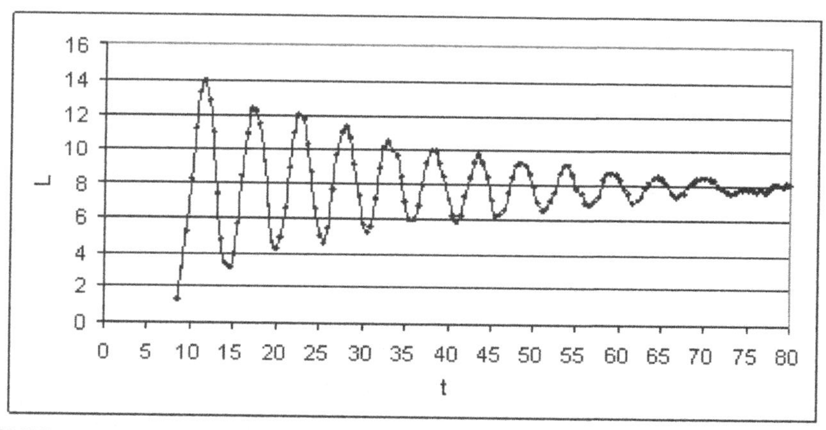

FIGURE 5.20 Fullerene oscillations in nanocontainer, where *L*—immersion depth into the nanotube (Å), and *t*—time (ps).

Change in the system temperature: nanotube (10.10), C_{60}, and hydrogen molecules are shown in Figure 5.21. Here the first temperature peak corresponds to the initial unsuccessful try of C_{60} to penetrate into the nanotube. Further, with the second attraction of C_{60} to the nanotube a sharp temperature growth in the system up to 97 K is observed, when the fullerene is immersed into the nanotube to the maximum depth—the second peak (10–15 ps). The second peak is especially interesting—it is double, that is, there is an insufficient temperature drop in its center, then the temperature elevates again.

The temperature drop is conditioned by a sharp drop of the fullerene rate when reaching the extreme point of hydrogen compression inside the nanotube. As a result, the system kinetic energy decreases and, consequently, the temperature drops. After passing the extreme point the fullerene rate increases due to the pressure of expanding hydrogen, and the system temperature elevates again. At the next peaks the similar situation is observed, but, in the course of time, that is, with the dying of oscillations of C_{60} in nanotube (10.10), double peaks disappear merging into one. Further oscillations of C_{60} determine temperature fluctuations as well, which decrease with the reduction of fullerene oscillation amplitude, that is, temperature peaks in Figure 5.21 become diffused.

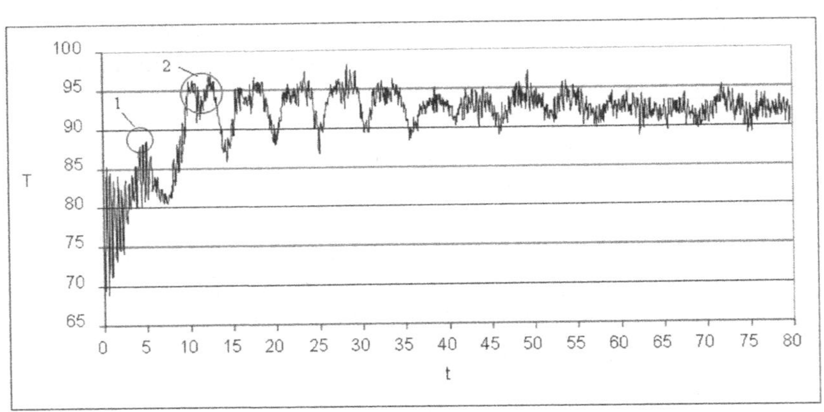

FIGURE 5.21 Change in the temperature of nanocontainer with hydrogen at its blocking, where *T*—temperature (K) and *t*—time (ps).

In Figure 5.22, there is a graph of hydrogen density change inside the nanotube (10.10) due to the oscillations of fullerene C_{60}. Hydrogen density at the first peak nearly reached the density of liquid hydrogen. However, fullerene C_{60} produced dying oscillations; therefore, the density fluctuations further became insignificant, conditioned by heat oscillations of fullerene and nanotube walls.

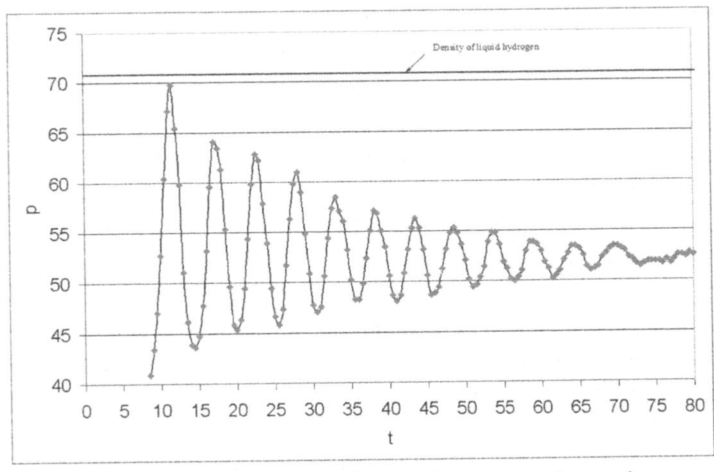

FIGURE 5.22 Change in the hydrogen density in nanocontainer, where *p*—hydrogen density inside the nanotube (kg/m^3) and *t*—time (ps).

When the temperature is close to the temperature of the fullerene outlet from nanotube (10.10), the partial removal of the fullerene from the nanotube edge is observed. At the same time, the insignificant emission of hydrogen from the nanotube occurs. After that C_{60} returns to the nanotube again. The outlet of C_{60} from nanotube (10.10) at the nanotube length of 32 Å takes place at 550 K.

When modeling the interactions of bigger fullerene C_{240}, the nanotubes with the parameters (15.15) and hydrogen molecules the following results were obtained. The system: hydrogen molecules, C_{240}, and nanotube (15.15)—more inertial than the aforesaid nanotube (10.10) and C_{60}, the oscillations take place in a more extended time interval. The period of oscillations of C_{240} is two times longer than that of C_{60}. The hydrogen desorption from the nanocontainer occurs at 620 K, when hydrogen extrudes C_{240} from the nanotube (15.15).

5.2.3 NANOCONTAINERS WITH A CHARGED BLOCKING ELEMENT

One of the ways to control the location of the blocking fullerene is the action of electrostatic filed on it. The location of the charged fullerene in the nanotube determines the stages of adsorption, storage, and desorption of the substances stored. A similar operation mechanism does not depend on thermodynamic conditions under which the system storage or discharge occur, as described above. The sample of the nanocontainer consisting of the nanotube with the parameters (10.10) with the defect (hole) in the wall, which is rather big for gas molecules' penetration into the nanotube internal space, and charged fullerene C_{60}^{5+}, is given in Figure 5.23. The nanotube (10.10) is of a single-wall armchair type with internal diameter 11.8 Å and external—15.2 Å.

Functioning of the nanocontainer for hydrogen storage is split into the following stages:

(a) preparation and adsorption stage: C_{60}^{5+} is deflected by an electric field to the right end of the nanotube for the adsorption of hydrogen molecules into the nanotube through the defect. The system temperature during modeling is 77 K, pressure—5 MPa.
(b) hydrogen compression stage: electric field intensity changes to the opposite. C_{60}^{5+} compresses the adsorbed hydrogen molecules in the left end of the nanotube. Now the hydrogen molecules cannot leave

the nanotube through the defect as they are hindered by the charged fullerene. Thermodynamic conditions of the system modeling are the same as at the adsorption stage.

(c) hydrogen storage stage: thermodynamic parameters of the system are brought to normal. The value of the electric field intensity is the same. The hydrogen pressure in the nanotube increases due to the temperature elevation, thus resulting in the deflection of the charged fullerene from the initial position. However, the fullerene deflection is insufficient for opening the defect and further hydrogen desorption.

(d) hydrogen desorption stage: vector of the electric field intensity changes its direction. The charged fullerene moves to the right part of the nanotube opening the defect through which hydrogen molecules are desorbed. Thermodynamic conditions of desorption—temperature: 300 K, external pressure: 0.1 MPa.

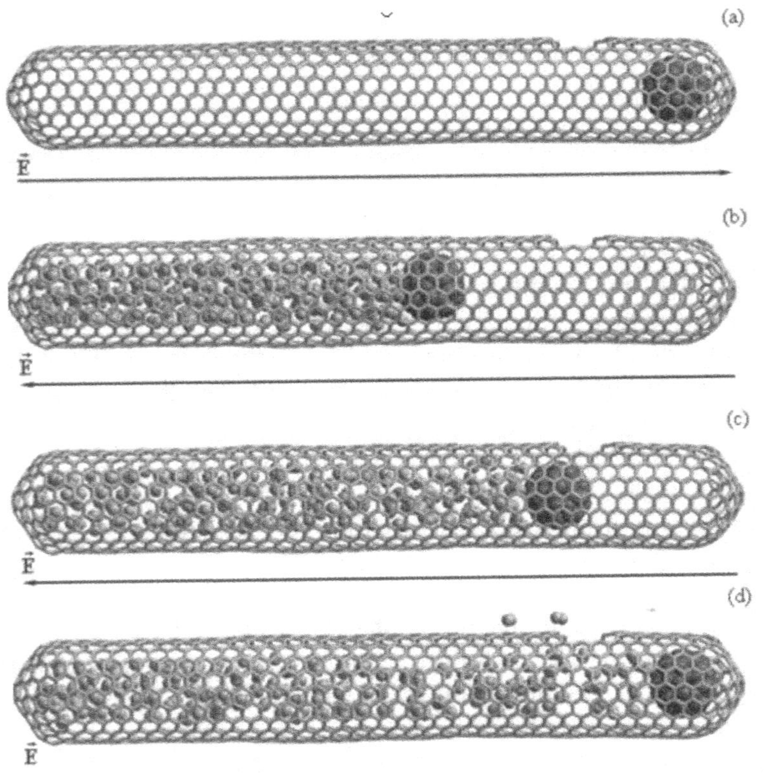

FIGURE 5.23 Functioning stages of the nanocontainer for hydrogen storage, consisting of the nanotube with parameters (10.10) and charged fullerene C_{60}^{5+}.

The value of external electric field intensity obtained with the help of modeling $\vec{\mathbf{E}}$, required to move the charged fullerene C_{60}^{5+} in the nanotube (10.10) and equaled to 1.044×10^9 V/m, is in good compliance with the modeling results described in Ref. 150, where the value 1×10^9 V/m was obtained. The value $\vec{\mathbf{E}}$ is determined by the necessity to overcome the attracting forces at the ends of the nanotube, in which the fullerene is located at the stages of adsorption preparation and desorption. The nanotube length is 96 Å. Coordinate L characterizes the distance from the left end of the nanotube to fullerene. The fullerene is in the right end of the nanotube. The hydrogen molecules penetrate into the nanotube through the defect and are adsorbed onto the internal surface of its walls.

The stage of hydrogen molecule compression in the nanotube took place at T = 77 K and P = 5 MPa. The compression as described above is carried out for hydrogen molecules not to desorb from the nanotube due to blocking the access to the defect moved by the charged fullerene. C_{60}^{5+} moves due to the change in the direction of electric field, which intensity $\vec{\mathbf{E}}$ stays constant.

In Figure 5.24, you can see the change in the coordinate of the charged fullerene C_{60}^{5+} in the nanotube with the parameters (10.10) at the stage of hydrogen compression. Approximately to 7 ps the coordinate L practically did not change, this is explained by the attraction of the right end of the nanotube end, where the charged fullerene was located at the previous stage, as well as relatively low electric field intensity. After C_{60}^{5+} overcome the attraction of the nanotube end, the dying oscillations started ending when by 70 ps the fullerene was at the distance 42 Å from the left end of the nanotube.

The change in the hydrogen density is characterized in Figure 5.25.

From the figure it is seen that when the fullerene is in the right part of the nanotube and does not cover the defect, the hydrogen density is constant and equals 30.9 kg/m^3. After passing the defect region, the compression of hydrogen molecules starts in the left part of the nanotube resulting in the sharp increase in hydrogen density with the peak (70.3 kg/m^3) nearly at liquid hydrogen density (70.8 kg/m^3). After passing the peak, the energy of compressed hydrogen is sufficient, overcoming the electric field action, to nearly restore the hydrogen density in the nanotube before the compression. Further asymptomatically dying fluctuations of hydrogen density conditioned both by the action of the expanding compressed hydrogen and electric field counteraction are observed. The final density is ~55 kg/m^3. Insignificant density fluctuations are explained by heat movements of the nanotube atoms changing its volume and oscillations of charged fullerene C_{60}^{5+}.

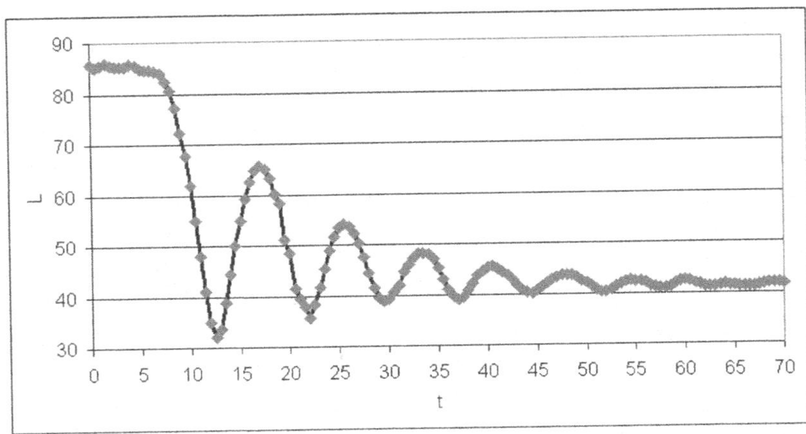

FIGURE 5.24 Movement of the fullerene C_{60}^{5+} along the axis of the nanotube (10.10) at the stage of hydrogen compression, where *L*—distance from the left end of the nanotube to the fullerene (Å) and *t*—time (ps).

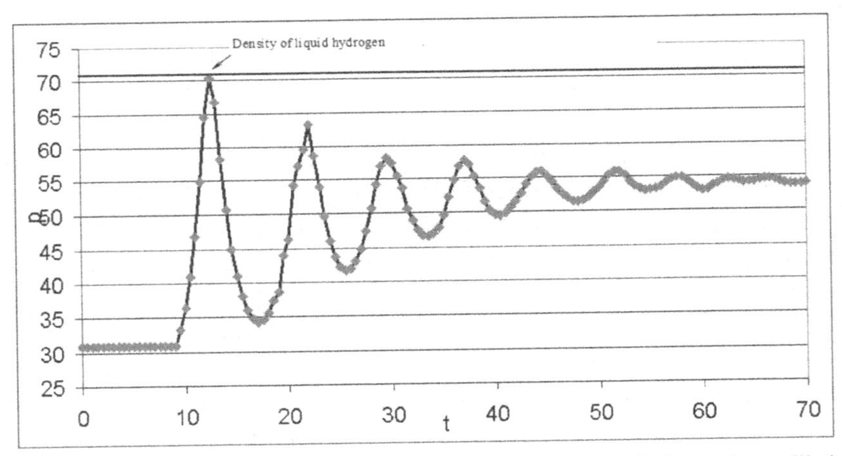

FIGURE 5.25 Changes in hydrogen density in the nanotube (10.10) due to the oscillations of the charged fullerene C_{60}^{5+} at the compression stage, *p*—hydrogen density inside the nanotube (kg/m) and *t*—time (ps).

5.2.4 NANOCONTAINERS OF INCREASED CAPACITY WITH A CHARGED BLOCKING ELEMENT

To increase the quantity of gas stored in nanocontainers it is necessary to use nanotubes of "bottle-like" shape. In Figure 5.26, you can see the operation stages of the container containing different single-wall armchair nanotubes:

(20.20), (10.10), and (8.8), combined by pentagonal and heptagonal rings. The internal and external diameters of the nanotubes in the nanocontainer: (20.20)—25.3 and 28.7 Å, (10.10)—11.8 and 15.2 Å, (8.8)—9.0 and 12.4 Å, respectively.

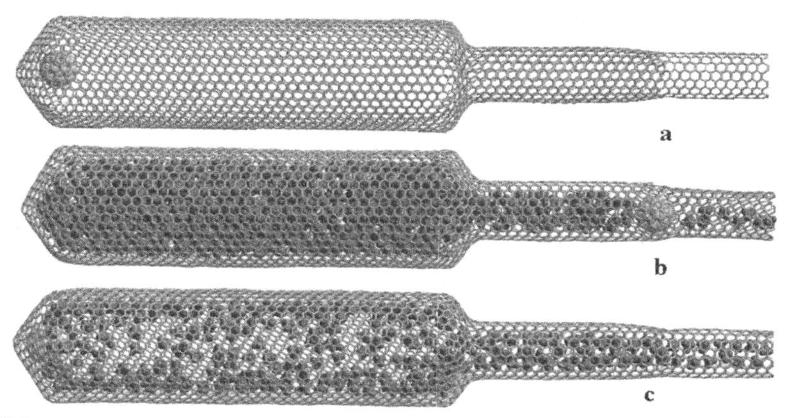

FIGURE 5.26 Functioning stages of the nanocontainer for natural gas storage: (a) empty nanocontainer ready to adsorb the natural gas molecules ($P = 10$ MPa); (b) K@C60 is moved to the narrow part of the nanocontainer by the electric field, methane molecules cannot leave the nanocontainer internal space ($P = 0.1$ MPa); and (c) K@C60 is moved to the thickened part of the nanocontainer by the electric field and methane molecules are freely desorbed from the nanocapsule internal space ($P = 10$ MPa).

The nanocontainer contains the endohedral complex K@C60, in which potassium atom has a singular positive charge. The nanocontainer diameter d_2 in the section (10.10) is large enough for the penetration of K@C60, but small for gas molecules to move between K@C60 and the nanocontainer wall. The nanocontainer diameter d_3 in the section (8.8) is already insufficient for the movement of K@C60; however, the internal space is sufficient to move methane molecules. The nanocontainer section (20.20) serves to accumulate and store gas molecules. The formation of similar structures is possible with a specific change in the nanotube structure while irradiating with ions or electrons. The movement of K@C60 in the nanocapsule is carried out under the electrostatic field action.

The nanocontainer operation can be split into several stages: adsorption, storage, and desorption. At the adsorption stage the endohedral complex K@ C60 is near nanocontainer base (left end of the nanocontainer). The methane molecules penetrate through the hole in the nanocontainer and adsorb on its walls. To store methane under the normal conditions it is necessary to close the nanocontainer inlet. With help of electrostatic field K@C60 moves to the inlet and blocks it so that methane molecules cannot leave the nanocontainer. When K@C60 reaches the nanocontainer section (8.8), it stops as its dimensions are smaller than the internal space of the section (8.8). Due to the methane pressure and capillary forces action K@C60 stays in the nanocontainer section representing the nanotube (10.10) even after the electric field disappears. Now the stage of methane storage stage begins. At this stage the external thermodynamic conditions are getting normal. Due to the action of the expanding methane inside the nanocontainer K@C60 is pressed to the beginning of the section (8.8) and blocks the methane molecules from exiting the nanocontainer, since the distance between the nanotube walls and endohedral complex is insufficient for the gas penetration between them.

The molecular-dynamic modeling at the methane storage stage in the nanocontainer within 10 ns at 300 K and external pressure 0.1 MPa demonstrated that there are no methane leakages from the nanocapsule observed.

Figure 5.27 demonstrates the dependencies of methane content in the nanocapsule on time at the adsorption stage.

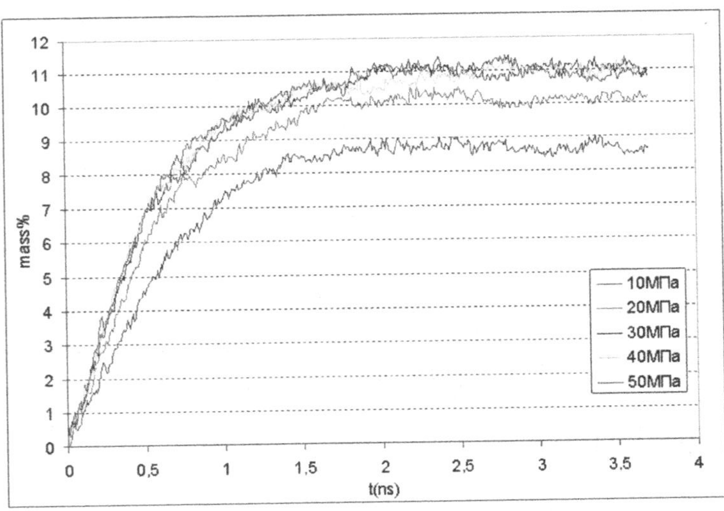

FIGURE 5.27 Adsorption of methane molecules by the nanocontainer under various pressures, where *t*—adsorption time and mass%—mass content of methane molecules inside the nanocontainer.

It is seen that at 300 K and external pressure 10 MPa the methane content in the nanocapsule is 9 mass%. Under the pressure 20 MPa the number of methane molecules is 10.2 mass%. The rest of the curves: P = 30 MPa, P = 40

MPa, and P = 50 MPa practically coincide along their length showing the methane content in the capsule—11 mass%. For all the curves the plateau by the second nanosecond is indicative, characterized by the filling of the nano-capsule internal space.

In Figure 5.28, you can see the nanocapsule complete operation cycle—adsorption, storage, and desorption of methane molecules.

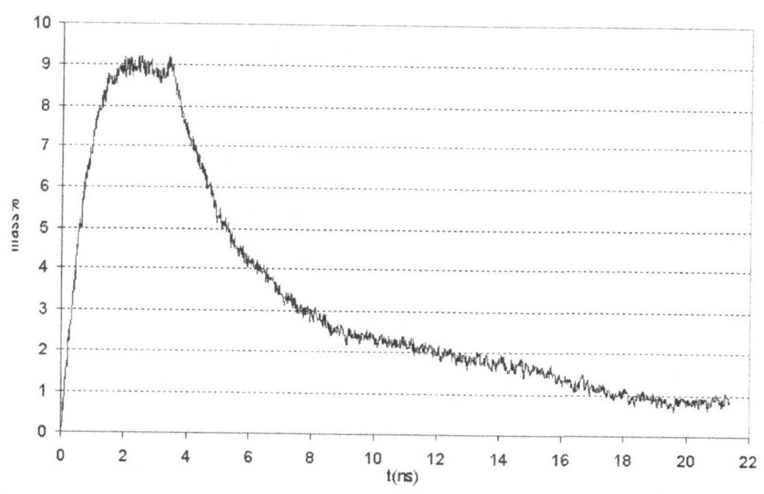

FIGURE 5.28 Stages of methane adsorption, storage, and desorption by the nanocapsule, where *t*—adsorption time and mass%—mass content of methane molecules inside the nanocontainer. There is methane adsorption up to 2.5 ps, then up to 3.7 ps—methane storage and its desorption up to 21 ps.

In the interval from 0 to 2.5 ps the methane adsorption into the nanocapsule internal space is observed. Then from 2.5 to 3.7 ps—methane storage. Obviously the storage is also possible in a wider time interval. Methane desorption when opening the nanocapsule takes rather long period from 3.7 to 21 ps with the determination of residual methane quantity (~1 mass%), which did not desorb from the nanocapsule. The residual gas in the nanocapsule is explained by the significant value of adsorbed potential of the nanocapsule internal space.

5.2.5 CONCLUDING REMARKS

We demonstrated the operation of the nanocapsule of a complex structural form suitable for effective methane storage. The nanocapsule under normal conditions retains 14.5 mass% of methane molecules, which were filled up under T = 300 K and P = 20 MPa. The nanocapsule opening and closing processes are performed by the displacement of the $K@C_{60}^{1+}$ endohedral complex under the action of the electric field. The calculations showed that the nanocapsules of such structure and principles of operation present a new and effective way of methane storage and can be used to store other gases.

The calculation results presented in this section demonstrate that carbon nanocontainers are perspective devices for gas storage. Under the normal storage conditions nanocontainers operating at the change in thermodynamic conditions are able to preserve without loss the amount of hydrogen (1.6 mass%), which was adsorbed under the charge conditions (T = 77 K, P = 10 MPa). Hydrogen was isolated from such nanocontainers at the temperature elevation.

Nanocontainers with charged blocking element (charged fullerene) store hydrogen irrespective of thermodynamic parameters. Only the voltage of external electrostatic field defines adsorption, storage, and desorption stages.

Under the normal external conditions “bottle-like” nanocontainers of increased capacity with the charged blocking element are able to preserve 9 mass% of methane adsorbed under the pressure of 10 MPa and at normal temperature.

Various constructions of nanocontainers for gas storage proposed by the authors show the application area and development of synthesis methods for the creation of highly effective adsorbents for gas storage. Let us compare the results obtained with the existing experimental material and developed techniques to evaluate the modeling reality and to create nanocontainers based on it. It is experimentally demonstrated that nanotubes are adsorbents for the storage of various gases. The calculation results presented highly precisely correspond to the experimental data for hydrogen adsorption under different thermodynamic conditions. It is shown that the structures based on existing “bucky shuttle” nanostructures are perspective adsorbents. To obtain nanocontainers (nanocapsules) of complex structural shapes we can use the already developed techniques for the formation of nanostructured carbon under the action of energy particles: electrons, gamma rays, protons, and ions. This results in the change of structure, morphology and electron, mechanical and chemical properties of carbon materials. The techniques developed allow creating defects (holes) in nanotubes with the accuracy up to 1 A, changing nanotube diameters, “welding” various nanotubes with the creation, for example, “Y” and “T-like” structures. The developed means and techniques of nanostructure directed change are the basis for the creation of future hardware base for the formation of nanocapsules of more complicated structural shapes.

In the conclusion, let us consider the important for practice issue on the hydrogen mass content in nanocapsules. We evaluate this parameter for the nanocontainer from the single-layer nanotube with the length l, diameter d, and wall thickness h. The hydrogen quantity in the nanotube is determined with the following correlation:

$$M_i = \rho_i l \pi d^2 / 4, \tag{5.3}$$

and the nanotube weight is

$$M_{nt} = \rho_{nt} l \pi d h, \tag{5.4}$$

where ρ_H, ρ_{tn} are densities of hydrogen and nanotube, respectively.

The mass content of hydrogen, taking into account (5.3) and (5.4) equals

$$W = M_H / M_{nt} = (\rho_H / 4\rho_{nt} h) \cdot d. \tag{5.5}$$

From the correlation (5.5) it is seen that with the same hydrogen density in the nanotube its mass content is proportional to the nanocontainer diameter. The comparison for different diameters of containers and nanotubes is given in Figure 5.29.

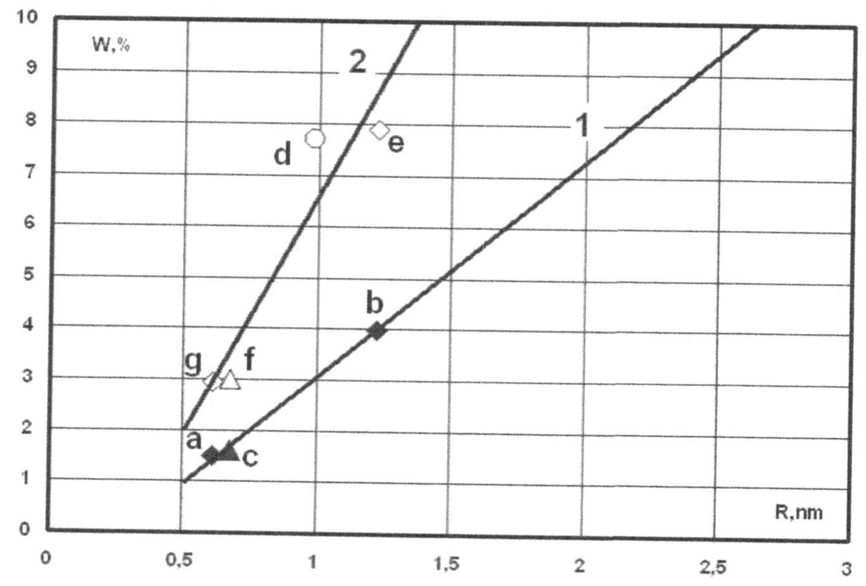

FIGURE 5.29 Dependence of hydrogen mass content on the nanotube diameter.

The research results of different authors are marked with dots. Dots g, f, and e correspond to the calculation conditions of dots a, c, and b when the hydrogen density equals the density of liquid hydrogen. Line 1, according to eq 5.5, approximates the results a, b, and c. Line 2—calculations for the systems filled with liquid hydrogen (g, f, d, and e). Line 2 corresponds to the maximum hydrogen mass content. For all dots located above it, hydrogen density exceeds the density of liquid hydrogen. Figure 5.29 demonstrates that nanocontainers with the radius 2.5 nm can provide hydrogen mass content up to 10%.

5.3 CALCULATION OF NANOPARTICLE ELASTICITY PROPERTIES

5.3.1 STRUCTURE AND FORMS OF NANOPARTICLES

After the relaxation process (the paragraph 4.1.1), the nanoparticles can have quite diverse shapes: globe-like, spherical centered, spherical eccentric, spherical icosahedral nanoparticles, and asymmetric nanoparticles (Fig. 5.30).

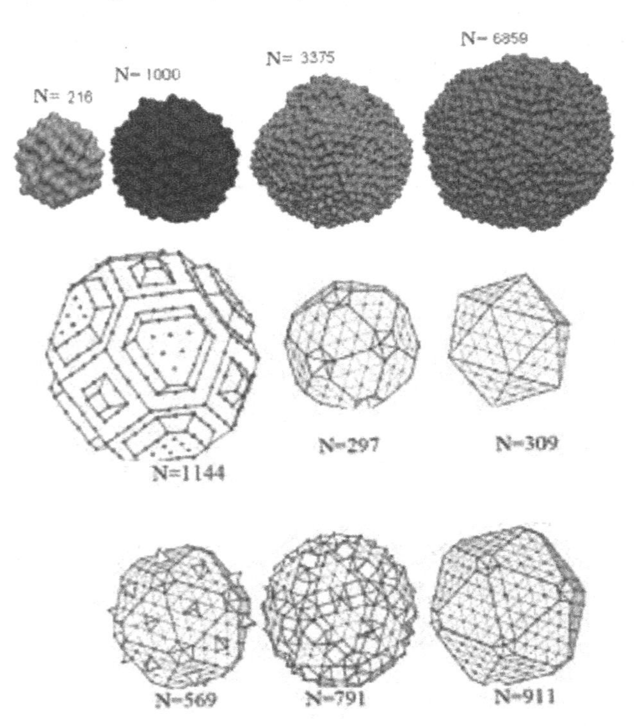

FIGURE 5.30 Nanoparticles of various forms, depending on the number (N) of atoms from which they consist.

In this case, the number of atoms N significantly determines the shape of a nanoparticle. Note, that symmetric nanoparticles are formed only at a certain number of atoms. As a rule, in the general case, the nanoparticle deviates from the symmetric shape in the form of irregular raised portions on the surface. Besides, there are several different equilibrium shapes for the same number of atoms. The plot of the nanoparticle potential energy changes in the relaxation process (Fig. 5.31) illustrates it.

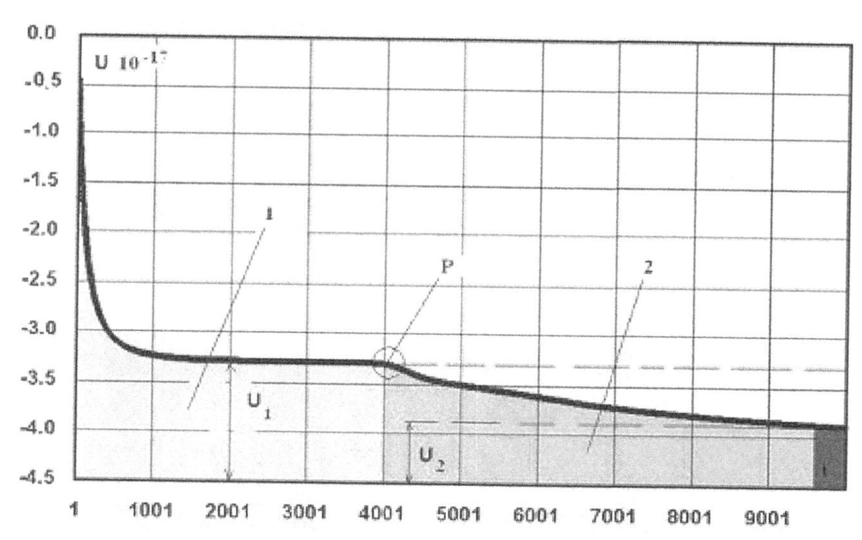

FIGURE 5.31 The plot of the potential energy change of the nanoparticle in the relaxation process. 1—a region of the stabilization of the first nanoparticle equilibrium shape; 2—a region of the stabilization of the second nanoparticle equilibrium shape; and P—a region of the transition of the first nanoparticle equilibrium shape into the second one.

As it follows from Figure 5.31, the curve has two areas: the area of the decrease of the potential energy and the area of its stabilization promoting the formation of the first nanoparticle equilibrium shape (1). Then, a repeated decrease in the nanoparticle potential energy and the stabilization area corresponding to the formation of the second nanoparticle equilibrium shape are observed (2). Between them, there is a region of the transition from the first shape to the second one (P). The second equilibrium shape is more stable due to the lesser nanoparticle potential energy. However, the first equilibrium shape also "exists" rather long in the calculation process. The change of the equilibrium shapes is especially characteristic of the nanoparticles with an "irregular" shape.

The internal structure of the nanoparticles is of importance since their atomic structure significantly differs from the crystalline structure of the bulk materials: the distance between the atoms and the angles change, and the surface formations of different types appear. In Figure 5.32, the change of the structure of a two-dimensional nanoparticle in the relaxation process is shown.

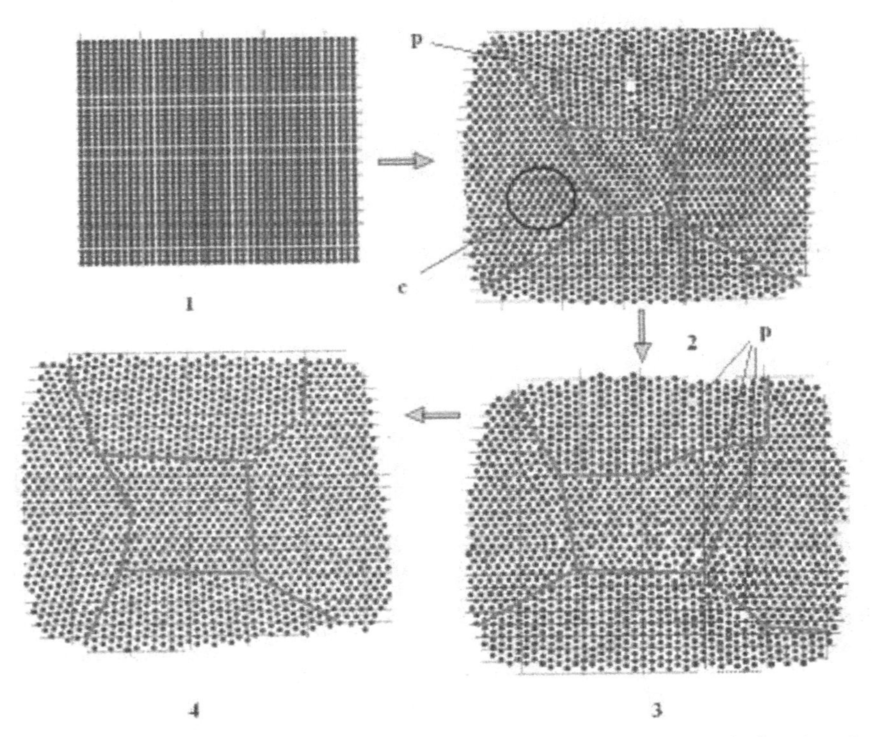

FIGURE 5.32 The change of the structure of a two-dimensional nanoparticle in the relaxation process: (1) the initial crystalline structure; (2), (3), and (4) the nanoparticles' structures which change in the relaxation process; p—pores; and c—the region of compression.

The figure shows how the initial nanoparticle crystalline structure (1) is successively rearranging with time in the relaxation process (positions 2, 3, and 4). Note that the resultant shape of the nanoparticle is not round, that is, it has "remembered" the initial atomic structure. It is also of interest that in the relaxation process, in the nanoparticle, the defects in the form of pores (designation "p" in the figure) and the density fluctuation regions (designation "c" in the figure) have been formed, which are absent in the final structure.

5.3.2 *METHOD OF CALCULATION OF ELASTICITY PROPERTIES OF THE NANOELEMENTS*

After construction of the form nuclear structure of the nanoelement, it is possible to calculate its characteristics. Let's consider a problem of calculation of the elastic parameters of nanoelement. For this purpose let us consider a nanoelement of volume Ω and with a surface S, which consists of N atoms (Fig. 5.33-1) and an "equivalent" elastic element that similar to the nanoelement in shape and sizes (Fig. 5.33-2) under a load of the same set of external balanced surface and volume forces, $\vec{\mathbf{F}}_s$ and $\vec{\mathbf{F}}_\Omega$, respectively.

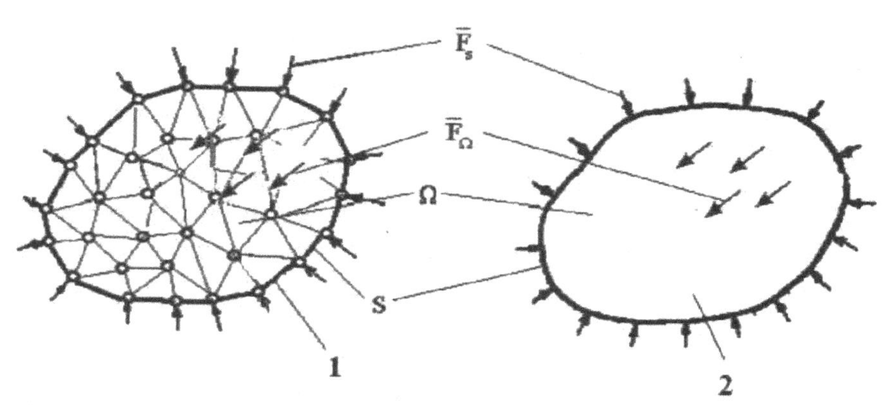

FIGURE 5.33 A nanoelement (1) and an "equivalent" elastic element (2).

Let us introduce the definition of the "equivalent" elastic element. Its material is considered uniform and isotropic; consequently, its elastic properties are determined by two constants elasic modulus and Poisson coefficient. It should be noted that in the general case the nanoelement properties can be inhomogeneous (the possible anisotropy of properties are not considered in the work). This is due to the change in the nanoelement atomic structure on its surface. Therefore, the notion of the elastic "equivalent" element is based on the similarity of the response of the nanoelement and the elastic element to the given set of loads. In other words, the change in the shape and sizes of the elastic "equivalent" element should be the same under the action of the same given set of loads.

The nanoelement deformation will be calculated by the MD method.

The solution of the task about the stress and deformation of "equivalent" element will allow calculating the displacement vector, deformation tensor, and strain tensor in the elastic "equivalent" element, which depend on the matrix of elasticity constants. For example, the comparison of the problem solutions performed by MD with the theory of elasticity is carried out using the displacement vectors at the points coinciding with the position of the atoms of the nanoelement

$$\vec{\mathbf{u}}_e, \vec{\mathbf{u}}_{md}, \tag{5.6}$$

where $\vec{\mathbf{u}}_e, \vec{\mathbf{u}}_{md}$ are the displacement vectors of the elastic "equivalent" element and nanoelement, respectively.

To fulfill condition (5.6), the variation of the elasticity constants of the elastic "equivalent" element is carried out in such a way that the total error determined by the displacement vector difference is minimal.

As a rule it is not possible to have an analytical solution for $\vec{\mathbf{u}}_e$, therefore let consider the numerical schemes of calculation of nanoelement properties.

To derive the finite element (FE) set of equations for the problem of a nanoparticle elastic deformation, let us use the principle of possible displacements and take into account surface and volume forces affecting a nanoparticle. The principle is formulated as follows[196]

$$\int_{\Omega} \delta\, \hat{\varepsilon} \circ \hat{\sigma}\, d\Omega - \int_{S} \delta \vec{\mathbf{u}} \circ \vec{\mathbf{F}}_s\, dS - \int_{\Omega} \delta \vec{\mathbf{u}} \circ \vec{\mathbf{F}}_{\Omega}\, d\Omega = 0\,. \tag{5.7}$$

Let us divide a nanoparticle occupying a calculation region Ω into a number of disjoint sub regions (FEs) and specify tensors of deformation $\{\varepsilon\}_k$, stresses $\{\sigma\}_k$, and a vector of displacements $\{u\}_k$ for each FE. Then the integrals over the volume Ω and the surface S of the nanoparticle can be presented in the form of a sum of integrals over all the FEs filling the given region.

$$\begin{aligned} &\sum_{k=1}^{N} \int_{\Omega_k} \delta\{\varepsilon\}_k^{\mathrm{T}} \{\sigma\}_k\, d\Omega_k - \sum_{k=1}^{N} \int_{S_k} \delta\{u\}_k^{T} \{F\}_k^{S}\, dS_k - \\ &- \sum_{k=1}^{N} \int_{\Omega_k} \delta\{u\}_k^{T} \{F\}_k^{\Omega}\, d\Omega_k = 0 \end{aligned}, \tag{5.8}$$

where Ω_k and S_k are a volume and a surface of a FE; $\{F\}_k^{\Omega}$ and $\{F\}_k^{S}$ are vectors-matrices of the surface and volume forces acting in the volume and on the surface of FEs, respectively; T is a symbol of transposition; δ is a symbol of variation.

Let us specify an approximation of the displacement vector $\{u\}_k$ in the form

$$\{\mathbf{u}\}_k = [\mathbf{N}]\{\boldsymbol{\delta}\}, \tag{5.9}$$

where $[\mathbf{N}]$ is a shape function; $\{\boldsymbol{\delta}\}$ is a vector of the displacement of the nodes of the FE model.

Deformations within a FE can be found from the relation

$$\{\boldsymbol{\varepsilon}\}_k = [\mathbf{B}]\{\boldsymbol{\delta}\}, \tag{5.10}$$

$$[\mathbf{B}] = [\mathbf{N}] \otimes \vec{\nabla} + \vec{\nabla} \otimes [\mathbf{N}], \tag{5.11}$$

$[\mathbf{B}]$ is a matrix of the interrelation of the displacements of the nodes $\{\boldsymbol{\delta}\}$ of FEs and the deformations $\{\boldsymbol{\varepsilon}\}_k$ within a FE. Let us consider the tensor of total deformations $\{\boldsymbol{\varepsilon}\}_k$ as equal to the tensor of elastic deformations $\{\boldsymbol{\varepsilon}\}_k^e$

$$\{\boldsymbol{\varepsilon}\}_k = \{\boldsymbol{\varepsilon}\}_k^e. \tag{5.12}$$

Using the interrelation of the elastic deformation components and stresses in its linear form

$$\{\boldsymbol{\varepsilon}\}_k^e = [\mathbf{D}]^{-1}\{\sigma\}_k, \tag{5.13}$$

we find the stress $\{\boldsymbol{\sigma}\}_k$ from eq 5.13, taking into account eqs 5.11 and 5.12

$$\{\boldsymbol{\sigma}\}_k = [\mathbf{D}][\mathbf{B}]\{\boldsymbol{\delta}\}, \tag{5.14}$$

where **[D]** is the matrix of the elasticity constants of the material.

Substitution of eqs 5.10, 5.13, and 5.14 in eq 5.8 gives the system of algebraic equations

$$[\mathbf{K}]\{\boldsymbol{\delta}\} = \{\mathbf{f}\}_S + \{\mathbf{f}\}_\Omega, \tag{5.15}$$

where

$$[\mathbf{K}] = \sum_{k=1}^{N} \int_{\Omega_k} [\mathbf{B}]^T [\mathbf{D}][\mathbf{B}] d\Omega_k \tag{5.16}$$

is a matrix of "rigidity" of the system;

$$\{\mathbf{f}\}_S = \sum_{k=1}^{N} \int_{S_k} [N]^T \{\mathbf{F}\}_k^S \, dS_k, \tag{5.17}$$

$$\{\mathbf{f}\}_\Omega = \sum_{k=1}^{N} \int_{\Omega_k} [N]^T \{\mathbf{F}\}_k^\Omega \, d\Omega_k , \tag{5.18}$$

are vectors of surface and volume forces reduced to the nodes of FEs, respectively.

From the solution of the system of equations (5.15), the displacements of the nodes of FEs are found.

$$\{\boldsymbol{\delta}\} = [\mathbf{K}]^{-1}(\{\mathbf{f}\}_S + \{\mathbf{f}\}_\Omega) = [\mathbf{K}]^{-1}\{\mathbf{F}\} \cdot \tag{5.19}$$

Let us calculate potential energy of a nanoparticle with the use of the stress and deformation known values

$$U_{np} = \frac{1}{2}\sum_{k=1}^{N} \int_{\Omega_k} \{\boldsymbol{\varepsilon}\}_k^{\mathrm{T}} \{\boldsymbol{\sigma}\}_k \, d\Omega_k \, . \tag{5.20}$$

Substitution of the stress and displacement values in eq 5.20 with the regard for eq 5.18 gives

$$\begin{aligned} U_{np} &= \frac{1}{2}\sum_{k=1}^{N} \int_{\Omega_k} ([\mathbf{B}]\{\boldsymbol{\delta}\})_k^{\mathrm{T}} ([\mathbf{D}][\mathbf{B}]\{\boldsymbol{\delta}\})_k \, d\Omega_k = \\ &= \frac{1}{2}\sum_{k=1}^{N} \int_{\Omega_k} \{\mathbf{F}\}_k^{\mathrm{T}} [\mathbf{K}]^{-1} \{\mathbf{F}\}_k \, d\Omega_k \end{aligned} \tag{5.21}$$

Equating the potential energy of a nanoparticle calculated from the solution obtained by MD with the potential energy of a nanoparticle calculated with the use of the finite element method (FEM) will give the equation for the determination of the elasticity constants of the material

$$\mathbf{U}_{md} = \frac{1}{2}\sum_{k=1}^{N} \int_{\Omega_k} \{\mathbf{F}\}_k^{\mathrm{T}} [\mathbf{K}]^{-1} \{\mathbf{F}\}_k \, d\Omega_k . \tag{5.22}$$

For constant forces, constants of elasticity and shape linear functions of FEs, we obtain

$$\mathbf{U}_{md} = \frac{1}{2}\sum_{k=1}^{N} \{\mathbf{F}\}_k^{\mathrm{T}} [\mathbf{K}]^{-1} \{\mathbf{F}\}_k \Omega_k . \tag{5.23}$$

Let us remove the modulus of elasticity E from the rigidity matrix

$$[\mathbf{K}] = E[\tilde{\mathbf{K}}], \tag{5.24}$$

where $[\tilde{\mathbf{K}}]$ is the rigidity matrix depending on Poisson coefficient ν.

Then

$$E = \frac{\frac{1}{2}\sum_{k=1}^{N} \{\mathbf{F}\}_k^{\mathrm{T}} [\tilde{\mathbf{K}}]^{-1} \{\mathbf{F}\}_k \Omega_k}{U_{md}}. \tag{5.25}$$

Formula (5.25) allows calculating the modulus of elasticity of the material for nanoparticles having arbitrary shapes and different sizes at known external forces. In this case, Poisson coefficient ν should be specified.

If Poisson coefficient ν is not known, it is necessary to solve two problems on the nanoparticle deformation at different external loads and to calculate two values of the potential energy for the above loads. In doing so, we obtain the system of two equations

$$E_1 = \frac{\frac{1}{2}\sum_{k=1}^{N} \{\mathbf{F}_1\}_k^{\mathrm{T}} [\tilde{\mathbf{K}}]^{-1} \{\mathbf{F}_1\}_k \Omega_k}{U_{1md}}, \tag{5.26}$$

$$E_2 = \frac{\frac{1}{2}\sum_{k=1}^{N} \{\mathbf{F}_2\}_k^{\mathrm{T}} [\tilde{\mathbf{K}}]^{-1} \{\mathbf{F}_2\}_k \Omega_k}{U_{2md}}. \tag{5.27}$$

Since $E_1 = E_2$, Poisson coefficient ν is found from the equation

$$\frac{\frac{1}{2}\sum_{k=1}^{N} \{\mathbf{F}_1\}_k^{\mathrm{T}} [\tilde{\mathbf{K}}]^{-1} \{\mathbf{F}_1\}_k \Omega_k}{U_{1md}} = \frac{\frac{1}{2}\sum_{k=1}^{N} \{\mathbf{F}_2\}_k^{\mathrm{T}} [\tilde{\mathbf{K}}]^{-1} \{\mathbf{F}_2\}_k \Omega_k}{U_{2md}}. \tag{5.28}$$

Equation 5.28 is non-linear, therefore its solution for determining Poisson coefficient will be found based on the method of successive approximations. Substituting the coefficient ν found into eq 5.27 or 5.28, we will determine Young's modulus.

5.3.3 CALCULATION OF THE ELASTICITY MODULE OF SPHERICAL NANOPARTICLE

Let us exemplify the computational investigation results on the influence of the size of a nanoparticle under axis tensile forces (Fig. 5.34) on its elasticity modulus magnitude. Calculation of this dependence is necessary for calculation of rigidity of the materials reinforced by the nanoparticles[9] and for creation of new techniques of experimental definition of elastic constants of the nanoparticles.[24–27]

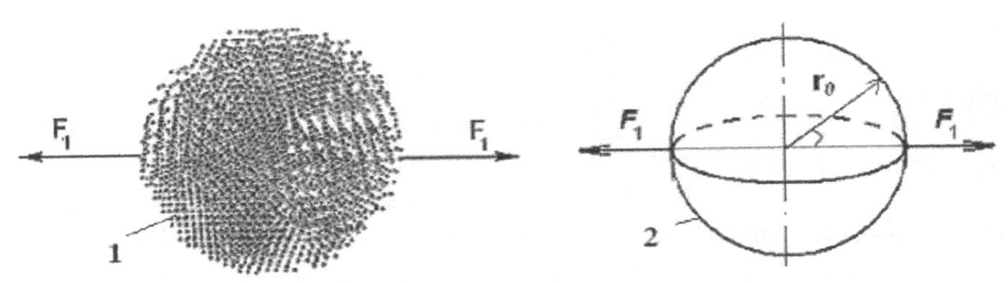

FIGURE 5.34 "Equilibrium" nanoparticle (1) and elastic "equivalent" element (a sphere) (2) stretched by point force.

At the first step of the calculation, an "equilibrium" nanoparticle is built up (the nanoparticle is formed in the process of minimization of its potential energy) that is free from external loads. Lennard-Johns potential[17] was used in the calculations. The calculations of the equilibrium configuration of nanoparticles show that the particles have the shape close to spherical. Therefore, a sphere was used as an elastic equivalent element. Then, tensile forces are applied to this nanoparticle as is shown in Figure 5.34 and its load deformation is investigated.

The problem of the calculation of the elastic "equivalent" element (a sphere) has an analytical solution at stretching by point forces:[159]

$$u = \frac{(1+v)F_1}{2\pi r_0 E}\sum_{n=0}^{\infty}(4n+3)\left\{(\cos\alpha)\frac{P_{2n+1}(\cos\alpha)}{2n+1} + \left[n(2n+1)\rho^{-2} - n(2n-1) - v(4n+1)\right] * \right.$$

$$\left. * \frac{P_{2n+1}(\cos\alpha)}{m_n}\right\}\rho^{2n+1}, \tag{5.29}$$

where r_0 is the sphere radius; $P_{2n}(\cos\alpha)$ is Lagrange polynomials; $\rho = r/r_0 < 1$; F_1 the value of the tensile force. The function m_n is determined from the expression

$$m_n = (1+v)(4n+1)n+n+1 \cdot \quad (5.30)$$

The value of the tensile force was $F_1 = 2.086*10^{-11}$ H and it was selected in such a way that the atom, to which it was applied, did not detach from the particle. It is required that the values of the applied tensile forces should be in the range, where the relation between the displacements at the force application point and the magnitude of the force is linear (Fig. 5.35).

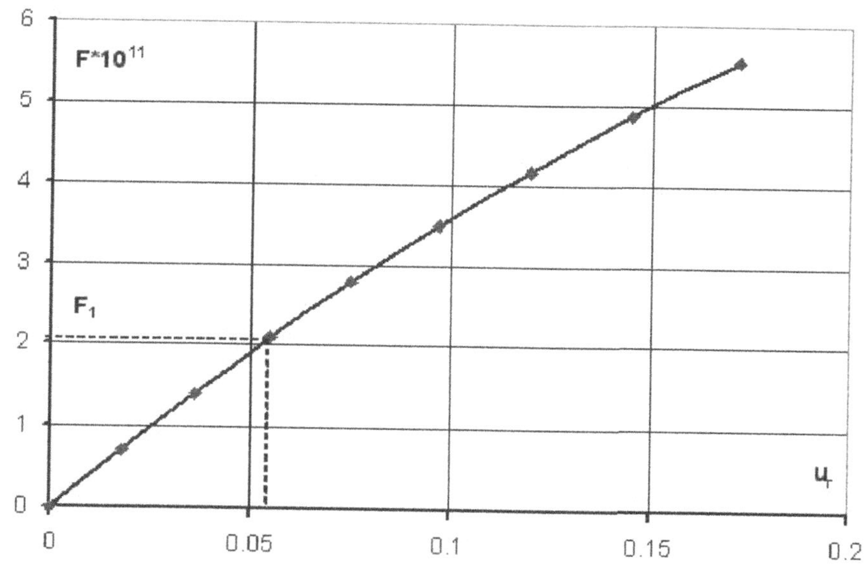

FIGURE 5.35 The force F (H) dependence on the displacements u_r (A) at the point of the force action on the zinc nanoparticle consisting of 1000 atoms.

In Figure 5.36a-1, a solid line designates the dependence of the displacements on the distance to the centre of the "equivalent" elastic element for Young's modulus $E = 9.5 \times 10^{10}$ Pa and Poisson coefficient $v = 0.21$. The same plot (Fig. 5.36a-2) shows the dependence of the radial displacement of atoms u on the distance to the center of mass of the nanoparticle r. It can be seen that the above dependences do not coincide. Therefore, the change of Young's modulus leads to the curves' merging (Fig. 5.36b).

The criterion of this is a root-mean-square error, the change of which depending on Young's modulus is shown in Figure 5.37. It is seen from the plot that the root-mean-square error has a clearly expressed minimum. The best coincidence of the displacement vectors corresponds to this minimum. It is the point where the elasticity modulus of nanoparticles is determined.

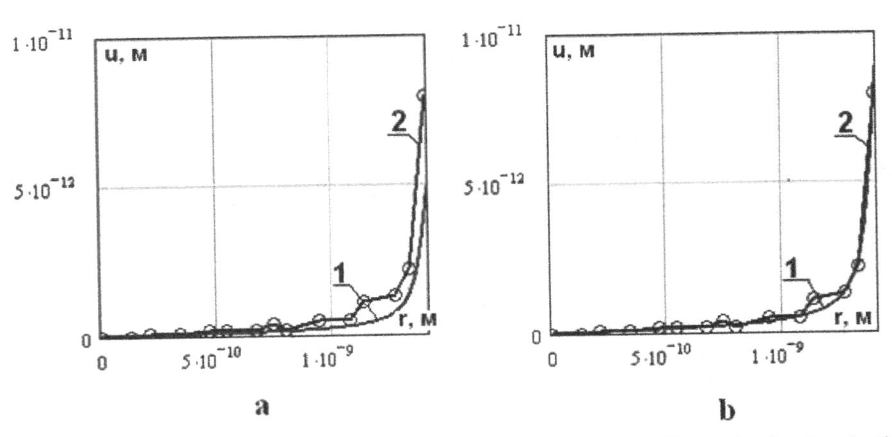

FIGURE 5.36 The dependence of the displacements u on the radius r: 1—for the elastic sphere, 2—for the nanoparticle consisting of 2197 atoms; (a) $E = 9.5 \times 10^{10}$ Pa, (b) $E = 5.25 \times 10^{10}$ Pa.

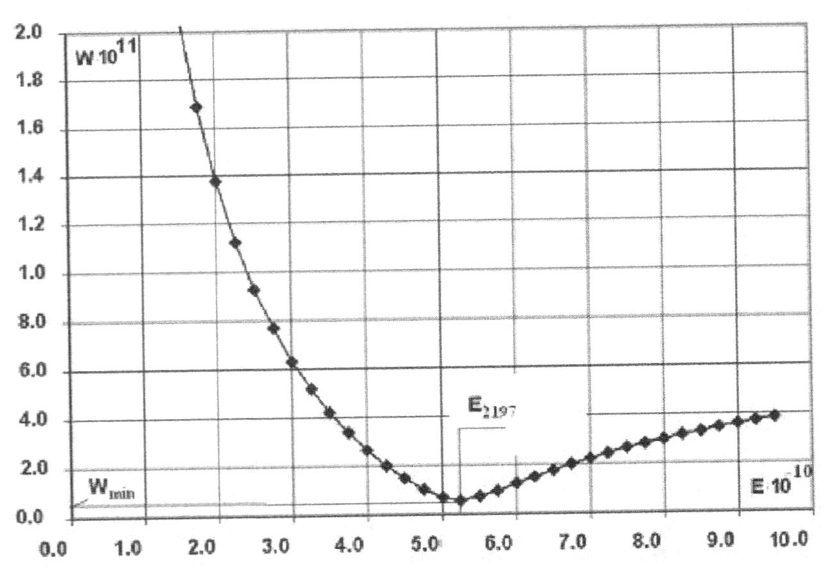

FIGURE 5.37 The dependence of the root-mean-square error W on Young's modulus E for the elastic "equivalent" element; E_{2197} is Young's modulus (Pa) of the nanoparticle containing 2197 atoms.

Carrying out the above procedure for nanoparticles of different diameters, we construct the dependence of Young's modulus on the nanoparticle diameter (Fig. 5.38).

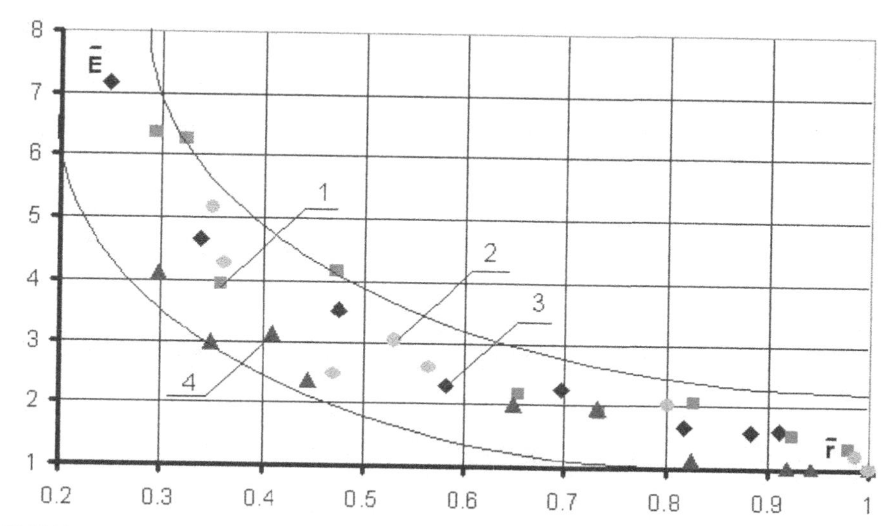

FIGURE 5.38 The dependence of relative Young's modulus $\bar{E}$ on the relative radius $\bar{r}$ for nanoparticles from different materials: 1—cesium, 2—calcium, 3—zinc, and 4—magnesium.

Figure 5.38 displays the dependence of relative Young's modulus $\bar{E}$ on the radius r for nanoparticles from different materials. The relation of Young's modulus to its asymptotic value at the maximal nanoparticle diameter was chosen as the relative modulus. In order to bring the carried-out calculations to a single scale, we divide the nanoparticle radius by its limit radius, for which the relative modulus is 1. In this case, all estimated points are grouped near a generalized curve and can be approximated by a single equation.

5.3.4 CALCULATION OF THE ELASTICITY MODULE OF NONSPHERICAL NANOPARTICLE

The task calculation of the elasticity module of nonspherical nanoparticle is divided on two stages: the development testing stage and the stage of the numerical researches and analysis of their results (Fig. 5.39). Development testing stage is carried out using an integrated system for determining the physical and mechanical properties of micro and nanomaterials NanoTest 600.

Nanoaerosol material particles evaporation is produced on the working surface of the glass substrate. Scanning and numeralization of surface points with spray-coated micro or nanoparticles are carried out. Indentation procedure is performed using Berkovich indenter that has the tip with the radius approximately 200 nm. Indentation in the particle is conducted with a force F that certainly is not resulted in exceeding of the transition threshold of elastic deformation in the plastic deformation of the testable material (thus, the deformation of the material is reversible). Value of indenter tip penetration depth into the particle is determined h_{exper}.

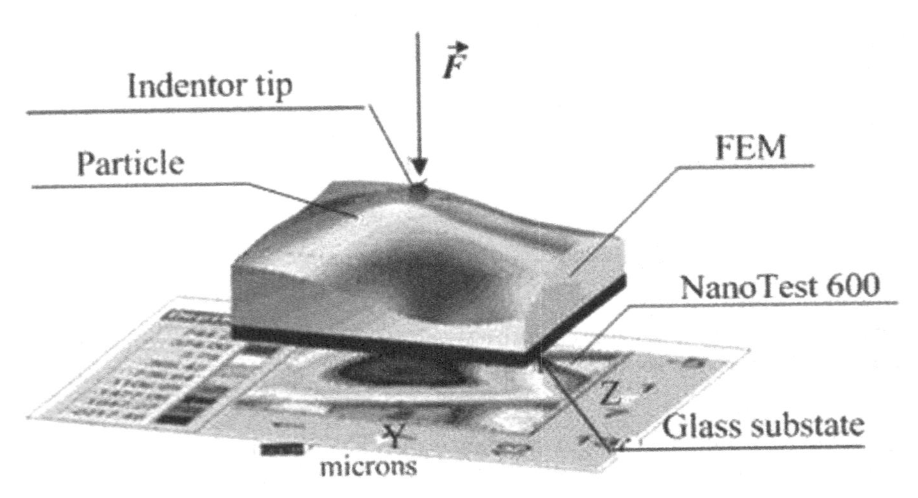

FIGURE 5.39 Modeling by finite element method of the aerosol particle surface created on the basis of scanning procedure results by system NanoTest 600 and indentation procedure with the force F.

Numerical researches stage and analysis of results is carried out. Calculation FE model of the field of the investigation by the FEM is created on the basis of particles surface numeralization data. Numerical solution of the contact introduction task of the indenter tip into the particles is realized (Fig. 5.40). The indenter penetration depth is defined h_{FEM}.

The calculation is performed using the known values of the elastic constants and the form of the indenter, the Poisson's ratio of the material particle, as well as the experimental indentation force F. Experimental curve of the dependence E–h' is plotted after series of numerical experiments, under a fixed value of force by varying Young's modulus in each calculating, where

$$\Delta h = h_{FEM} - h_{exper} \tag{5.31}$$

On the basis of obtained dependence by means of approximation under $\Delta h = 0$ the value of the Young's modulus E of investigated particle is defined.

High-precision scanning of glass surface (substrate) with spray-coated micro and nanoparticles is carried out with minimal scanning load 0.01 mN and established scanning step 50 nm, by means of the piezo profilometer of system NanoTest 600 with a resolution of the axes *Y*, *Z* 2 nm, the axis *X* 0.1 nm.

An experimental indentation in the investigated particle with a force $F = 1$ mN is carried out. The depth of the indentation $h_{exper} = 85.421$ nm is defined. A series of numerical solutions by the FEM is carried out of the contact task of indenter tip introduction into the top of the investigated particle by means of varying Young's modulus of the particle, the Poisson's ratio is set to $P = 0.33$. Young's modulus and Poisson's ratio of Berkovich indenter are 1141 GPa and 0.07.

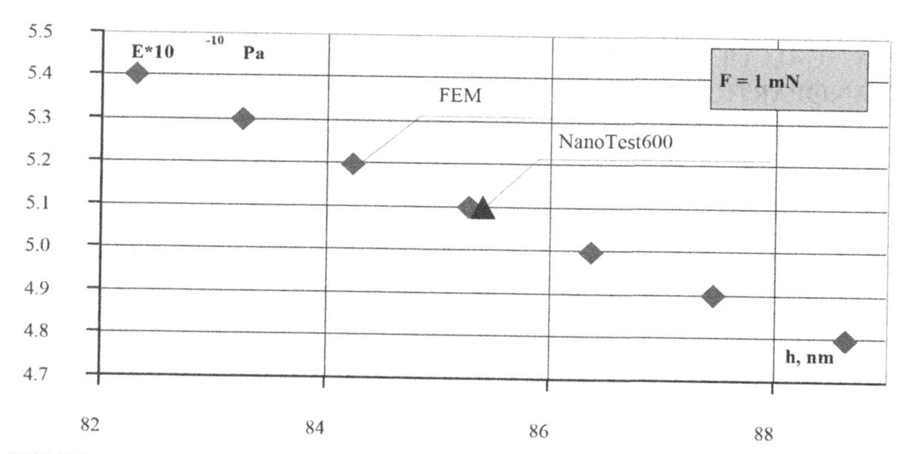

FIGURE 5.40 The dependence of the Young's modulus *E*, Pa from indenter penetration depth *h*, microns for the force $F = 1$ mN (NanoTest 600, FEM).

The dependence of the indenter penetration depth into the particle from the Young's modulus values is plotted (Fig. 5.41).

Young's modulus $E_1 = 5.1 \times 10^{10}$ Pa is determined for investigated particle by means of *approximation of obtained* dependence under $\Delta h = 0$ (Fig. 5.41).

Thus, elastic modulus is determined for micro and nanoparticles.

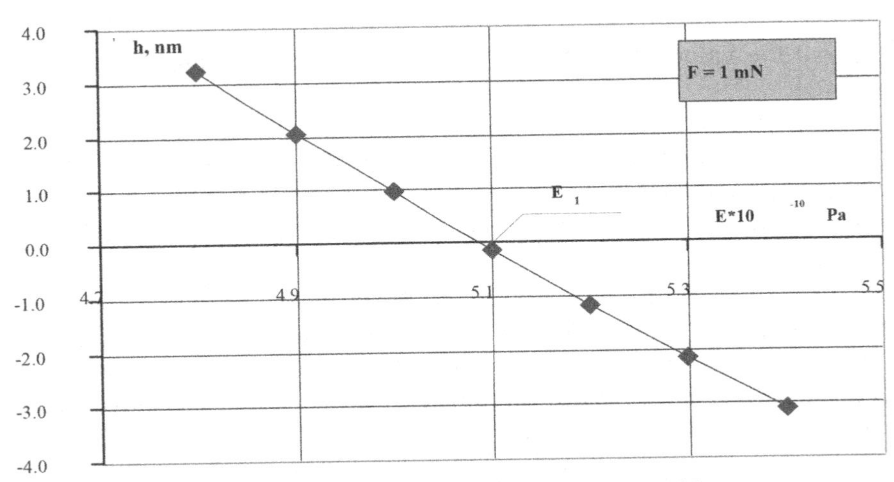

FIGURE 5.41 The dependence $E(\Delta h)$ for indentation force $F = 1$ mN.

5.4 CALCULATION OF MICROCRACKS PROPERTIES WITH NANOPARTICLES

One of the main factors determining the durability of machine parts is the abrasion resistance of the materials from which the parts are made. Increased wear resistance is based on the study of the physics of the phenomena occurring on the surfaces of parts in friction and, above all, the evolution of the structure of the surface layers of the contacting materials.

For process control structure formation in the surface layer and reduce the work of friction lubricants are applied, including additives that provide the desired surface modification of the layer structure. At the moment, there are plenty of liquids and plastic lubricants, relating to the type of metal cladding. As the active ingredient powders of various metals, their oxides and alloys with different dispersion, including nanopowders are often used.

It should be noted that despite the significant number of publications[160] devoted to the influence metal lubricants on the friction and wear processes of structural materials, the lubricating mechanism of action of nanopowders is studied insufficiently. In this connection, a computer simulation of complex multi-phase nanosystems will complement experimental studies detailing the processes of interaction in the two-phase medium consisting of components: surface, nanoparticles, and a liquid lubricating medium.

The chapter presents in-depth analysis and modeling of processes of interaction of nanoparticles with microcracks on a solid surface. The results of numerical calculations and analyzes of the interaction of aluminum and copper nanoparticles with microcracks in the aluminum and silumin are presented.

5.4.1 FORMULATION OF THE PROBLEM

Consider nano system including (Fig. 5.42a) a solid body (1) with a crack (2), inside which there are one or more nanoparticles (3).

System modeling includes three sequential tasks: the construction of the initial configuration of nanosystems; the relaxation to equilibrium state nanosystems; and loading nanosystems external forces before failure (Fig. 5.42b).

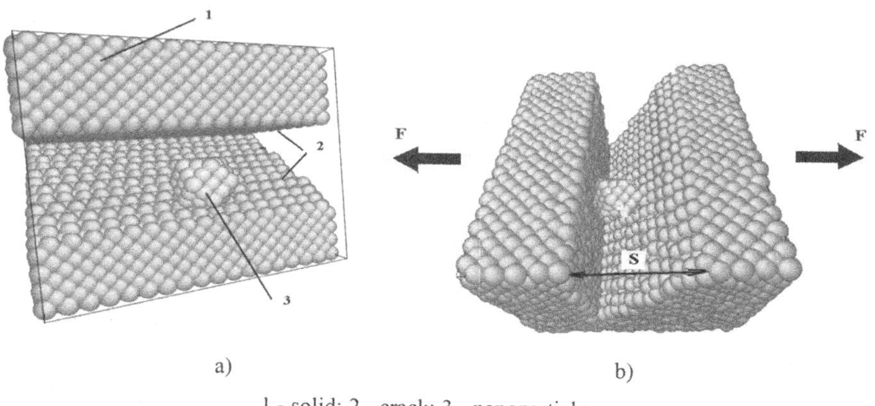

FIGURE 5.42 The settlement scheme of the task: 1—solid; 2—crack; and 3—nanoparticle.

To simulate this nanosystems using the method of MD[71] and a numerical scheme Verlet for the integration of MD equations. MD equations for the nano systems containing nanoparticles are given in Chapter 2. At the initial time nanosystems it is at rest. Initial temperature of the system defines velocities with which the atoms oscillate about the equilibrium position.

For calculations table interatomic interaction potential of atoms with a radius of interaction of 6.68 Å is used. The potential is calculated by the embedded atom. The potential is presented graphically in Figure 5.43.

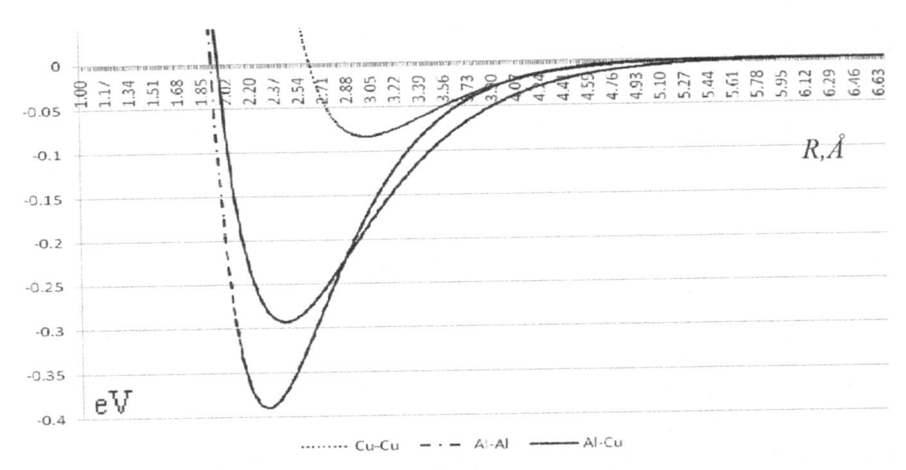

FIGURE 5.43 The potential of interatomic interaction Al–Al, Cu–Cu, Al–Cu.

The calculations were performed using problem-oriented program complex Nano Crack, including as a computing module Large-scale atomic/molecular massively parallel simulator (LAMMPS) program developed by Sandia National Laboratories. To visualize the data AtomEye program and Visual molecular dynamics (VMD) program were used.

5.4.2 RESULTS OF CALCULATIONS

Let us consider the results of numerical calculations and analysis of the interaction of aluminum and copper nanoparticles with microcracks in the aluminum and silumin. At the first stage will examine the interaction of a single nanoparticle with a crack, and then examine the interaction of "nanoparticle system–crack." The calculations were performed at $T = 400$ K.

Figure 5.44 shows the interaction of copper nanoparticle with surface crack in aluminum at a time 20 ps. Calculations showed that the aluminum atoms, actively interacting with the atoms of copper, crawl on the nanoparticle, durably fixing it on the wall of the microcracks. Further nanoparticle is not moved. It can be seen that in this case "healing" of crack do not occur.

Consider the interaction of nanoparticles system with a crack (Fig. 5.45). In this case, nanosystems include nine nanoparticles of copper and crack in the aluminum. Each aluminum nanoparticles comprise 55 atoms. The results of calculations showing the time variation of the structure of the nanosystems are shown in Figure 5.46.

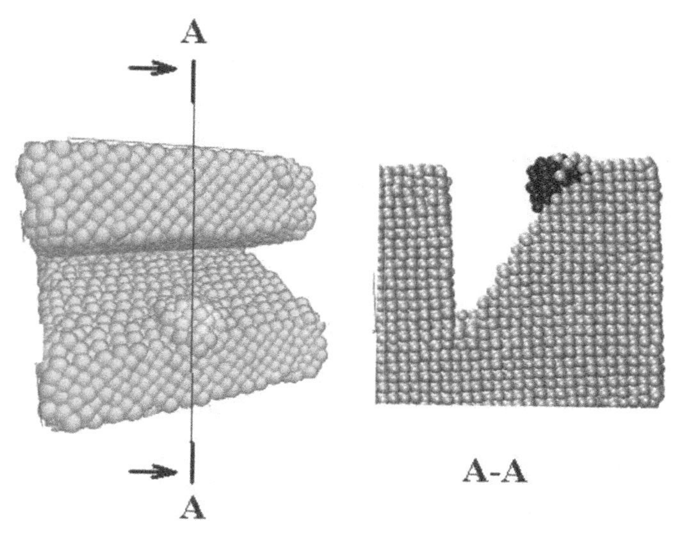

FIGURE 5.44 Copper nanoparticle interacting with a crack in aluminum: A–A—cross-sectional of model; $t = 20$ ps.

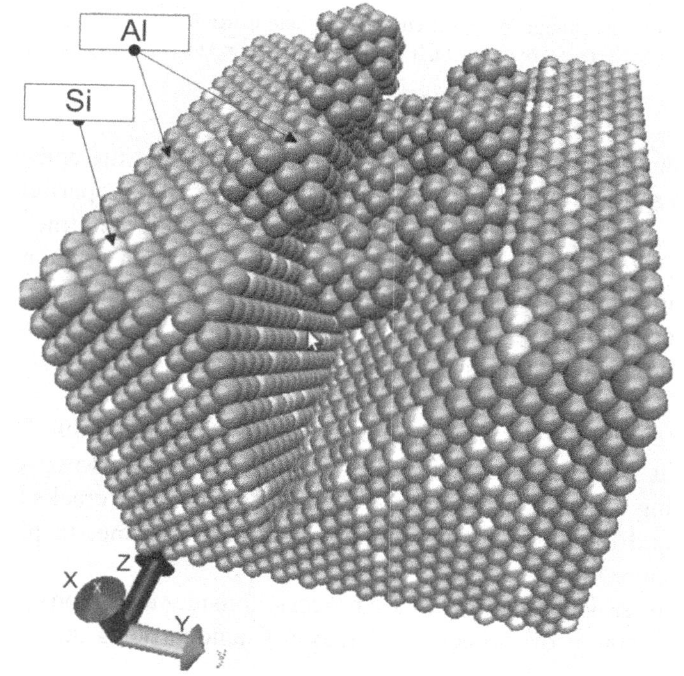

FIGURE 5.45 The initial state of the model "nanoparticles–crack."

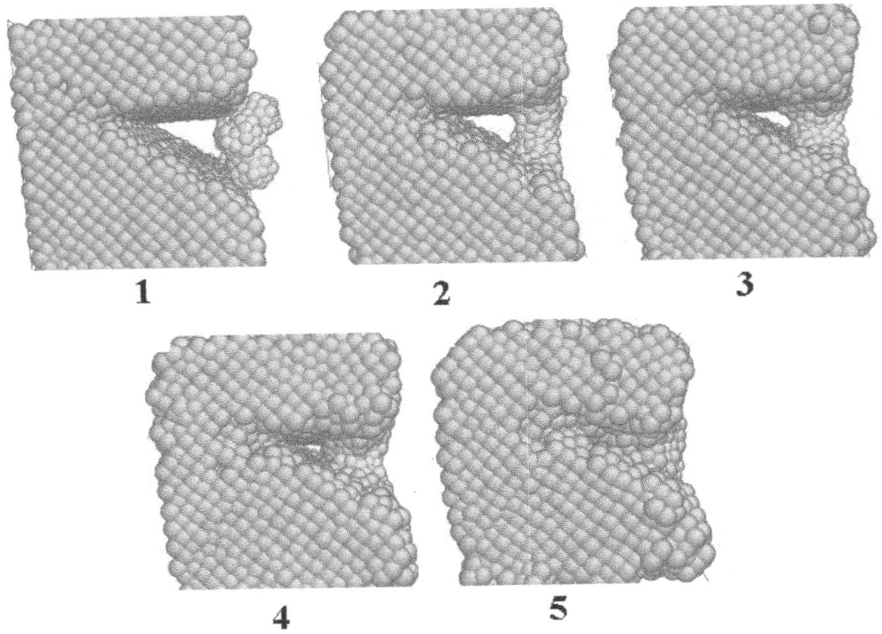

FIGURE 5.46 The change in the structure of the "nanoparticle—crack in aluminum" model in time: (1) $t = 6$; (2) $(t) = 36$; (3) $t = 66$; (4) $t = 96$; and (5) $t = 160$. Time is given in picosecond (ps).

The figure shows that the nanoparticles are placed in the cracks begin to move into a crack to its base. Gradually a "bridge" of nanoparticles formed connecting wall cracks. Next, a "bridge" of nanoparticles contracts the walls of the crack. The result is a gradual complete overgrowth of the microcrack and a composite material of Cu and Al in place of the crack is formed.

Another situation is observed in the interaction of nanoparticles with a crack in the system silumin consisting of 87% aluminum and 13% silicon (Fig. 5.47).

In this case, like the above, the nanoparticles are pulled into the fracture and form a "bridge" connecting the walls of the crack. However, the full overgrowing of the crack is not observed, the walls of the cracks do not pull together, and the "bridge" has remained stable over time. In place of the fracture a pore forms.

For a better understanding of the process, consider the graphs of displacement and velocity of the center of mass of nanoparticles inside the crack (Fig. 5.48).

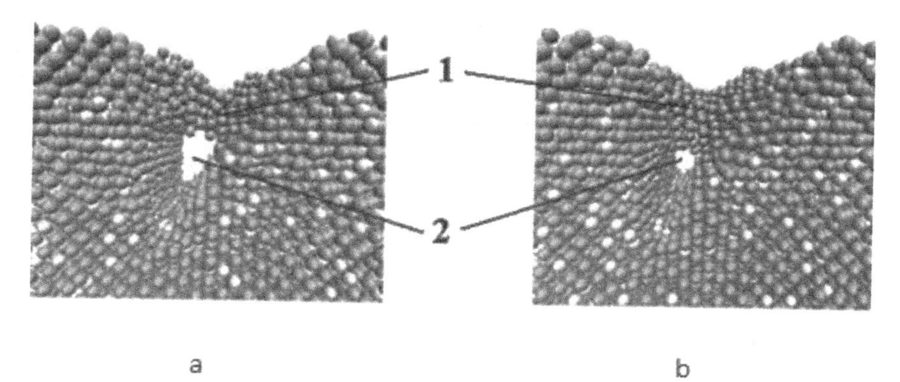

FIGURE 5.47 The change of the structure of the model "nanoparticles—a crack in the siuminum" in time: 1—nanoparticles bridge; 2—pore; time: (a) $t = 10$ ps; (b) $t = 250$ ps.

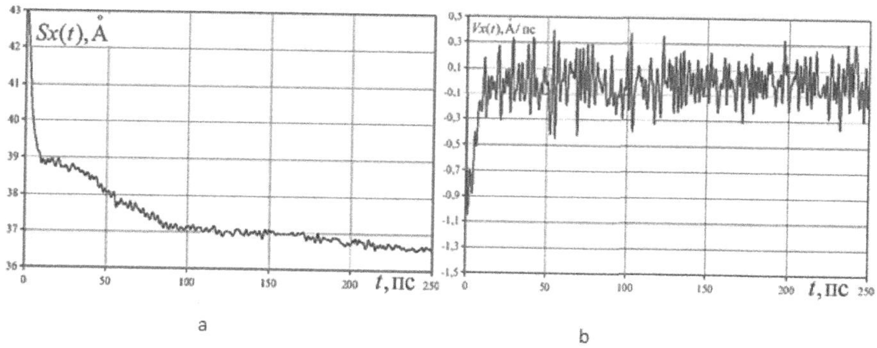

FIGURE 5.48 Change in time displacement (a) and the velocity (b) of the center of mass of nanoparticles for the model "nanoparticles—a crack in the silumin."

As the chart shows the initial time to $t = (5 \div 7)$ ps, there is a rapid movement of the nanoparticles into the microcracks. Then the speed is reduced, and at time $t = 100$ ps, nanoparticle movement practically is terminated. Atoms of the nanoparticles have temperature fluctuations around the equilibrium position, and the whole system of nanoparticles does not move, that corresponds to the final formation of the "bridge" structure.

It is important to estimate the strength of the formed nanosystems. For this purpose, calculations are made of the crack opening under the action tensile forces (Fig. 5.45b). As an example, consider the deformation and destruction of the structure shown in Figure 5.45. At each step, the force F stepwise increased by a certain value. Then executes the calculation of deformation of nanosystems, and for each value of the force F determines the value of the crack opening ΔS. The loading of nanosystems was carried until its destruction. The calculation executed for the crack with and without of nanoparticles. Figure 5.49 shows the result of calculation as a function of disclosure of crack (ΔS) on the applied forces (F).

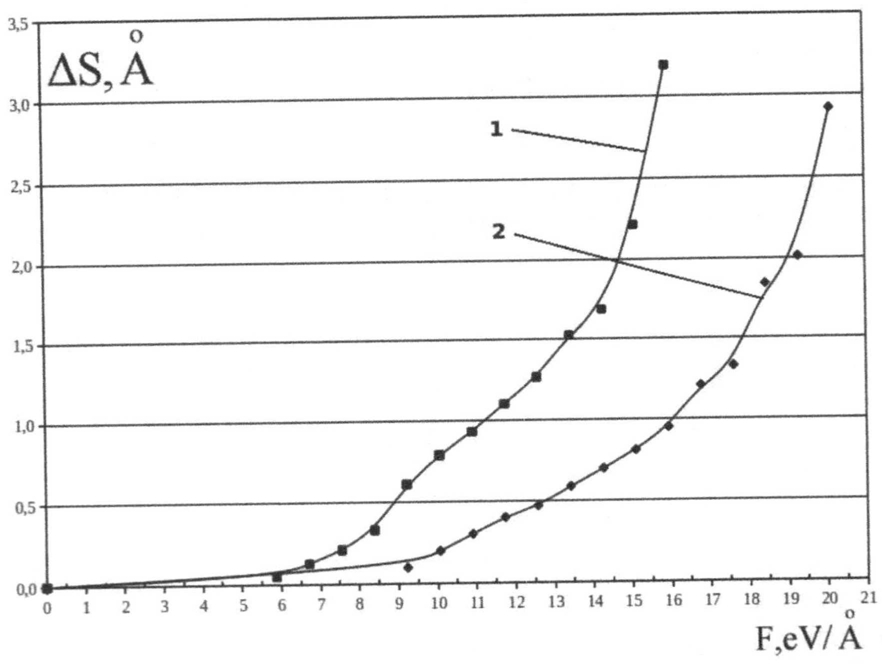

FIGURE 5.49 ΔS crack opening size depending on the value of the applied force F: 1—crack without nanoparticles; 2—crack with nanoparticles.

The graph shows that the failure load for crack with the nanoparticles is 30% more load at which a crack is destroyed without nanoparticles. Note that the critical value of crack opening ΔS for cracks with nanoparticles and without them is practically the same (about 3 Å). Taking into account that the fracture toughness is linearly dependent on the load applied[72] can be considered as a first approximation, that the introduction of the nanoparticles into the crack to increase the fracture toughness at 30%. In general, the presence of nanoparticles in microcrack stabilizes and strengthens it.

5.5 CALCULATION OF THERMAL CONDUCTIVITY OF NANOSYSTEMS

5.5.1 FORMULATION OF THE PROBLEM

As a method of modeling, MD was used. In the calculations, the formalism of Green–Kubo formalism (Green–Kubo) was used. This method connects the autocorrelation function of the heat flow with the thermal conductivity. Thermal conductivity in the Green–Kubo model is calculated by the following formula[55,82]

$$k = \lim_{\tau\to\infty}\lim_{L\to\infty}\frac{1}{k_B T^2 L^d}\int_0^{\tau}\langle J(t)J(0)\rangle dt, \qquad (5.32)$$

where k —thermal conductivity of the d-dimensional system of linear size L, T—temperature, k_B —the Boltzmann constant, J—heat flow component.

Autocorrelation functions of the right side in formula (5.32) are estimated in equilibrium, without a temperature gradient. Autocorrelation is a statistical relationship between random variables from one series, but taken with a shift, for example, for a random process—with a time shift. The autocorrelation function may be defined as

$$\Psi(\tau) = \int f(t)f(t-\tau)dt \cdot \qquad (5.33)$$

Total heat flow in the system is calculated as

$$J(t) = \int j(x,t)dx \cdot \qquad (5.34)$$

There are various forms of writing the eq 5.32. In this work, formula (5.35) to calculate the coefficient of thermal conductivity is used.

$$k = \frac{1}{Vk_B T^2}\int_0^{\infty}\langle J_x(0)J_x(t)\rangle dt = \frac{1}{3Vk_B T^2}\int_0^{\infty}\langle \mathrm{J}(0)\cdot \mathrm{J}(t)\rangle dt, \qquad (5.35)$$

where V is the system volume.

5.5.2 RESULTS OF CALCULATIONS

Consider the results obtained in simulating the formation of quantum dots and calculation of coefficient of thermal conductivity by the example of the formation of quantum dots based on chromium in silicon. The modeling was carried out for the modified embedded atom method (MEAM) potential of the atom interraction.[99]

After the deposition of chromium atoms onto the substrate surface and short-term annealing, three-dimensional islands are formed on the substrate surface. This corresponds to the experimental data. The dimensions and number of islands depend on the thickness of the deposited chromium layer. Thus, Figure 5.50 shows a system with one island.

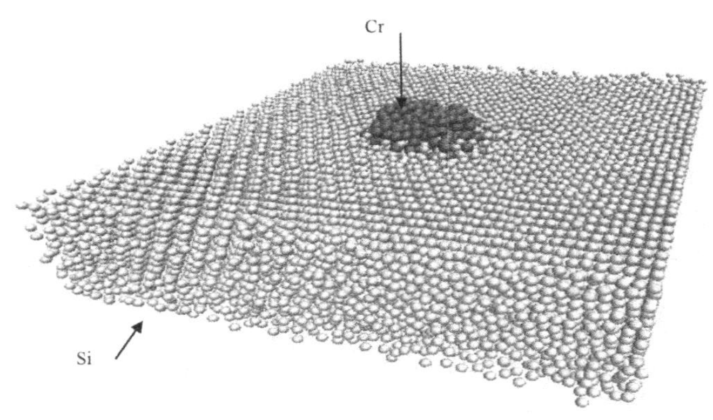

FIGURE 5.50 Simulated system. Here, the chromium atoms are dark, the silicon atoms are light.

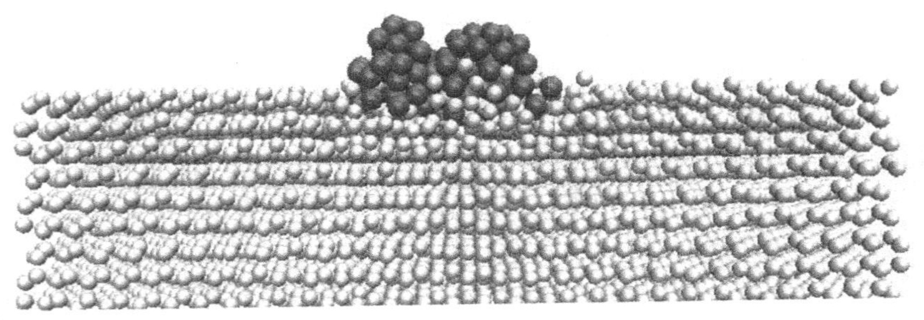

FIGURE 5.51 Diffusion process.

The process of the formation of nanostructures on the substrate surface is greatly influenced by lattice mismatch of deposited material and substrate. The diffusion of chromium atoms into the substrate only in the surface layers was observed. The presence of diffusion in the surface layers is attributed to the structure and properties of the elements participating in simulation. Figure 5.51 illustrates the process of diffusion.

Figure 5.52 illustrates the plot of the radial distribution function for the system obtained using the MEAM potential. Proceeding from the good compliance of the profiles of the distribution curves, it is possible to conclude that simulation results qualitatively correspond to experimental data, and that the C40 structure of $CrSi_2$ is formed in simulations.

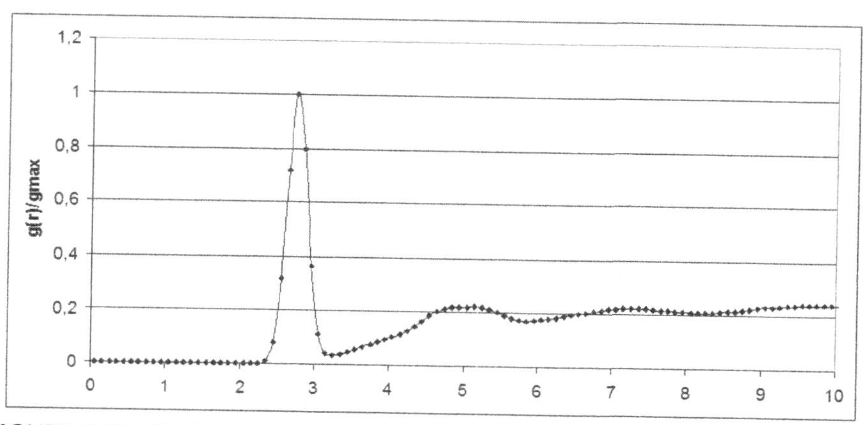

FIGURE 5.52 Radial pair distribution functions for the system obtained as a result of simulation with the MEAM potential.

Thus, the mathematical model presented in this chapter made it possible to evaluate the structure and properties of metal-silicon quantum dots. So, we can use this model to determine the heat-physical properties of nanomaterials, such as thermal conductivity. For example, the thermal conductivity for the chromium-silicon system at temperature 1000 K is 208,069 W/(mK).

KEYWORDS

- **numerical simulation**
- **nanostreams properties**
- **absorption properties**
- **elasticity properties**
- **crack resistance**
- **thermal conductivity**

BIBLIOGRAPHY

1. Abe, K.; Kondoh, T.; Nagano, Y. A New Turbulence Model for Predicting Fluid Flow and Heat Transfer in Separating and Reattaching Flows-i. Flow Field Calculations. *Int. J. Heat Mass Transfer.* **1994,** *37,* 139–151.
2. Anderson, H. S. Molecular Dynamics Simulation at Constant Pressure and/or Temperature. *J. Chem. Phys.* **1980,** *72,* 2384–2396.
3. Adachi, M., Lockwood, D. J., Eds. Self-Organized Nanoscale Materials; In *Series: Nanostructure Science and Technology;* Springer: New York, 2006; p 317.
4. Andrievski, R. A.; Ragulia, A. V. *Nanostructural Materials;* Academia: Moscow, 2005; p 192.
5. Andreeva, A. V. *Fundamentals of Physics-Chemistry and Technology of Composites;* IPRZhR: Moscow, 2001; p 192.
6. Allen, M. P.; Tildesley, D. J. *Computer Simulation of Liquids;* Oxford Science Publications: New York, 1987; p 400.
7. Alikin, V. N.; Vakhrouchev, A. V.; Golubchikov, V. B.; Lipanov, A. M.; Serebrennikov, S. Y. *Development and Investigation of the Aerosol Nanotechnology;* Mashinostroenie: Moscow, 2010; p 196.
8. Alferov Zh, I.; Andreev, V. M.; Garbuzov, D. Z. The Influence of Heterostructures Parameters in the AlAs–GaAs System on the Over Current of Lasers and Obtain Continuous Lasing at Room Temperature. *Fiz. Tekh. Poluprovodn.* **1970,** *4,* 1826–1829.
9. Andereck, C. D.; Liu, S. S.; Swinney, H. L. Flow Regimes in a Circular Couette System with Independently Rotating Cylinders. *J. Fluid Mech.* **1986,** *164,* 155–183.
10. Atempa-Rosiles, P., et al. Simulation of Turbulent Flow of a Rotating Cylinder Electrode and Evaluation of its Effect on the Surface of Steel API 5L X-56 during the Rate of Corrosion in Brine Added with Kerosene and H2S. *J. Elec. Sci.* **2014,** *9,* 4805–4815.
11. Averill, A. F.; Mahmood, H. S. Determination of Tertiary Current Distribution in Electrodeposition Cell–Part 1. Computational Techniques. *Trans. Inst. Metal Finish.* **1997,** *75,* 228–233.
12. Bader, R. F. W. *Atoms in Molecules. A Quantum Theory;* Clarendon Press: Oxford, 1990; p 434.
13. Bischof, C.; Bucker, M. Computing Derivatives of Computer Programs. In *Modern Methods and Algorithms of Quantum Chemistry;* Grotendorst, J., Ed.; NIC Series, John von Neumann Institute for Computing: Julich, Germany, 2000; Vol. 1, pp 287–299.
14. Berdichevsky, V. L. *Variational Principles of Continuum Mechanics;* Springer: Heidelberg, London, 1983; p 447.
15. Baraton, M. I., Ed. *Synthesis, Functionalization and Surface Treatment of Nanoparticles;* University of Limoges: Limoges, France, 2003; p 323.
16. Brooks, B. R.; Bruccoleri, R. E.; Olafson, B. D.; States, D. J.; Swaminathan, S.; Karplus, M. CHARMM: A Program for Macromolecular Energy Minimization, and Dynamics Calculations. *J. Comput. Chem.* **1983,** *4 (2), 187–217.*
17. Barajas Barraza, R. E.; Guirado Lopez, R. A. Clustering of H2 Molecules Encapsulated in Fullerene Structures. *Phys. Rev. B Condens. Matter.* **2002,** *66* (15), 1–12.

18. Becke, A. D. A New Mixing of Hartree–Fock and Local Density Functional Theories. *J. Chem. Phys.* **1993,** *98* (2), 1372–1377.
19. Becke, A. D. Density–Functional Thermochemistry I. The Effect of the Exchange–Only Gradient Correction. *J. Chem. Phys.* **1992,** *96* (3), 2155–2160.
20. Burkert, U.; Allinger, N. L. *Molecular Mechanics;* American Chemical Society: Washington, D.C., 1982.
21. Benoit, R.; Klaus, S. Computational Studies of Membrane Channels. *Structure.* **2004,** *12,* 1343–1351.
22. Berlin A. A.; Balabaev N. K. Simulation of the Properties of Solids and Liquids by Computer Simulations. *Soros Obraz. Zh.* **1997,** *11,* 85–92.
23. Bellani, V.; Guizzetti, G.; Marabelli, F., et al. Theory and Experiment on the Optical Properties of CrSi2. *Phys. Rev. B.* **1992,** *46,* 9380–9389.
24. Bockris, J.; Reddy, A. *Modern Electro-Chemistry. V. 2, A Fundamentals of Electrodics;* Kluwer Academic Publishers: New York, 2002; p 817.
25. Chikara, H.; Ryozi, U.; Akira, T. *Ultra-Fine Particles: Exploratory Science and Technology: Materials Science and Process Tech Series;* Editors Noyes Publication: Westwood, NJ, 1996; p 467.
26. Chushak, Y.; Bartell, L. S. Molecular Dynamics Simulations of the Freezing of Gold Nanoparticles. *Eur. Phys. J. D.* **2001,** *16,* 43–46.
27. Cagın, T.; Che, J.; Qi, Y.; Zhou, Y.; Demiralp, E.; Gao, G.; Goddard III, W. Computational Materials Chemistry at the Nanoscale. *J. Nanopart. Res.* **1999,** *1,* 51–69.
28. Case, D. A.; Cheatham, T. E.; Darden, T.; Gohlke, H.; Luo, R.; Merz, K. M.; Onufriev, A.; Simmerling, C.; Wang, B.; Woods, R. The Amber Biomolecular Simulation Programs. *J. Comp Chem.* **2005,** *26,* 1668–1688.
29. Curtin, W. A.; Miller, R. E. Atomistic/Continuum Coupling Methods in Multi-Scale Materials Modeling Modeling Simul. *Mater. Sci. Eng.* **2003,** *11,* 33–68.
30. Chernov, A. A.; Givargizov, E. I. Modern Crystallography. In *The Formation of Crystals;* Vainshtein, B. K., Ed.; Nauka: Moscow, 1980; Vol. 3, p 401.
31. Coe-Sullivan, S.; Stecke,l J. S.; Woo, W. K.; Bawendi, M. G.; Bulovic, V. Large-Area Ordered Quantum Dot Monolayers Via Phase Separation during Spin-Casting. *Adv. Funct. Mater.* **2005,** *15* (7), 1117–1124.
32. Drexler, E.; Peterson, C.; Pergamit, G. *Unbounding the Future: The Nanotechnology Revolution;* William Morrow & Company, Inc.: New York, 1991; p 158.
33. Drexler, K. E. *Engines of Creation: The Coming Era of Nanotechnology;* Anchor Press/Doubleday: New York, 1986; p 271.
34. Drexler, K. *Eric Nanosystems: Molecular Machinery, Manufacturing, and Computation;* Wiley: New York, 1992; p 576.
35. Duan, H. L.; Wang, J.; Huang, Z. P.; Karihaloo, B. L. Size-Dependent Effective Elastic Constants of Solids Containing Nano-Inhomogeneities with Interface Stress. *J. Mech. Phys. Solids.* **2005,** *53* (7), 1574–1596.
36. Duan, H. L., et al. A Point-Charge Force Field for Molecular Mechanics Simulations of Proteins Based on Condensed-Phase Quantum Mechanical Calculations. *J. Comput. Chem.* **2003,** *24* (16), 1999–2012.
37. Dubrovskii, V. G.; Sibirev, N. V.; Cirlin, G. E. Shape Modification of III-V Nanowires: The Role of Nucleation on Sidewalls. *Phys. Rev. E.* **2008,** *77* (3), 0316061–0316067.
38. Dubrovskii, V. G. *Theoretical Foundations of the Technology of Semiconductor Nanostructures: Textbook;* St. Petersburg State University: St. Petersburg, Russia, 2006.

39. Dubrovskii, V. G.; Sibirev, N. V.; Tsyrlin, G. E. Nucleation on the Lateral Surface and its Influence on the Shape of Filamentary Nanowires. *Fiz. Tekh. Poluprovodn.* **2007,** *41* (10), 1257–1264.
40. Dongqing Li. The Small Flow Becomes Main Stream. *Microfluid Nanofluid.* **2004,** *1,* 1–12.
41. Dingreville, R.; Qu, J.; Cherkaoui, M. Surface Free Energy and its Effect on the Elastic Behavior of Nano-Sized Particles, Wires and Films. *J. Mech. Phys. Solids.* **2004,** *53* (8), 1827–1854.
42. Eijkel, J. C. T.; van den Berg, A. Nanofluidics: What is it and what can we Expect from it?. *Microfuid Nanofuid.* **2005,** *1,* 249–267.
43. Frenkel, D.; Smit, B. *Understanding Molecular Simulation: From Algorithms to Applications;* Academic Press: New York, 2002; p 638.
44. Friedlander, S. K. Polymer-Like Behavior of Inorganic Nanoparticle Chain Aggregates. *J. Nanopart. Res.* **1999,** *1,* 9–15.
45. Feynman, R. There's Plenty of Room at the Bottom. *Eng. Sci. Magazine.* **1959,** *23,* 22–36.
46. Fischer, D. *Theoretical Investigation of Nanoscale Solid State and Cluster Structures on Surfaces, Dissertation;* Universitat Konstanz Fachbereich Physik: Konstanz, Germany, 2002; p 144.
47. Foulkes, W. M. C.; Mitas, L.; Needs, R. J.; Rajagopal, G. Quantum Monte Carlo Simulations of Solids, *Rev. Modern Phys.* **2001,** *73,* 33–54.
48. Frisch, M. J.; Trucks, G. W.; Schlegel, H. B.; Scuseria, G. E.; Robb, M. A.; Cheeseman, J. R., et al. *Gaussian 98 (Revision A.1);* Gaussian, Inc.: Pittsburgh, PA, 1998.
49. Gabe, D. R.; Wilcox, G. D. The Rotating Cylinder Electrode: Its Continued Development and Application. *J. Appl. Elec.* **1998,** *28,* 759–780.
50. Gauss, J. Molecular Properties. *Modern Methods and Algorithms of Quantum Chemistry, Proceedings...;* 2nd ed.; Grotendorst, J., Ed.; NIC Series, John von Neumann Institute for Computing: Julich, 2000; Vol. 3, pp 541–592.
51. Galfetti, L.; De Luca, L. T.; Severini, F.; Meda, L.; Marra, G.; Marchetti, M.; Regi, M.; Bellucci, S. Nanoparticles for Solid Rocket Propulsion. *J. Phys. Condens. Matter.* **2006,** *18,* 1991–2005.
52. Gusev, A. I.; Rempel, A. A. *Nanocrystalline Materials;* Physical Mathematical Literature: Moscow, 2001; p 224.
53. Gubanov, V. A.; Zsukov, V. P.; Litinsky, A. O. *Empirical Molecular Orbital Methods in Quantum Chemistry;* Science: Moscow, 1976; p 219.
54. Gramer, C. J. Essentials of Computational Chemistry. In *Theories and Models;* 2nd ed.; Wiley: Hoboken, NJ, 2009; p 596.
55. Green, M. S. 'Markoff Random Processes and the Statistical Mechanics of Time-Dependent Phenomena. II Irreversible Processes in Fluids'. *J. Chem. Phys.* **1954,** *22,* 398–413.
56. Grzegorczyk, M.; Rybaczuk, M.; Maruszewski, K. Ballistic Aggregation: An Alternative Approach to Modeling of Silica Sol–Gel Structures. *Chaos Solitons Fract.* **2004,** *19,* 1003–1011.
57. Galkin, N. G.; Dozsa, L.; Turchin, T. V., et al. Properties of CrSi2 Nanocrystallites Grown in a Silicon Matrix. *J. Phys. Condens. Matter.* **2007,** *19* (50), 506204–506216.
58. Galkin, N. G.; Goroshko, D. L.; Dotsenko, S. A.; Turchin, T. V. Self-Organization of CrSi2 Nanoislands on Si(111) and Growth of Monocrystalline Silicon with Buried

Multilayers of CrSi2 Nanocrystallites. *J. Nanosci. Nanotechnol. Amer. Sci. Publ.* **2008,** *8* (2), 557–563.

59. Galkin, N. G.; Turchin, T. V.; Goroshko, D. L. The Influence of the Thickness of Chromium Layer on the Morphology and Optical Properties of Heterostructures Si (111)/ Nanocrystallites CrSi2 (111). *Fiz. Tverd. Tela.* **2008,** *50* (2), 345–353.
60. Gergel, V. A.; Zeleny, A. P.; Yakupov, M. N. A Mathematical Simulation of the Effect of the Bistability of Current Characteristics in Nanosized Multiple-Layer Heavily Doped Heterostructures. *Semiconductors.* **2007,** *41* (3), 314–319.
61. Golovnev, I. F.; Basova, T. V.; Aleksandrova, N. K.; Igumenov, I. K. Computer Simulation of the Synthesis of Nanoscopic Heterostructures. *Fiz. Mezomekh.* **2001,** *4* (6), 17–26.
62. Guo, L. J.; Zhao, G. F.; Gu, Y. Z.; Lia, X.; Zeng, Z. Density-Functional Investigation of Metal– Silicon Cage Clusters MSin (M = Sc, Ti, V, Cr, Mn, Fe, Co, Ni, Cu, Zn; n = 8–16). *Phys. Rev. B.* **2008,** *77,* 1954171–195178.
63. Hummer, G.; Rasaiah, J. C.; Noworyta, J. P. Water Conduction through the Hydrophobic Channel of a Carbon Nanotube. *Nature.* **2001,** *414,* 188–190.
64. Halgren, T. A. Merk Molecular Force Field. 1. Basis, form, Scope, Parametrization and Performance of MMFF94. *J. Comp. Chem.* **1996,** *17,* 490–519.
65. Holian, B. L. Formulating Mesodynamics for Polycrystalline Materials. *Europhys. Lett.* **2003,** *64,* 330–336.
66. Hari, S. N. *Handbook of Nanostructured Materials and Nanotechnology;* Academic Press: New York, 2000; Vol. 1–5.
67. Hoare, M. R. Structure and Dynamics of Simple Microclusters. *Adv. Chem. Phys.* **1979,** *40,* 49–135.
68. Haile, M. J. Molecular Dynamics Simulation–Elementary Methods. Wiley Interscience: New York, 1992; p 386.
69. Han, S. S.; Lee, H. M. Adsorption Properties of Hydrogen on (10.0) Single-Walled Carbon Nanotube through Density Functional Theory. *Carbon.* **2004,** *42,* 2169–2177.
70. Humphrey, W.; Dalke, A.; Schulten, K. VMD-Visual Molecular Dynamics. *J. Mol. Graphics.* **1996,** *14* (1), 33–38.
71. Heerman, W. D. *Computer Simulation Methods in Theoretical Physics;* Springer-Verlag: Berlin, 1986; p 145.
72. Hellan, K. *Introduction to Fracture Mechanics;* Mc Graw-Hill Book Company: New York, 1984; p 364.
73. Imry, Y. *Introduction to Mesoscopic Physics;* University Press: Oxford, 2002; p 304.
74. Jagadeesh, P.; Murali, K. Application of Low-Re Turbulence Models for Flow Simulations Past Underwater Vehicle Hull Forms. *J. Nav. Architect. Mar. Eng.* **2005,** *1,* 41–54.
75. Jan, C. T. E.; Albert van den, B. Nanofluidics: What is it and what can we Expect from it? *Microfluid Nanofluid.* **2005,** *1,* 249–267.
76. Joseph, S.; Aluru, N. R. Why are Carbon Nanotubes Fast Transporters of Water? *Nano Lett.* **2008,** *8,* 452–458.
77. Van Kampen, N. G. *Stochastic Processes in Physics and Chemistry;* Elsevier: Amsterdam, 1981; p 463.
78. Kalra, A.; Garde, S.; Hummer, G. Osmotic Water Transport through Carbon Nanotube Membranes. *Proc. Natl. Acad. Sci. USA.* **2003,** *100* (18), 10175–10180.
79. Koch, C. C. *Nanostructured Materials-Processing, Properties, and Potential Applications;* William Andrew Publishing: Norwich, Norfolk, 2002.

80. Kendall, M.; Stuard, A. *Statistic Conclusions and Connections;* Science: Moscow, 1973; p 899.
81. Kang, J. W.; Hwang, H. J. Atomistic Modeling on Structures of Ultrathin Copper Nanotubes. *Nanotech.* **2003,** *3,* 195–198.
82. Kubo, R. 'Statistical Mechanical Theory of Irreversible Processes. I. General Theory and Simple Applications to Magnetic and Conduction Problems'. *J. Phys. Soc. Japan.* **1957,** *12,* 570–586.
83. Kang, Z. C.; Wang, Z. L. On Accretion of Nanosize Carbon Spheres. *J. Physical Chem.* **1996,** *100,* 5163–5165.
84. Kim, D.; Lu, W. Self-Organized Nanostructures in Multi-Phase Epilayers. *Nanotechnology.* **2004,** 15, 667–674.
85. Kim, L. A.; Anikeeva, P. O.; Coe-Sullivan, S.; Steckel, J. S.; Bulovic, V. Contact Printing of Quantum Dot Light-Emitting Devices. *Nano Lett.* **2008,** *8* (12), 4513–4517.
86. Korn, G. A.; Korn, M. T. *Mathematical Handbook;* McGraw-Hill Book Company: New York, 1968; p 1130.
87. Kenny, B. L.; Donald, B. B., Eds. *Reviews in Computational Chemistry;* Wiley-VCH: New York, 2000; Vol. 16, p 210.
88. Kosevich, A. M. *The Crystal Lattice: Phonons, Solitons, Dislocations, Superlattices;* 2nd ed.; Wiley-VCH: Germany, 2006; p 345.
89. Kaplan I. G. *Introduction into the Theory of Intermolecular Interactions;* Nauka Press: Moscow, 1983; pp 220–221.
90. Lindahl, E.; Hess, B.; van der Spoel, D. Gromacs 3.0: A Package for Molecular Simulation and Trajectory Analysis. *J. Mol. Mod.* **2001,** *7,* pp 306–317.
91. Lachawiec, A. J., et al. Hydrogen Storage in Nanostructured Carbons by Spillover: Bridge-Building Enhancement. *Langmuir.* **2005,** *21,* 11418–11424.
92. Lee, J. W., et al. Methane Adsorption on Multi-Walled Carbon Nanotube at (303.15, 313.15, and 323.15) *K. J. Chem. Eng. Data.* **2006,** *51* (3), 963–967.
93. Lee Yoon, S. *Self–Assembly and Nanotechnology: A Force Balance Approach;* Wiley: New York, 2008; p 344.
94. Liu, W. K.; Karpov, E. G.; Zhang, S.; Park, H. S. An Introduction to Computational Nano Mechanics and Materials. *Comput. Methods Appl. Mech. Eng.* **2004,** *193,* 1529–1578.
95. Liu, W. K.; Karpov, E. G.; Zhang, S.; Park, H. S. *Nano Mechanics and Materials. Theory, Multiscale Methods and Applications;* John Wileys & Sons: New York, 2006; p 320.
96. Leach, A. R. *Molecular Modelling: Principles and Applications;* Pearson Education Limited: Edinburgh, Scotland, 2001; p 773.
97. Laurendeau, N. M. *Statistical Thermodynamics: Fundamentals and Applications;* Cambridge University Press: New York, 2005; p 448.
98. Marcos, G. L.; Preece, J. A.; Stoddard, J. F. The Art and Science of Self-Assembling Molecular Machines. *Nanotechnology.* **1996,** *7,* 183–192.
99. Marx, D.; Hutter, J. *Ab Initio Molecular Dynamics: Theory and Advanced Methods;* Cambridge University Press: New York, 2009; p 567.
100. Metropolis, N.; Rosenbluth, A.; Rosenbluth, M.; Teller, A.; Teller, E. Equation of State Calculations by Fast Computing Machines. *J. Chem. Phys.* **1953,** *21,* 1087–1092.
101. MacKerell, Jr. A. D.; Banavali, N.; Foloppe, N. Development and Current Status of the CHARMM Force Field for Nucleic Acids. *Biopolymers.* **2001,** *56,* 257–265.
102. Melikhov, I. V.; Bozhevol'nov, V. E. Variability and Self-Organization in Nanosystems. *JNR.* **2003,** *5,* 465–472.

103. Miller, R. E.; Tadmor, E. B. Hybrid Continuum Mechanics and Atomistic Methods for Simulating Materials Deformation and Failure. *MRS. Bull.* **2007,** *32,* 920–926.
104. Miller Ronald, E.; Tadmor, E. B. A Unified Framework and Performance Benchmark of Fourteen Multiscale Atomistic/Continuum Coupling Methods. *Modelling Simul. Mater. Sci. Eng.* 2009, 17 (5), 053001.
105. Mercle, C. R. Computational Nanotechnology. *Nanotechnology.* **1991,** *2,* 134–141.
106. Merkle, C. R. "It's a Small, Small, Small, Small World", *MIT Technol. Rev.* **1997,** *100,* 25–32.
107. Mattia, D.; Gogotsi, Y. Review: Static and Dynamic Behavior of Liquids inside Carbon Nanotubes. *Microfluid Nanofluid.* **2008,** *5,* 289–305.
108. Morris, D. G. *Mechanical Behaviour of Nanostructured Materials;* Trans Tech Publication: Uetikon-Zurich, Switzerland, 1998; p 100.
109. Noy, A.; Park, H. G.; Fornasiero, F.; Holt, J. K.; Grigoropoulos, C. P.; Bakajin, O. Nanofluidics in Carbon Nanotubes. *Nanotoday.* **2007,** *2,* 22–29.
110. Nose, S. A Molecular Dynamics Method for Simulation in the Canonical Ensemble. *Mol. Phys.* **1984,** *52,* 255–278.
111. Newman, J.; Thomas-Alyea, K. E. *Electrochemical Systems;* 3rd ed.; John Wiley & Sons: Hoboken, NJ, 2004; p 647.
112. Nastovjak, A. G.; Neizvestny, I. G.; Shwartz, N. L. Effect of Growth Conditions and Catalyst Material on Nanowhisker Morphology: Monte Carlo Simulation. *Solid State Phenomena.* **2010,** *156–158,* 235–240.
113. Oku, T.; Kuno, M.; Narita, I. Hydrogen Storage in Boron Nitride Nanomaterials Studied by TG/DTA and Cluster Calculation. *J. Phys. Chem. Solids.* **2004,** *65* (2–3), 549–552.
114. Oku, T.; Kuno, M. Synthesis, Argon/Hydrogen Storage and Magnetic Properties of Boron Nitride Nanotubes and Nanocapsules. *Diamond Relat. Mater.* **2003,** *12* (3–7), 840–845.
115. Ozawa, E. Properties, Production Methods, Use and Application of Ultra Dispersed Powders. *J. Japan Soc. Technol. Plast.* **1986,** *27,* 1166–1172.
116. Perez, T.; Nav, J. Numerical Simulation of the Primary, Secondary and Tertiary Current Distributions on the Cathode of a Rotating Cylinder Electrode Cell. Influence of Using Plates and a Concentric Cylinder as Counter Electrodes. *J. Elec. Chem.* **2014,** *719,* 106–112.
117. Palmer, L. C.; Velichko, Y. S.; de la Cruz, M. O.; Stupp, S. I. Supramolecular Self-Assembly Codes for Functional Structures. *Phil. Trans. R. Soc. A.* **2007,** *365,* 1417–1433.
118. Phillips, J.C., et al. Scalable Molecular Dynamics with NAMD. *Comput. Chem.* **2005,** *26,* 1781–1802.
119. Plimpton S. J. Fast Parallel Algorithms for Short-Range Molecular Dynamics. *J. Comput. Phys.* **1995,** *117,* 1–19.
120. Pozdeev, A. A.; Niyshin Yu, I.; Trusov, P. V. *Residual Stresses: Theory and Applications;* Science: Moscow, 1982; p 112.
121. Park, H. S.; Liu, W. K. An Introduction and Tutorial on Multiple-Scale Analysis in Solids. *Comput. Methods Appl. Mech. Eng.* **2004,** *193,* 1733–1772.
122. Parr, R. G.; Yang, W. *Density-Functional Theory of Atoms and Molecules;* Oxford University Press: Oxford, 1989; p 331.
123. Pelesko John, A. *Self Assembly: The Science of Things that Put Themselves Together;* Chapman & Hall/CRC: Boka Raton, FL, 2007; p 307.
124. Pippan, R., Gumbsch, P., Eds. *Multiscale Modelling of Plasticity and Fracture by Means of Dislocation Mechanics;* Springer: New York, 2010; p 394.

125. Qing-Qing, N.; Yaqin, F.; Masaharu, I. Evaluation of Elastic Modulus of Nano Particles in PMMA/Silica Nanocomposites. *J. Soc. Materials Sci.* **2004,** *53* (9), 956–961.
126. Ren, Y. X., et al. State of Hydrogen Molecules Confined in C60 Fullerene and Carbon Nanocapsule Structures. *Carbon.* **2006,** *44,* 397–406.
127. Ruoff, R. S.; Pugno, N. M. In *Strength of Nanostructures,* Proceedings of the 21st International Congress of Theoretical and Applied Mechanics, Warsaw, Poland, Aug 15–21, 2004; pp 303–311.
128. Ramachandran, K. I.; Deepa, G.; Namboori, K. *Computational Chemistry and Molecular Modelling;* Springer-Verlag: Berlin, Heidelberg, 2010; p 397.
129. Reet, M. *Nanoconstruction in Science and Technology: Introduction to Nanocomputations;* R&C Dynamics: Izhevsk, Moscow, 2005; p 160.
130. Tuzun Robert, E.; Noid Donald, W.; Sumpter Bobby, G.; Merkle Ralph, C. Dynamics of Fluid Flow Inside Carbon Nanotubes. *Nanotechnology.* **1996,** *7,* 241–246.
131. Steinbach, I. Phase-Field Models in Materials Science. Modelling Simul. *Mater. Sci. Eng.* 2009, 17 (7), 073001.
132. Schuler, L. D.; Xavier, D.; Wilfred F. van G. An Improved GROMOS96 Force Field for Aliphatic Hydrocarbons in the Condensed Phase. *J. Comput. Chem.* 2001; *22* (11), 1205–1218.
133. Simakin, A. V.; Voronov, A. V.; Shafeev, G. A. In *The Formation of Nanoparticles by Laser Ablation of Solids in Liquids*, Proceedings of the Institute of General Physics of the Academy of Science of the USSR, 2004; Prokhorov, A. M., Ed.; 2004; pp 83–107.
134. Sagoff, J. Nano-Boric Acid Makes Motor Oil More Slippery. *Argonne News.* **2007,** *60* (16), 1–2.
135. Steinhauser, O. M. *Computational Multiscale Modelling of Fluids and Solids. Theory and Application;* Springer-Verlag: Berlin, Heidelberg, 2008; p 427.
136. Simonyan, V. V.; Diep, P.; Johnson, J. K. Molecular Simulation of Hydrogen Adsorption in Charged Single-Walled Carbon Nanotubes. *J. Phys. Chem. B.* **1999,** *111,* 9778–9783.
137. Shigenobu, O.; Yoshitaka, U.; Masanori, K. First-Principles Approaches to Intrinsic Strength and Deformation of Materials: Perfect Crystals, Nano-Structures, Surfaces and Interfaces. *Modelling Simul. Mater. Sci. Eng.* **2009,** *17* (1), 013001.
138. Sholl, D. S.; Stecker, J. A. *Density Functional Theory. A Practical Introduction;* Wiley: Hoboken, NJ, 2009; p 328.
139. Sturgess, H. A. The Choice of Classic Intervals. *J. Amer. Statist. Ass.* **1926,** *21,* 47–55.
140. Schaefer Henry, F. *Modern Theoretical Chemistry: Methods of Electronic Structure Theory;* 3rd ed.; Springer: New York, 1977; Vol. 3, p 467.
141. Stojak, J. L.; Fransaer, J.; Talbot, J. B.; Review of Electrocodeposition. In *Advances in Electrochemical Science and Engineering;* Alkire, R. C., Kolb, D. M., Eds.; Wiley-VCH Verlag: Weinheim, 2002; Vol. 9, pp 193–225.
142. Stojak, J. L.; Talbot, J. B. Effect of Particles on Polarization during Electrocodeposition using a Rotating Cylinder Electrode. *J. Appl. Elec.* **2001,** *31,* 559–564.
143. Self-Organizing of Microparticles Piezoelectric Materials. *News of Chemistry,* date-news.php.htm
144. Schwab, I. V.; Dudnikova, G. I.; Jeger, D. Modeling of Nonlinear Multilayer Heterostructures for High-Speed Optical Processes. *Vychisl. Tekhnol.* **2000,** *5* (4), 104–110.
145. Severyukhina, O. Y.; Vakhrushev, A. V.; Galkin, N. G.; Severyukhin, A. V. Simulation of Formation of Multilayer Nanoheterostructures with Variable Chemical Bonds. *Khim. Fiz. Mezoskopiya.* **2011,** *13* (1), 53–58.

146. Schubert, L.; Werner, P.; Zakharov, N. D. Silicon Nanowhiskers Grown on 111 Si Substrates by Molecular-Beam Epitaxy. *Appl. Phys. Lett.* **2004,** *84,* 4968–4971.
147. Schmauder, S.; Mishnaevsky, L. Jr. *Micromechanics and Nanosimulation of Metals and Composites. Advanced Methods and Theoretical Concepts;* Springer Verlag: Berlin, Heidelberg, 2009; p 420.
148. Tanaka, H.; El-Merraoui, M.; Steele, W. A.; Kaneko, K. Methane Adsorption on Single-Walled Carbon Nanotube: A Density Functional Theory Model. *Chem. Phys. Lett.* **2002,** 352 (5–6), pp 334–341.
149. Turker, L.; Erkos, S. AM1 Treatment of Endohedrally Hydrogen Doped Fullerene; nH2@C60. *J. Mol. Struct. Theochem.* **2003,** *638* (1–3), 37–40.
150. Tomanek, D., et al. Nanocapsules Containing Charged Particles, their Uses and Methods of Forming Same. U.S. Patent 6,473,351, October 29, 2002.
151. Thompson, R. A. Mechanics of Powder Pressing. I.-III Model for Powder Densification. *Amer. Ceram. Soc. Bull.* **1981,** *60,* 237–243.
152. Tadmor, E. B.; Miller, R. E. *Modeling Materials. Continuum, Atomistic and Multiscale Techniques;* Cambridge University Press: New York, 2011; p 759.
153. Tsyrlin, G. E.; Dubrovsky, V. G.; Sibirev, N. V. The Diffusion Mechanism Underlying the Growth of GaAs and AlGaAs Nanowhiskers in the Method of Molecular-Beam Epitaxy. *Fiz. Tekh. Poluprovodn.* **2005,** 39 (5), 587–594.
154. Vakhrushev, A. V.; Lipanov, A. M. A Numerical Analysis of the Rupture of Powder Materials under the Power Impact Influence. *Comput. Struct.* **1992,** *44* (1/2), 481–486.
155. Vakhrouchev, A. V. Simulation of Nano-Elements Interactions and Self-Assembling. *Model. Simul. Mater. Sci. Eng.* **2006,** *14,* 975–991.
156. Vakhrouchev, A. V. Modelling of the Process of Formation and Use of Powder Nanocomposites. In *Composites with Micro and Nano-Structures. Computational Modeling and Experiments;* Computational Methods in Applied Sciences Series; Springer Science: Barcelona, Spain, 2008; Vol. 9, pp 107—136.
157. Vakhrushev, A. V.; Lipanov, A. M. Numerical Analysis of the Atomic Structure and Shape of Metal Nanoparticles. *Comp. Math. Math. Phys.* **2007,** 47 (10), 1702–1711.
158. Vakhrouchev, A. V.; Suyetin, M. V. Methane Storage in Bottle-Like Nanocapsules. *Nanotechnology.* **2009,** *40,* 125602.
159. Vakhrouchev, A. V. Modelling of the Nanosystems Formation by the Molecular Dynamics, Mesodynamics and Continuum Mechanics Methods. *Multidiscipl. Model. Mater. and Struct.* **2009,** 5 (2), 99–118.
160. Vakhrushev, A. V. *Theoretical Bases of Nanotechnology Application to Thermal Engines and Equipment;* Institute of Applied Mechanics Ural Branch of the Russian Academy of Sciences: Izhevsk, Russia, 2008; p 212.
161. Vakhrouchev, A. V.; Lipanov, A. M.; Suyetin, M. V. Simulation of the Processes of the Hydrogen and Hydrocarbon Accumulation by Nanostructures. Regular & Chaotic Dynamics: Moscow, Izhevsk, 2008; p 120.
162. Vakhrushev, A. V.; Fedotov, A. Y.; Vakhrushev, A. A.; Golubchikov, V. B.; Golubchikov, E. V. The Plant Nutrition from the Gas Medium in Greenhouses: Multilevel Simulation and Experimental Investigation. In *"Plant Science";* Nabin, K., Sudam, C. S., Eds.; InTech: Rijeka, Croatia, 2012; pp 65–104.
163. Vakhrushev, A.V.; Fedotov, A. Y.; Golubchikov, V. B. In *Research and Forecasting of Properties of Metallic Nanocomposites and Nanoaerosol Systems,* Proceeding of the 6th International Conference of Nanotechnology, Rom, Italy, Nov 7–9, 2015; pp 113–140.

164. Vakhrushev, A. V.; Molchanov, E. K. In *Nanocomposite Coating Created by Electrocodeposition Method,* Proceeding of the 6th International Conference of Nanotechnology, Rom, Italy, Nov 7–9, 2015; pp 31–40.
165. Volkova, E. I.; Vakhrushev, A.V.; Suyetin, M. V. Triptycene-Modified Linkers of MOFs for Methane Sorption Enhancement: A Molecular Simulation Study. *Chem. Phys.* **2015,** *459,* 14–18.
166. Vakhrushev, A. V.; Molchanov, E. K. Hydrodynamic Modeling of Electrocodeposition on a Rotating Cylinder Electrode. *Key Eng. Mater.* **2015,** *654,* 29–33.
167. Volkova, E. I.; Vakhrushev, A. V.; Suyetin, M. V. Improved Design of Metal-Organic Frameworks for Efficient Hydrogen Storage at Ambient Temperature: A Multiscale Theoretical Investigation. *Int. J. Hydrogen Energy.* **2014,** *39* (16), 8347–8350.
168. Vakhrushev, A. V.; Severyukhina, O. Y.; Severyukhin, A. V.; Vakhrushev, A. A. Galkin N. G. Simulation of the Processes of Formation of Quantum Dots on the Basis of Silicides of Transition Metals. *Nanomech. Sci. Tech.* **2012,** *3* (1), 51–75.
169. Vakhrushev, A. V.; Molchanov, E. K. Modeling Electrochemical Deposition of Al_2O_3 Nanoparticles into a Cu Matrix. *Nanomech. Sci. Technol.* **2012,** *3* (4), 353–371.
170. Vakhrushev, A. V.; Severyukhina, O. Y.; Severyukhin, A. V. Modeling of the Initial Stage of Formation of Nanowhiskers on an Activated Substrate. Part 1. Theory Foundations. *Nanomech. Sci. Technol.* **2012,** 3 (3), 193–209.
171. Vakhrushev, A. V.; Severyukhina, O. Y.; Severyukhin, A. V. Modeling of the Initial Stage of Formation of Nanowhiskers on an Activated Substrate. Part 2. Numerical Investigation of the Structure and Properties of Au–Si Nanowhiskers on a Silicon Substrate. *Nanomech. Sci. Technol.* **2012,** 3 (3), 211–237.
172. Vakhrushev, A. V.; Vakhrusheva, L. L. In *The Finite Element Analysis of the Powder Materials Compression,* Proceeding of the 4th International Conference of Numerical Methods in Industrial Forming Processes-NUMIFORM'92, Valbonne, France, 1992; pp 887–892.
173. Venikov, V. A. *The Theory of Similarity and Modeling;* Vysshaya Shkola Press: Moscow, 1976; p 479.
174. Verlet, L. Computer "Experiments" on Classical Fluids. I. Thermodynamical Properties of Lennard-Jones molecules. *Phys. Rev.* **1967,** 159 (1), 98–103.
175. Verlet, L. Computer "Experiments" on Classical Fluids. II. Equilibrium Correlation Functions. *Phys. Rev.* **1967,** *165,* 201–214.
176. Wang, Q.; Johnson, K. Molecular Simulation of Hydrogen Adsorption in Single-Walled Carbon Nanotubes and Idealized Carbon Slit Pores. *J. Chem. Phys.* **1999,** 110 (11), 577–586.
177. Wang, Q.; Johnson, K. Optimization of Carbon Nanotube Arrays for Hydrogen Adsorption. *J. Phys. Chem. B.* **1999,** *103,* 4809–4813.
178. Weinan, E.; Engquist, B.; Li, X.; Ren, W.; Vanden-Eijnden, E. Heterogeneous Multiscale Methods: A Review. *Commun. Comput. Phys.* **2007,** *2,* 367–450.
179. Weinan, E. *Principles of Multiscale Modeling;* Cambridge University Press: Cambridge, 2011; p 466.
180. Wagner, G. J.; Liu, W. K. Coupling of Atomistic and Continuum Simulations Using a Bridging Scale Decomposition. *J. Comput. Phys.* **2003,** *190,* 249–274.
181. Wagner, R. S.; Ellis, W. C.; Vapor–Liquid–Solid Mechanism of Single Crystal Growth. *Appl. Phys. Lett.* **1964,** *4,* 89–90.

182. Werder, T.; Walther, J. H.; Jaffe, R. L.; Halicioglu, T.; Koumoutsakos, P. On the Water-Carbon Interaction for Use in Molecular Dynamics Simulations of Graphite and Carbon Nanotubes. *J. Phys. Chem. B.* **2003,** *107,* 1345–1352.
183. Whitby, M.; Quirke, N. Fluid Flow in Carbon Nanotubes and Nanopipes. *Nat. Nanotechnol.* **2007,** *2,* 87–94.
184. Cheng-I, W.; Wang-Long, L.; Chi-Chuan, H. Gaseous Flow in Microtubes at Arbitrary Knudsen Numbers. *Nanotechnology.* **1999,** *10,* 373–379.
185. McWeeny, R. *Methods of Molecular Quantum Mechanics;* 2nd ed.; Academic Press: London, 1996; 591.
186. Walling, M. A.; Novak, S. Quantum Dots for Live Cell and *In Vivo* Imaging. *Int. J. Mol. Sci.* **2009,** *10* (2), 441–491.
187. Wang, L. W.; Cartoixa, X. Microscopic Dielectric Response Functions in Semiconductor Quantum Dots. *Phys. Rev. Lett.* **2005,** *94,* 236804–236808.
188. Yang F. H., et al. Adsorption of Spillover Hydrogen Atoms on Single-Wall Carbon Nanotubes. *J. Phys. Chem. B.* **2006,** *110,* 6236–6244.
189. Ye, X., et al. A Nanocontainer for the Storage of Hydrogen. *Carbon.* **2007,** *45,* 315–320.
190. Yoon, M.; Berber, S.; Tomanek, D. Energetics and Packing of Fullerenes in Nanotube Peapods. *Phys. Rev. B.* **2005,** *71,* 155406–1–155406–4.
191. Yoon S. L. *Self–Assembly and Nanotechnology;* Wiley: Hoboken, NJ, 2008; p 344.
192. Yitao, D.; Chun, T.; Wanlin, G. Simulation Studies of a "Nanogun" Based on Carbon Nanotubes. *Nano Res.* **2008,** *1,* 176 – 183.
193. Yesilata, B. Nonlinear Dynamics of a Highly Viscous and Elastic Fluid in Pipe Flow. *Fluid Dynamics Res.* **2002,** *31* (16), 41–64.
194. Yaghi, O. M.; O'Keeffe, M.; Ockwig, N. W.; Chae, H. K.; Eddaoudi, M.; Kim, J. Synthesis and the Design of New Materials. *Nature.* **2003,** *423,* 705–714.
195. Zhang, X.; Wang, W. Adsorption of Linear Ethane Molecules in Single Walled Carbon Nanotube Arrays by Molecular Simulation. *PCCP.* **2002,** *4,* 3048–3054.
196. Zienkiewicz, O. C. *The Finite Element Method in Engineering Science;* McGraw-Hill: New York, 1971; 521.
197. Zhu, F.; Tajkhorshid, E.; Schulten, K. Pressure-Induced Water Transport in Membrane Channels Studied by Molecular Dynamics. *Biophys. J.* **2002,** *83,* 154–160.
198. Zhu, F.; Schulten, K. Water and Proton Conduction Through Carbon Nanotubes as Models for Biological Channels. *Biophys. J.* **2003,** *85,* 236–244.
199. Zhu, F.; Tajkhorshid, E.; Schulten, K. Collective Diusion Model for Water Permeation Through Microscopic Channels. *Phys. Rev. Lett.* **2004,** *93,* 224–501.
200. Zhou, S. Y.; Xie, Q.; Yan, W. J.; Chen, Q. First-Principle Study on the Electronic Structure of Stressed CrSi2. *Sci. China Ser. G-Phys. Mech. Astron.* **2009,** *52* (1), 76–81.
201. Zhang, F. Molecular Dynamics Studies of Chainlike Molecules Confined in a Carbon Nanotube. *J. Chem. Phys.* **1999,** *111,* 9082–9085.

EPILOGUE

In conclusion, we emphasize that the theoretical chapters, presented in the book, give an initial idea of the methods of calculation of nanosystems. Author can recommend the following books[99,135,179] for a more in-depth study of theoretical foundations.

There is a frequently asked question: "Why do you check your ideas with the help of computer simulation, not in the experiments?." Modeling is necessary for an understanding of the processes taking place in nanosystems. Here the most important thing—the right to formulate a mathematical model that adequately describes the investigated processes. Simulation shows whether it is possible in principle to create a technology that allows you to predict how processes occur, and to calculate its parameters. This helps us to experiment more precisely, only those conditions that are promising.

In our practice, there are already examples where the method of "simulated-used" has led to successful results. For example, the preparation of nanoparticles in pulsed power installations, the development of nano aerosol technology feeding plants from the gas phase, the formation of the space charge capacitors based on micro- and nanoparticles.

A few years ago, when we started to get the first results on the modeling of hydrocarbon accumulation processes using nanocontainers, at international conferences, experimenters were told that it is impossible to get nanocontainers. Then they recognized that it is possible to create individual copies, only for experimentation. Now nanocontainers made of nanostructured carbon, which is irradiated with a beam of energetic particles—electrons, gamma rays, protons, and ions. As a result, changing the structure, morphology, as well as electronic, mechanical, and chemical properties of carbon materials. The parameters of this change can be controlled and obtain nanocontainers with specific properties. It is likely that in a few years, it will be possible to mass synthesis of nanocapsules' desired size and shape.

The main thing is that the author believes it is necessary for the reader to learn the need for multiscale modeling of nanosystems, as only such an approach will identify the connection structure of nanosystems with its macro characteristics.

INDEX

F

G

H

I

M

T

U

V

Printed in the United States
by Baker & Taylor Publisher Services